Introduction to Communication Science and Systems

Applications of Communications Theory
Series Editor: R. W. Lucky, *Bell Laboratories*

INTRODUCTION TO COMMUNICATION SCIENCE AND SYSTEMS
John R. Pierce and Edward C. Posner

A Continuation Order Plan is available for this series. A continuation order will bring delivery of each new volume immediately upon publication. Volumes are billed only upon actual shipment. For further information please contact the publisher.

Introduction to Communication Science and Systems

John R. Pierce
and
Edward C. Posner

California Institute of Technology
Pasadena, California

PLENUM PRESS · NEW YORK AND LONDON

Library of Congress Cataloging in Publication Data

Pierce, John Robinson, 1910-
  Introduction to communication science and systems.

  Includes bibliographical references and index.
  1. Telecommunication. I. Posner, Edward C., 1933-   joint author. II. Title.
TK5101.P53   621.38   80-14877
ISBN 0-306-40492-3

© 1980 Plenum Press, New York
A Division of Plenum Publishing Corporation
227 West 17th Street, New York, N.Y. 10011

Printed in the United States of America

To the students of Caltech —
who made this a better book

John R. Pierce
Edward C. Posner

September 1979

Preface

There are many valuable and useful books on electrical communication (References 1–5 are some examples), but they have certain disadvantages for the beginner. The more advanced books present some things in a basic way, but they are very narrow for an introduction to communication. The introductory books are broader but still narrow by our standards. Further, they often pick things out of thin air rather than derive them.

This book is aimed at giving the beginner a basic understanding of a wide range of topics which are essential in communication systems. These include antennas and transmission, thermal noise and its consequences, Fourier transforms, modulation and noise, sampling and pulse code modulation, autocorrelation and power spectrum, optimum filtering, gaussian noise and errors in digital transmission, data transmission, limits on data rate including information theory and quantum limits, and source encoding. We have not included communications traffic, switching, and multiplexing, nor protocols for digital and computer communications. For these, Reference 6 is excellent. In general, our book does not discuss the circuits used for communication or the physics of radio propagation. We assume that these will be taught in specialized courses, but such courses are not prerequisites for this one.

Chapter 1 introduces the transmission formula or antenna equation and antenna directivity. Only a very basic sophomore physics knowledge of electromagnetic theory is assumed. The radar equation is also treated. The transmission formula occurs throughout the rest of the book in examples and problems because it is often the main determinant of communication system performance.

In Chapter 2, we study noise and its measurement. Here we use an average or mean-square criterion, and defer probability until Chapter 8. Some of the details on noise in optical communication are drawn from a later chapter, Chapter 11. Chapters 1 and 2 enable the student to calculate signal-to-noise ratios, and so begin to analyze the performance of entire communication systems.

Chapter 3 introduces transfer functions and frequency spectra, but at a level above the electrical circuit. No prior knowledge of Fourier series, Fourier transforms, or circuit theory is assumed, and only the most elementary facts are used about differential equations, from sophomore mathematics. We also introduce impulses or delta functions. The two-dimensional Fourier transform is defined and is used to derive rigorously the antenna equation of Chapter 1. We also discuss short-term energy spectra or sonograms. This chapter is basic for all that follows.

In Chapter 4, we learn about all the classical radio engineering methods of modulation, including amplitude, frequency, and phase modulation. The results of Chapter 3 allow us to find the power spectra of these forms of modulation. In conjunction with Chapters 1 and 2, we find signal-to-noise ratios for some real communication systems, and compare them in this aspect of their performance.

Chapter 5 provides the link between the analog world of voltages and the digital world of bits through the study of sampling and the sinc function. The error due to sampling is introduced as one part of the overall system error in digital communication. All the necessary properties of sampling and sinc functions are derived.

Chapter 6 introduces the autocorrelation function and the concept of a stationary signal or noise. Here we use what we learned about Fourier transforms in Chapter 3. These concepts form the basis of our analysis in subsequent chapters of the error performance of communication systems. But, probability is not introduced as a tool for signal and noise analysis until Chapter 8.

Chapter 7 develops and uses simple optimization techniques from the calculus of variations to find pulse shapes that maximize the signal-to-noise ratio in the transmission of random binary data. We use the fact that a signal corresponding to random binary data is stationary as defined in Chapter 6. We also derive the concept of matched filtering for optimum detection. As an application of optimization, we derive the basic conditions for diversity reception and antenna arraying.

In Chapter 8, we introduce probability for the first time. No prior experience is assumed. We derive expressions for shot noise, thermal noise, and gaussian noise in general. We include the central limit theorem in order to explain why gaussian noise is so prevalent in communications. This derivation is based on what we have learned about Fourier transforms. For the first time in this book, we are able to relate signal-to-noise ratios to error probabilities. It is the error probability that the user of digital communication is concerned with.

Chapter 9 continues to develop probability, extending the concepts to include entire waveforms such as signals or noises. We extend Chapter 6 to probabilistic signals and noise by studying stationary random processes and their autocorrelation. The two-dimensional gaussian is treated, and so the two-dimensional Fourier transform comes into play. As an application, we study the detectability of gaussian signals in noise using a spectrum analyzer. We also derive properties of narrowband gaussian noise, which occurs almost universally in communication receivers as at least one component of noise.

Chapter 10 goes into more detail concerning digital data transmission systems. It shows how to shape the spectrum of a data signal by simple encoding in order to adapt the data to a channel, such as a telephone circuit, that does not pass dc. We also describe a simple encoding to randomize the data source in order to make synchronization of the data symbols easier. In addition, this chapter studies the errors associated with phase modulation used for digital data communication. The bandwidth occupancy and power spectrum of phase modulation are derived as well.

Chapter 11 goes beyond the concept of the probability of making an error in a single received data bit. Here we study information theory. We derive Shannon's expression for the ideal energy per bit needed for reliable communication on a gaussian channel, making use of the concept of signal space. This relies on what we have learned about sampling, correlation, and gaussian noise. We study ways to approach Shannon's limit by coding, and learn about the efficiency of multilevel signaling compared with Shannon's limit. We use Fourier transforms, optimum filtering, and gaussian noise theory to derive, via the quantum mechanical uncertainty principle, the quantum noise we first introduced in Chapter 2.

In Chapter 12, we study error-correcting coding used to approach Shannon's limit. In particular, we derive Viterbi decoding of convolutional codes. We study the losses inevitable in data communication if we detect bits one by one instead of using optimum detection involving many bits at once. Here we use the concept of mutual information, which permits us to define the capacity of any channel. The capacity of the photon-counting optical channel is then derived by using a conceptual coding scheme. Jamming, snooping, and spread-spectrum communication are also considered, and we find the optimum communication possible, according to information theory, in such adverse circumstances.

Previous chapters treated the communication link. Chapter 13, the final chapter, deals with the data source. Ways are given for removing redundancy, both within and outside of the context of information theory and mutual information. Huffman coding and rate distortion theory allow some gains, but we will see that the major gains are attained by getting at the underlying mechanism of signal generation. The vocoder is described as an instance of this. Multidimensional scaling is explained as a means for getting at the message behind the message in other situations. We close by giving some thoughts on how to design entire communication systems.

This book, then, aims at basic ideas, simple derivations, and tractable but meaningful examples. The examples are intended to confirm that what is being learned has relevance and application to communication systems. The aim is to introduce and direct, to give a sense of meaning and purpose, not to complete. To communication science and systems there is no end, but a good beginning can help the student or practitioner on his or her way.

The book may serve as a three-quarter or two-semester text for juniors, seniors, or graduate students in electrical engineering who want to understand the fundamental ideas of communication, either to enhance their value as electrical engineers or as a preparation for advanced courses leading to a career in research, development, or engineering in any area of communications. Engineers in other fields who understand the basics of electricity and magnetism and a very little electronics will also find this book useful as an introduction to the technology of one of the world's most significant and exciting industries. For such engineers, the book may also be useful for self-study.

Physical science students may find this book useful. Physicists, earth scientists, and astronomers who must detect signals in their work can use this book to learn about the limits of detectability of signals, whether they be laser signals, seismological signals, or the signals of radio astronomy. This applies as well for students of neurological signals, psychophysics, and acoustics. Pure and applied mathematicians working in information theory and coding theory will learn how their work fits into a broader framework of communications technology, and they may find new and exciting areas to work in.

Proofs of key results are provided wherever practical. The proofs have been chosen for simplicity and insight. They may not always be the best proofs, but they do not require much previous knowledge of mathematics, and they present important concepts.

Problems are provided at the ends of the sections. Chiefly, these are problems with a moral. Some illustrate material in the text with specific numbers plausible for real-life systems. Others extend the material presented in the text. A small book of solutions is available to instructors.

One of the authors (JRP) is gratified that members of his Caltech classes, as projects in the third quarter of a three-quarter course based on drafts of this book, gathered particular data from the literature, from Caltech's Jet Propulsion Laboratory, and from other sources, and sketched the design and performance of plausible communication systems for distances ranging from a few hundred feet to the distance from Earth to Alpha Centauri. Many others have helped improve this book by commenting on portions of it. A few students who used this book for a reading course were especially helpful, and so were the graders of the course for which this book was designed. Special thanks should be given to Patricia Neill, who typed the multiple versions of the class notes on which this book is based as well as the final manuscript delivered to the publisher.

Pasadena, California   John R. Pierce
                                                                         Edward C. Posner

Contents

Antennas and Transmission

Communication is usually limited by the amount of power that we can transfer from the transmitter to the receiver, because this, in part, determines the ratio of received signal power to noise power.

In radio communication, the power received depends not only on the transmitted power, but also on the characteristics of the transmitting and receiving antennas. In general, the larger the transmitting antenna (measured in wavelengths), the more directive the beam of radio waves it sends out, and the greater the power density at the receiving antenna. And, the larger the receiving antenna, the greater the total power that it picks up.

In this chapter we shall learn how to deal quantitatively with the antenna parameters that determine the ratio of received power to transmitted power.

Figure 1.1 shows a microwave antenna with a *cassegrainian* feed system. The microwave radiation is fed to the antenna surface by reflection from a hyperboloidal subreflector. The antenna surface itself is paraboloidal. Elementary geometry shows that the path length, and hence the time it takes a radio wave to travel from the source to a very distant point, is the same no matter what part of the main paraboloidal surface or subreflector the wave is reflected from. This is the condition for focusing the radiation into a narrow beam. Reference 1 tells much more about reflector antennas and their properties.

Antennas for communication or radar must often transmit and receive at the same time. The 64-m paraboloidal antennas with cassegrainian feed systems that are used for deep space communication routinely receive signals 26 orders of magnitude weaker than the signal they transmit. This means that we may deal with very weak signals in communication.

We have seen that antennas are a very important part of a communication system. They are always the most visible parts of microwave radio

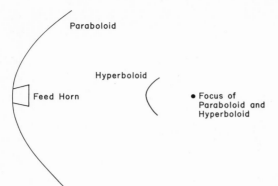

Figure 1.1. Cassegrainian reflector antenna.

systems that operate at frequencies from 1 to 30 GHz. The antennas for space communication are usually paraboloidal. Some microwave antennas have the form of a horn with a dielectric lens at the large end (Fig. 1.2) or a horn with a paraboloidal reflector at the large end (Fig. 1.3). The latter type of antenna is used in transcontinental microwave transmission for telephone and TV. The lens or the curved reflector makes the time it takes a radio wave to travel from the source to a very distant point the same no matter what part of the antenna aperture the wave emerges from. As we have noted, this is the condition for focusing.

Suppose that we have two radio antennas, a transmitting antenna and a receiving antenna. What fraction of the transmitted power does the receiving antenna pick up? What is the relation between this fraction (the path loss) and the directivity or radiation pattern of the antenna? These questions are of primary importance in radio systems, whether they be transcontinental microwave systems with antennas on hilltop towers at spacings of some 30 miles, or satellite systems with an earth–satellite distance of over 20,000 miles, or links to the moon, Mars, Jupiter, or the stars.

This chapter is concerned with path loss for line-of-sight paths. The phenomenon of fading in atmospheric transmission is not considered, nor are problems of generating radio frequency power. In this book, a resistance of 1 ohm is usually assumed in relating power to mean-squared voltage. Our expressions will be dimensionally correct if we interpret V^2 as power.

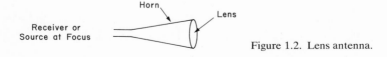

Figure 1.2. Lens antenna.

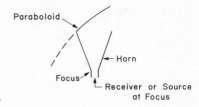

Figure 1.3. Horn reflector antenna.

1.1. Path Loss from Satellites

A synchronous satellite is 42,000 km or 26,000 miles from the center of the earth (at an altitude of 22,000 miles from the nearest point on earth). What fraction of the transmitted power can be picked up by a circular antenna of diameter D? This depends on the radiation pattern of the antenna on board the satellite. It is an important question because more received power means that more telephone conversations or TV programs or other communication can be transmitted over the path. Let us find what fraction of the transmitted power can be received.

If the satellite transmitting antenna sends out power isotropically or equally in all directions (as the Telstar antenna pretty much did), the transmitted power will be distributed equally over 4π steradians.† A circular antenna of diameter D at a distance r covers $\pi D^2/4r^2$ steradians and gathers together and in phase at the focus all of a plane radio wave that it intercepts. Thus, in this case of isotropic radiation, the ratio of received power P_R to transmitted power P_T will be

$$P_R/P_T = (\pi D^2/4r^2)/4\pi$$

$$P_R/P_T = D^2/16r^2 \tag{1.1-1}$$

But, satellite antennas no longer radiate power isotropically. They send out power preferentially toward the receiving antennas on earth. If the beam of radio waves from the satellite falls uniformly on the disk of the earth and none is lost into space, the ratio of the power received to the power transmitted is very nearly

$$P_R/P_T = D^2/D_e^2 \tag{1.1-2}$$

where D_e is the diameter of the earth.

The most recent satellite antennas are more directive than this. If the radio waves from the satellite antenna fall on an area A of the earth's surface, the ratio of received to transmitted power is

$$P_R/P_T = \pi D^2/4A \tag{1.1-3}$$

† An infinitesimal area dA normal to the radius at a distance r has a solid angle dA/r^2 steradians.

It is a basic law of radio transmission that the path loss (ratio of power received to power transmitted) is the same if the transmitting antenna and the receiving antenna are interchanged. This path loss is commonly expressed in decibels, or dB. Path loss in dB is

$$10 \log_{10} (P_R/P_T)$$

In the next section we shall consider the effectiveness of antennas in focusing the radiated energy into a tight beam. This is an important factor in determining received power, together with the collecting area of the receiving antenna.

Problems

1. How much is 26 orders of magnitude in dB?
2. Assume that the receiving antenna on earth has a diameter of 150 ft. What is the path loss in dB if the satellite (1) radiates isotropically, (2) just covers the disk of the earth ("global beam"), (3) just covers the 3.6×10^6 miles^2 area of the continental United States ("spot beam")?
3. What is the gain in dB (i.e., advantage over isotropic radiation) in (1), (2), and (3) of Problem 2?

1.2. Directivity of Microwave Antennas—Rough Treatment

A good deal is said about waves in this book, but no derivation is given of the properties of any physical waves, electromagnetic or acoustic. This is not a book about the physical properties of waves. Reference 2 is useful in learning about these properties.

The wavelength λ of a sinusoidal wave is the velocity v divided by the frequency f. We do need to know that beyond a few of these wavelengths from a small source, waves travel in straight lines with a constant velocity. For electromagnetic waves this velocity v is

$$v = c \doteq 3 \times 10^8 \, \text{m/sec}$$

This is the velocity of light. If a plane wave, or one which varies only in the direction of propagation, is sinusoidal and has a frequency f or a radian frequency $\omega = 2\pi f$, then along a straight line in the direction of propagation the electric or magnetic field, E or H, varies with distance x as

$$E = E_0 e^{-j(\omega/c)x} = E_0^{-j(2\pi/\lambda)x}$$
$$H = H_0 e^{-j(\omega/c)x} = H_0 e^{-j(2\pi/\lambda)x}$$

Here E_0 and H_0 are the peak amplitudes of the electric and magnetic fields. The power density carried by an electromagnetic wave is proportional to EE^* or HH^* or EH^*. (Throughout this book, $*$ denotes the complex conjugate.)

A wave is *isotropic* if at a given distance from the source, or on a sphere concentric with the source, the power density is the same in all directions or at all positions on the surface of the sphere. The power density decreases inversely as the square of the radius because the total power flowing through a sphere surrounding the source does not depend on the radius. This is the law of conservation of energy. Isotropic electromagnetic waves are impossible to generate, but we sometimes use hypothetical isotropic electromagnetic waves as a standard of comparison, as we did in Section 1.1. Within a small volume sufficiently far from its source, an isotropic electromagnetic wave, or any other electromagnetic wave, looks like a plane wave.

The electric or magnetic fields of waves from two or more sources add as vectors to give the total electric or magnetic field. Diffraction theory tells us that a wave itself can be represented as an array of small sources over the *wave front*, which is the surface of constant phase. If we want to find the field produced at some distant point by a plane wave emanating from an aperture, we look back at the aperture and add vectorially and with appropriate phase, or as complex vectors, contributions from each little area $dx \, dy$ of the aperture.

Electric and magnetic fields are vectors. However, if the aperture is far enough away and if the electric field points in the same direction in each little area $dx \, dy$, then the electric fields produced at the distant point will all point in the same direction, and the amplitudes will add just as if the fields were scalars, not vectors. Scalar diffraction theory takes no account of vectors but merely adds scalar intensities, taking account, of course, of phase. This is a good approximation in dealing with electromagnetic waves under the conditions given above. We shall explore this in Section 3.10, where we derive the exact formula for transmission of electromagnetic waves between antennas. In this section, we will derive an approximate version of the transmission formula by taking into account quantitatively the relative phases of waves from different portions of the antenna aperture.

Figure 1.4 illustrates a paraboloidal microwave antenna, with a focal point feed. A microwave source at the focus of a paraboloidal reflector illuminates the reflector more or less uniformly. The microwaves are reflected as shown.

If we draw any line from the focus to the surface of the paraboloid and thence parallel to the axis (the line between the focus and apex), the distance to a plane $P-P'$ which is normal to the axis is always the same.

Thus the number of wavelengths that a ray travels from the focus to the plane P–P' is the same for all rays or paths. Over the whole aperture of parabolic dish (the part of P–P' within the diameter D) the microwaves from the source at the focus will be in phase.

According to diffraction theory, then we look back at the aperture from a distant point, the waves over the plane of the aperture constitute sources. If we move off the axis on a line from the center of the antenna which makes a small angle of θ radians with respect to the axis, the distances to the two edges of the dish differ approximately by a length

$$D \sin \theta \doteq D\theta \qquad (1.2\text{-}1)$$

Here D is the antenna diameter. When $D\theta$ is a half a wavelength, the signals from the two edges cancel because they are exactly out of phase, and the radio signal is substantially less than it is on the axis. Thus we can say roughly that the angular half-width or directivity of the radio beam sent out by the antenna is given by

$$D\theta = \lambda/2$$

$$\theta = \lambda/2D \qquad (1.2\text{-}2)$$

Here λ is the wavelength. Figure 1.5 shows an antenna pattern which displays this phenomenon. These patterns were taken simultaneously with an extragalactic point radio source (quasar) on a 64-m NASA deep space communication antenna using a dual-frequency feed. The crude equation (1.2-2) suggests beamwidths of 0.058° and 0.016°, respectively. The values which can be read off the figure are about 0.08° and 0.02°.

Assume that all the radiated energy lies in a cone of half-angle θ given by (1.2-2). Consider a receiving antenna of diameter D_R at a distance L from a transmitting antenna of diameter D_T. Here D_R is small enough so that the entire receiving aperture is in the cone, and so that the electric field is approximately constant over the receiving aperture. The receiving antenna accepts a fraction P_R/P_T of the radiated power which is the fraction of the area of the cone of radiation at distance L that the

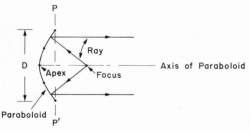

Figure 1.4. Paraboloidal antenna geometry.

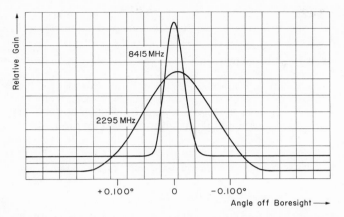

Figure 1.5. Antenna patterns at 13 cm (S-band) and 3.5 cm (X-band).

receiving antenna intercepts. This is just the ratio of the squares of the diameters of the two circles:

$$\frac{P_R}{P_T} = \frac{D_R{}^2}{[L(\lambda/D_T)]^2} \quad \text{(approximate)}$$

$$\frac{P_R}{P_T} = \frac{D_R{}^2 D_T{}^2}{L^2\lambda^2} \quad \text{(approximate)} \tag{1.2-3}$$

or

$$\frac{P_R}{P_T} = \frac{16 A_R A_T}{\pi^2 L^2 \lambda^2} \quad \text{(approximate)} \tag{1.2-4}$$

Here A_T and A_R are the areas, i.e., the areas of the apertures, of the circular transmitting and receiving antennas. For uniform in-plane illumination, the true formula does not depend on the fact that the cross section is circular, even though this crude derivation does. The exact answer will be given in the next section, where we introduce the concept of effective area. We will work out some of the implication of this in Section 1.5.

Problems

1. Calculate the beam half-widths in radians for the following cases:

	D	λ
(1)	10 ft	7.5 cm
(2)	200 ft	1 cm
(3)	15 ft	6000 Å

2. For the cases in Problem 1, calculate the beam half-diameters in feet and miles at the following distances: 20 miles, 20,000 miles, 200,000 miles, 100,000,000 miles, 4 light years.

3. A domestic satellite system for Japan will provide national coverage if its footprint on earth has an approximately elliptical shape properly oriented with major axis 2400 km and minor axis 800 km. A synchronous satellite at altitude 40,000 km is used, with a frequency of 4 GHz. What would you guess the appropriate shape, dimensions, and relative orientation of the satellite antenna should be to produce the desired pattern or footprint on earth?

1.3. The Transmission Formula and Effective Area

Equation (1.2-2) is not a measure of beam half-angle that leads to a correct constant in (1.2-4). An exact argument to be given in Section 3.10 leads to an expression called *Friis's transmission formula* or antenna equation:

$$\frac{P_R}{P_T} = \frac{A_T A_R}{\lambda^2 L^2} \tag{1.3-1}$$

This assumes that the antennas are *pointed at each other for maximum received power* and are far from each other. A_T and A_R are the *effective areas* of the transmitting and receiving antennas and λ is the wavelength.

If, when we use the antenna as a transmitting antenna, the illumination (field strength) is uniform over an antenna of geometrical area A, the effective area of the antenna is equal to the geometrical area A. This is shown in Section 3.10. What happens in the case of a nonuniform illumination?

Assume that the electric (or magnetic) field strength over the area of the antenna is

$$E(x, y)$$

The power P_T which flows out over the aperture is

$$P_T = (1/K) \iint E^2(x, y)\, dx\, dy$$

The integration is carried out over the area of the aperture. K is a constant called the *impedance of vacuum*:

$$K = 120\pi \doteq 377 \text{ ohms}$$

The electric field E_R at the aperture of the receiving antenna is the sum of contributions from all areas of the transmitting antenna

$$E_R = B \iint E(x, y)\, dx\, dy$$

Here B is a constant. The formula for received power must be analogous to that for transmitted power:

$$P_R = \frac{E_R^2 A_R}{K} = \frac{B^2}{K}\left(\iint E(x, y)\, dx\, dy\right)^2 A_R$$

We see that

$$\frac{P_R}{P_T} = B^2 A_R \frac{\left(\iint E(x, y)\, dx\, dy\right)^2}{\iint E^2(x, y)\, dx\, dy}$$

The mean or average field $\bar{E}$ at the transmitting antenna is

$$\bar{E} = \frac{1}{A_{TG}} \iint E(x, y)\, dx\, dy$$

where A_{TG} is the (actual) *geometrical* area of the transmitting antenna. The mean-square field $\overline{E^2}$ at the transmitting antenna is

$$\overline{E^2} = \frac{1}{A_{TG}} \iint E^2(x, y)\, dx\, dy$$

Hence

$$P_R/P_T = B^2 A_R A_{TG}[(\bar{E})^2/\overline{E^2}]$$

If the illumination over the transmitting antenna is of constant amplitude and in phase, we will have

$$(\bar{E})^2 = \overline{E^2}$$

This tells us what B^2 must be, and (1.3-1) holds with

$$B^2 = \frac{P_R}{P_T} \frac{1}{A_R A_T}$$

The formula for effective area to be used in (1.3-1) is then

$$A_T = A_{TG}[(\bar{E})^2/\overline{E^2}] \tag{1.3-2}$$

Duality between transmitting and receiving leads to the similar formula

$$A_R = A_{RG}[(\bar{E})^2/\overline{E^2}] \tag{1.3-3}$$

In the case of A_{RG}, the geometric area of the receiving antenna, we mean the field over the aperture when the antenna is used for transmitting.

We will note that

$$(\bar{E})^2/\overline{E^2} \leq 1 \tag{1.3-4}$$

This will be derived in Section 9.1 when we study random waveforms. Thus the effective area cannot be greater than the geometrical area. The ratio $(\bar{E})^2/\overline{E^2}$ is called the *aperture efficiency*.

Sometimes only a part of the power directed toward a parabolic reflector by a feed actually strikes the dish and is reflected to form a part of the antenna beam. When this is so, the aperture efficiency we should use in the transmission formula (1.3-1) is

$$\alpha[(\bar{E})^2/\overline{E^2}]$$

Here α is the fraction of the power from the feed that actually strikes the reflector.

We are also interested in the effective areas of antennas that are not simple reflectors. We may assume a receiving antenna of effective area A_R, and calculate the received power P_R. We will now define the effective area A_T of the nonreflector antenna so as to make (1.3-1) hold:

$$A_T = \frac{\lambda^2 L^2}{A_R} \frac{P_R}{P_T} \qquad (1.3\text{-}5)$$

To find P_R for given P_T we will need to find the electric field strength at the receiving antenna using elementary electromagnetic theory.

We may expect that the effective area of a given nonreflector antenna can depend on wavelength. This is not true of reflector antennas whose diameter is large compared with wavelength, if we do not vary the wavelength so much that surface accuracy effects become important.

The next section relates the tightness of the antenna beam to the hypothetical isotropic radiator as a reference standard.

Problems

1. A type of antenna called a horn reflector antenna has a nearly square aperture of width W. The field strength at the aperture is constant in one direction, which we shall call the y direction, and varies in the other, x, direction as

$$\cos(\pi x/W)$$

where x is the distance from the center line of the antenna. What is the aperture efficiency?

2. A microwave source pointed at a circular parabolic reflector produces a field strength which varies as

$$\exp(-r^2/a^2):$$

We can adjust a so that it is greater or less than $D/2$, where D is the diameter of the parabolic dish.

    (1) What fraction of the power from the source falls on the dish?
    (2) What is $(\bar{E})^2/\overline{E^2}$ for the power that does fall on the dish?
    (3) What is the ratio of the effective area of the antenna to its actual area?
    (4) What is the largest value of this ratio as a function of D/a, and for the largest value, what fraction of the energy falls on the dish?

3. What is the effective area of an isotropic source?

4. The field strength of a dipole antenna such as a small loop, with or without a ferrite rod through it, varies as

$$(1/r)\cos\phi:$$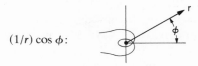

Here ϕ is angle with respect to the plane of the loop. What is the effective area of a dipole? (Assume the receiving antenna is located to maximize received power.)

5. Derive (1.3-1) for arbitrary reflector antennas which have all dimensions much larger than the wavelength, knowing it for circular apertures. We are to show that for uniform illumination, effective area is the ordinary geometric area.

1.4. Gain above Isotropic Radiator

We can only approximate isotropic radiation of electromagnetic waves. Nonetheless, the directivity of an antenna is sometimes specified in terms of gain (in power received from the antenna) over the power that would be received if the antenna radiated isotropically.

If we used a receiving antenna of effective area A_R, the fraction of the power received from an isotropic antenna at a distance L would be

$$P_R/P_T = A_R/4\pi L^2 \tag{1.4-1}$$

We have from the last section that for a transmitting antenna of effective area A

$$P_R/P_T = A_R A/\lambda^2 L^2 \tag{1.3-1}$$

Thus the gain G of an antenna of effective area A must be

$$G = 4\pi A/\lambda^2 \tag{1.4-2}$$

In dB the gain is

$$10\log_{10} G \text{ dB} \tag{1.4-3}$$

If we have a transmitting antenna of gain G_T and a receiving antenna of gain G_R separated by a distance L, the ratio of received power to transmitted power will be

$$\frac{P_R}{P_T} = \frac{G_T G_R \lambda^2}{16\pi^2 L^2} \qquad (1.4\text{-}4)$$

The idea of the gain of an antenna with respect to a hypothetical isotropic antenna is quite natural in dealing with antennas of rather low directivity such as those used in the early days of radio. When we use such antennas, the received power does, indeed, increase with wavelength of operation, as indicated by equation (1.4-4).

Antenna gain also provides a way of dealing with directive array antennas such as those used in VHF television reception; Section 7.8 discusses arrays.

There is a serious problem in thinking only in terms of gain in the case of paraboloidal reflectors or other antennas which have a well-defined geometrical aperture. When the geometrical areas of the transmitting and receiving antennas are held constant, the received power varies roughly (because efficiency changes somewhat with λ) as $1/\lambda^2$, in accord with (1.3-1), rather than with λ^2, in apparent accord with (1.4-4). The reason is, of course, that the directivity of an antenna of constant aperture increases as the wavelength is decreased (or as the frequency is increased). For a fixed antenna construction cost, then, it may pay to choose as *high* a frequency as technology will permit. We shall study this in Section 12.6 in connection with the photon channel.

But, antenna gain seems to be permanently enshrined in a figure of merit for communication satellite ground stations, which is the gain divided by noise temperature:

$$\text{figure of merit} = G/T_n \qquad (1.4\text{-}5)$$

Here G is the power gain (as a factor, not dB) of the receiving antenna, and T_n is the noise temperature of the receiving system (see Section 2.2). We will note here that the noise power in a receiver of specified bandwidth and gain is proportional to T_n, which is a measure of the noisiness of the receiver.

We find the effective area corresponding to a cone of constant-density radiation in the next section.

Problems

1. An antenna has a diameter of 3 ft and an aperture efficiency of 0.65. What is the gain at wavelengths of 1 cm, 3 cm, 7.5 cm, 15 cm, and 6000 Å?

2. What is the antenna gain of a dipole? (See Problems 3 and 4 of Section 1.3.)

3. Find the effect on received power of doubling the communication frequency but keeping the transmitted power the same in each of the four cases below. Assume that the efficiency of the reflector antennas is independent of frequency over the octave range in question. The term "omni antenna" refers to an omnidirectional or perfect isotropic radiator.
   (a) Reflector antenna transmitting and reflector antenna receiving.
   (b) Reflector antenna transmitting and omni antenna receiving.
   (c) Omni antenna transmitting and reflector antenna receiving.
   (d) Omni antenna transmitting and omni antenna receiving.

1.5. Effective Area for a Cone of Constant-Density Radiation

Section 1.2 gave a preliminary discussion of antenna directivity. There are many possible definitions of directivity. We might use the half-angle at which the square of the field (proportional to power density) is $\frac{1}{2}$ that on the axis. Or, we might use the half-angle of a cone containing half the radiated power. We can define a different sort of directivity that is easy to calculate: the half-angle of a cone that would contain all the power radiated if the field were zero outside of the cone and were equally strong for all angles within the cone.

First we calculate the solid angle ψ of a cone of half-angle θ:

$$\psi = \frac{1}{r^2} \int_0^\theta (2\pi r \sin\phi)(r\,d\phi) = \frac{1}{r^2} 2\pi r^2 (-\cos\phi)\Big|_0^\theta = 2\pi(1 - \cos\theta)$$

or

$$\psi = 4\pi \sin^2(\theta/2) \text{ steradians} \qquad (1.5\text{-}1)$$

When $\theta = \pi$, $\psi = 2\pi$, as it should.

If all the power transmitted had a constant density per steradian in a cone of half-angle θ, what would the angle θ be? The received power is

$$P_R = \frac{A_T A_R P_T}{\lambda^2 L^2} = \frac{A_R P_T}{L^2[4\pi \sin^2(\theta/2)]}$$

Here A_T is the effective area of a very special antenna whose radiation pattern is constant power density within a cone of half-angle θ. We see that

$$\sin^2\frac{\theta}{2} = \frac{\lambda^2}{4\pi A_T}$$

$$\theta = 2\sin^{-1}\frac{\lambda}{2\pi^{\frac{1}{2}}(A_T)^{\frac{1}{2}}} \qquad (1.5\text{-}2)$$

We can also solve for A_T:

$$A_T = \frac{\lambda^2}{4\pi \sin^2(\theta/2)} \qquad (1.5\text{-}3)$$

This gives the effective area corresponding to a cone of half-angle θ of constant-radiation density. When $\theta = \pi$ (a complete sphere), we obtain, as we should, the area of an isotropic radiator, which we worked out in Problem 3 of Section 1.3:

$$A_T = \frac{\lambda^2}{4\pi}$$

An isotropic radiator does not exist for electromagnetic waves. Nonetheless, it is frequently used as a standard, and, as we noted in Section 1.4, antenna gain is measured with respect to an isotropic radiator.

For a circular antenna of diameter D,

$$A_T = \frac{\pi D^2}{4} \frac{(\bar{E})^2}{\overline{E^2}}$$

$$\theta = 2 \sin^{-1}\left[\frac{\lambda}{\pi D}\left(\frac{(\bar{E})^2}{\overline{E^2}}\right)^{-\frac{1}{2}} \right] \tag{1.5-4}$$

If the angle is very small

$$\theta \doteq \frac{2\lambda}{\pi D}\left(\frac{(\bar{E})^2}{\overline{E^2}}\right)^{-\frac{1}{2}} \tag{1.5-5}$$

We compare this with the expression in Section 1.2:

$$\theta = \lambda/2D \tag{1.2-2}$$

We note that in (1.5-4) and (1.5-5) we have used $(\bar{E})^2/\overline{E^2}$ rather than the overall efficiency. The directivity depends on the power actually radiated from the dish, disregarding transmitter power which does not reach the dish but misses it.

So far we have assumed distant antennas. The next section considers how far from an antenna we must be to make this assumption valid.

Problems

1. What is the ratio of the angle given by (1.2-2) to the angle given by (1.5-5)?

2. Let $\lambda = 10$ cm and $D = 3$ m. Let $(\bar{E})^2/\overline{E^2} = 0.7$. What is θ in radians?

1.6. Near and Far Fields

The transmission formula of Section 1.3,

$$P_R/P_T = A_T A_R/\lambda^2 L^2 \tag{1.3-1}$$

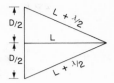

Figure 1.6. Path lengths differing by half a wavelength.

holds when the antennas are "far" apart. How far? Here there are two considerations.

One is that a parabola focuses or images a microwave source at infinity. The aperture of the antenna is illuminated all in phase. If we look at the aperture from too short a distance, the edges are appreciably farther away than the center is. When the edges are a half-wavelength farther away than the center is, we are in real trouble.

Figure 1.6 casts some light on this. We see that this real trouble occurs when

$$(L + \lambda/2)^2 = (D/2)^2 + L^2$$

$$L^2 + L\lambda + \lambda^2/4 = D^2/4 + L^2$$

$$L = (D^2 - \lambda^2)/4\lambda$$

Usually, $D^2 \gg \lambda^2$ and we can write

$$L = D^2/4\lambda \qquad (1.6\text{-}1)$$

If we are to measure the effective area or the gain of an antenna by sending a signal to it from a distant source, the source must be several times the distance given by (1.6-1). Very large antennas are sometimes calibrated using point source radio stars of known strength, such as quasars, which are certainly far enough away.

Let us turn to another but secondary aspect of (1.3-1). If our calculations show that P_R/P_T is greater than unity, something is certainly wrong with the formula, but not necessarily with the system. When the value of P_R/P_T as computed from (1.3-1) is unity or greater, we may actually be receiving a large fraction of the transmitted power. Rudolf Kompfer has called a pair of such antennas a "Hertzian cable." It is a wireless device for transmitting power with a very low loss. It has, in fact, been proposed that antennas for which (1.3-1) gives a value of P_R/P_T greater than unity be used to transmit electric power, generated by solar radiation, from a satellite to the earth using microwave frequencies.

The concave mirrors at the ends of a gas laser likewise form an antenna system for which P_R/P_T as computed by (1.3-1) is much greater than unity. These mirrors trap the laser radiation in a beam between the mirrors.

So far we have been looking at the situation in which one antenna is transmitting and another is receiving. In the next section, we study the situation in which the goal is not communication from one antenna to another, but rather the probing of a distant object by means of electromagnetic waves reflected from it. This is called radar.

Problems

1. Assume that in making an accurate measurement of an antenna the source must be far enough away so that the distance from the source to an edge of the aperture plane is only $\frac{1}{8}$ wavelength longer than the distance from the source to the center of the aperture plane. What must the distance of the source be for the following cases in kilometers and miles?

Diameter	meters:	3	100	30	3
	feet:	9.84	328	98.4	9.84
Wavelength, cm:		7.5	7.5	1	6×10^{-7} (6000 Å)

2. Assume that the antennas of a hertzian cable have diameters of 1 m. What is the largest allowable separation in meters for wavelengths of 10 cm, 1 cm, 1 mm, 6000 Å? HINT: pretend (1.3-1) holds.

1.7. Passive Reflection—The Radar Equation

In a typical radar we send out a short microwave pulse and measure the time and the energy of the reflection from some distant object. The time T which elapses between the sending of the pulse and the return of the reflection is related to the distance L of the object by

$$T = 2L/c$$

Here c is the velocity of light.

What about the power or energy of the received pulse? In some instances this pulse represents the energy reflected from a single object, such as an airplane or missile we are trying to *track*. In other instances the returned pulse comes from an area of diffusely reflecting terrain (larger than the antenna spot size or footprint) that we are trying to *map*. Expressions for the ratio of received to transmitted power are quite different for the two different cases.

Let us first consider the simpler case, that of the power reflected by a single object.

Let the effective area of the transmitting antenna be A. Let it be aimed at an object a distance L away. Let the cross-sectional area of this object be A_0. Then according to our transmission formula, the fraction of the transmitted power which is intercepted by the object is

$$\frac{AA_0}{\lambda^2 L^2}$$

Here λ is the wavelength of the radar. Now suppose that the object reflects the incident power equally in all directions—that is, scatters it isotropically. The fraction of the scattered power which will be received by the transmitting antenna, now used as a receiving antenna (this is a "monostatic" radar), will be, from (1.4-1),

$$\frac{A}{4\pi L^2} \tag{1.7-1}$$

Thus the ratio of the received power to the transmitted power will be

$$\frac{P_R}{P_T} = \frac{A^2 A_0}{4\pi\lambda^2 L^4} \tag{1.7-2}$$

Equation (1.7-2) is the *radar equation* for reflections from a single object. We should note how rapidly the received power increases with the effective area of the antenna (as A^2) and especially how rapidly the received power decreases with the distance (as $1/L^4$).

Of course, not all objects scatter isotropically. A smooth reflecting sphere does. A corner reflector reflects strongly toward the transmitter. A wedge or cone pointed at the transmitter reflects very little back to the transmitter. Sometimes we want a large reflection; a corner reflector is good for this. Sometimes, as in the case of the Echo passive communication satellite, we want a reflection in many directions; a sphere was good for this. Sometimes we want to avoid reflecting energy back to the radar, as in the case of escaping radar detection. Reference 3 discusses these basics of radar, especially Chapters 1–3.

Let us now consider the more complicated case in which a radar antenna is pointed at a diffuse (rough) reflecting surface which is much larger than the radar beam. All areas of a perfectly diffuse surface by definition appear equally bright from each point of view, no matter what the direction of illumination. In this case, *all* the transmitter power reaches the surface, and *some* of the reflected or scattered power is picked up by the antenna. Because the surface is rough, reflections from various parts of the surface will arrive at the receiver in different random and independent phases. As a consequence, the *powers* reflected from different areas will add. This will become more rigorous in Section 7.7 (when we discuss the phenomenon of speckle) and in Section 8.4 (when we discuss the variance of a random variable).

If the beam of incident power P_i strikes normal to a totally reflecting totally diffuse surface, the amount of power ΔP_r reflected back in a small solid angle of $\Delta\psi$ steradians around the axis of the beam is

$$\Delta P_r = \frac{\Delta\psi}{\pi} P_i \qquad (1.7\text{-}3)$$

The expression comes from the conservation of power, which is scattered back from one side only of the surface (none goes through it). The derivation (1.7-3) is not completely routine. We shall assume it here.

   If the diffusely reflecting surface absorbs some of the incident power, and if the radar beam is directed at the surface from an angle θ with respect to the normal, we can write the power reflected back in a narrow angle $\Delta\psi$ around the axis of the radar beam as

$$\Delta P_r = R(\theta)\frac{\Delta\psi}{\pi} P_i \qquad (1.7\text{-}4)$$

Here $R(\theta)$ is a diffuse reflection coefficient or backscatter function. It is identically equal to 1 for a perfectly diffuse surface.

   We will assume that the radar beam is reasonably narrow so that the distance L to any part of the surface that it strikes is a constant. The beam makes an angle θ with respect to the normal to the surface, and this angle is constant for all the illuminated points.

   At the reflecting surface we will measure distance normal to the axis of the radar beam by distances x and y; they are not distances *on* the surface (unless the beam is normal to the surface). We will express the power density p in the beam as

$$p = p_0 f((x/L), (y/L)) \qquad (1.7\text{-}5)$$

Here p_0 is the power density on the axis of the beam and hence

$$f(0, 0) = 1 \qquad (1.7\text{-}6)$$

We will assume that the beam is weaker off axis.

   How much power is incident on a little area $dx\,dy$ at (x, y)? An amount

$$p_0 L^2 f(u, v)\,du\,dv \qquad (1.7\text{-}7)$$

Here

$$u = x/L, \qquad v = y/L$$
$$dx = L\,du, \qquad dy = L\,dv \qquad (1.7\text{-}8)$$

The total transmitted power P_T must be

$$P_T = p_0 L^2 \int_{-\infty}^{\infty} \int_{-\infty}^{\infty} f(u, v)\,du\,dv \qquad (1.7\text{-}9)$$

   Equation (1.7-4) can be used to find the fraction of this power that will be reflected and picked up by the antenna of area A acting as a receiving antenna. At a point (x, y) off axis, the effective area of the antenna as a receiving antenna will be reduced by a factor $f(u, v)$. Thus the effective solid angle, $\Delta\psi$, of the receiving antenna for a source at a distance L will be

$$\Delta\psi = \frac{Af(u, v)}{L^2} \tag{1.7-10}$$

From (1.7-4), (1.7-7), and (1.7-10) we see that the total received power over the incident beam

$$P_R = \int \Delta P_r$$

will be

$$P_R = \frac{R(\theta)Ap_0}{\pi} \int_{-\infty}^{\infty} \int_{-\infty}^{\infty} [f(u, v)]^2 \, du \, dv \tag{1.7-11}$$

This is because powers from different areas add.

From (1.7-7) and (1.7-11) we see that

$$\frac{P_R}{P_T} = \frac{A}{\pi L^2} R(\theta)F \tag{1.7-12}$$

$$F = \frac{\displaystyle\int_{-\infty}^{\infty} \int_{-\infty}^{\infty} [f(u, v)]^2 \, du \, dv}{\displaystyle\int_{-\infty}^{\infty} \int_{-\infty}^{\infty} f(u, v) \, du \, dv} \tag{1.7-13}$$

Thus, the received power depends only on the effective area A of the radar antenna, the distance L, the reflection coefficient $R(\theta)$, which is a function of the angle of the beam with respect to the normal to the surface, and the factor F, which depends only on the antenna pattern.

Further, the received power depends on A/L^2 rather on A^2/L^4 as in (1.7-2), and in all real cases

$$A/L^2 \ll 1$$

Thus, the return from a large diffuse surface will be much greater than that from a single small object. This is not surprising.

The factor F in (1.7-12) depends on the shape of the antenna pattern. If $f(u, v)$ were unity within some given area and zero outside, F would be unity. For any nonuniform power distribution in the radar beam in which the power density decreases away from the axis, F will be less than unity. We can see this by noting that $f(u, v)$ has been assumed to be unity on axis and less than unity off axis. Thus, off axis the square of the function is everywhere less than the function.

As an example, let us assume that

$$f(u, v) = \exp(-u^2) \exp(-v^2) = \exp[-(u^2 + v^2)] \quad (1.7\text{-}14)$$

If we let

$$r^2 = u^2 + v^2 \quad (1.7\text{-}15)$$

then the appropriate unit of area is

$$2\pi r \, dr \quad (1.7\text{-}16)$$

and

$$F = \int_0^\infty 2r \exp(-2r^2) \, dr \Big/ \int_0^\infty 2r \exp(-r^2) \, dr$$

$$F = 1/2 \quad (1.7\text{-}17)$$

The last few sections have been space oriented. Now let us turn to earthbound systems. In the next section, we study what the effect on communication may be if the medium is not free space but rather is an absorber of electromagnetic radiation.

Problems

1. A spherical satellite has a diameter of 100 ft and is 1500 miles from the transmitting and receiving antennas. Each of these has a diameter of 100 ft and an aperture efficiency of 70%. The wavelength is 7.5 cm. What is the path loss in dB?

2. In passive satellite communication such as the Echo balloon, the transmitting and receiving antennas of effective areas A_T and A_R are separate and at distances L_T and L_R from the satellite or target of area A_0. Generalize the radar equation for such a system. (This is called a "bistatic" radar.)

3. A distant corner reflector of apparent area A_0 acts as a transmitting antenna of area A_0 pointed back at the radar antenna. What is the radar equation for this case?

4. What is the ratio of energy received from a corner reflector to energy received from an isotropic scatterer? How large is this ratio in dB for the dimensions given in Problem 1?

5. Equation (1.7-3) comes from the following considerations: (a) at any angle from the surface, any part of a diffusely reflecting surface seems equally bright, that is, emits the same power per unit solid angle, but (b) when seen slantwise, any small area of the surface appears smaller than when looked at normal to the surface. Derive (1.7-3) from these assumptions.

6. Give a heuristic justification of the fact that received power depends on $1/L^2$ (rather than on $1/L^4$) in equation (1.7-12) for the received power in a radar illuminating a large diffuse area. Why is the received power proportional to antenna area A rather than A^2?

7. A perfectly diffuse sphere is very distant from a radar, so that it fills only a small part of the beam. Show that the received power is $1/\pi$ times what would be received from an isotropically scattering sphere of the same radius, as given by (1.7-2). (We say that it has *radar cross section* $1/\pi$ times as great.) Give an explanation of why the radar return might be less. Where roughly might we put the receiving antenna of a *bistatic* radar to maximize the received power?

1.8. Earthbound Systems

So far we have considered microwave propagation through free space. By free space, we mean space that is really free—of rain or gases that may absorb microwaves, of layers of different atmospheric density that may reflect or refract radio waves, or of hills or trees that may stop or absorb microwaves. In such free space, the transmission formula, (1.3-1), holds.

But, there are many earthbound microwave systems. Some consist of a single hop, as from a TV studio to a mountain-top transmitter. Others span the country. A signal is received, amplified, and retransmitted by microwave *repeaters* 100 times in crossing the continent.

Microwave transmitting and receiving antennas near the earth's surface must be in line of sight of one another. Over fairly level terrain, microwave antennas are put on towers on hilltops. The towers may be around 100 ft tall and spaced about 30 miles apart. In mountainous regions, somewhat longer paths (around 50 miles) may be used.

When the ground is very smooth, a wave reflected from the ground may reach a receiving antenna out of phase with the transmitted wave and thus cause *fading*. This happens particularly in very flat desert areas. One cure is to put one antenna (transmitting or receiving) close to the ground and the other on a high tower, so that the reflected and direct paths do not differ much in length or position and remain stably in phase despite fluctuations in atmospheric density. Other ways of coping with fading will be discussed in Section 7.7. Reference 4 contains a good discussion of propagation and fading in various propagation media that earthbound transmission systems must traverse.

Sometimes refraction by layers in the atmosphere creates two paths between a transmitting and receiving antenna, and the signals via the two paths may be out of phase. This may cause fades of several tens of dB. The remedy is to switch to a different frequency (*frequency diversity*) or to an antenna at a different height on the tower (*spatial diversity*).

Fading is generally worse over long paths than over short paths. Figure 1.7, taken from Reference 4, shows the median duration of fast fades as a function of depth of fade for 4 GHz and a path length of 30–35 miles. In designing earthbound microwave systems a fading margin of around

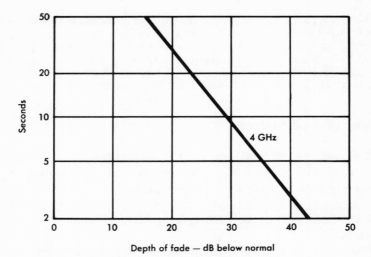

Figure 1.7. Median duration of fast fading.

30 dB is commonly allowed on each hop. But, in a long system with many hops it is uncommon for several paths to fade deeply at once.

The absorption of microwaves by rain increases with increasing frequency. It is small (in dB/km) at 4 GHz and large at 18 GHz, as we can see from Fig. 1.8, also taken from Ref. 4. At 4 GHz fading dominates. At 18 GHz attenuation due to rain dominates, and repeaters (which pick the signal up, amplify it and send it on) must be spaced more closely.

The transmission formula tells us that the ratio of received power to transmitted power is proportional to $1/L^2$, where L is the distance between the transmitting antenna and the receiving antenna. Thus the path loss in dB is

$$\text{path loss} = K + 20 \log_{10} L \text{ dB} \qquad (1.8\text{-}1)$$

Here K is a constant which depends on antenna size and on wavelength. Equation (1.8-1) should be contrasted with the path loss or attenuation for transmission systems in which a signal is sent over a transmission line such as a pair of wires or a coaxial cable, through a waveguide, or as light through an optical fiber. In these cases

$$\text{path loss} = aL \text{ dB} = A \text{ dB} \qquad (1.8\text{-}2)$$

Here a is an attenuation constant that gives the attenuation in dB per unit length, and the total attenuation A is simply this constant a times the length L of the cable, fiber, or waveguide.

The difference between attenuation or path loss for the microwave system (1.8-1) and attenuation for the transmission-line system (1.8-2)

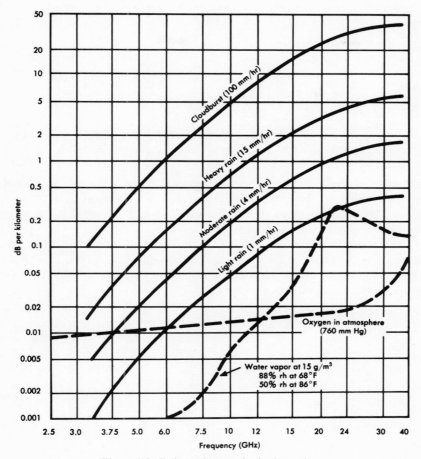

Figure 1.8. Estimated atmospheric absorption.

can be best appreciated through an example. Assume that in each case the path loss for a given distance (say, 10 miles) is 60 dB. If we double the distance, in the case of the microwave system the path loss will become 66 dB, while in the case of the transmission line system the path loss will become 120 dB. We shall explore the implications of this phenomenon for jamming and snooping in Section 12.7.

In the microwave case increasing the power 6 dB or four times makes it possible (in free space) to receive the same signal strength twice as far away, and another 6 dB will double that distance again. In the transmission line case increasing the power does not increase the distance we can send a signal very much. Hence transmission line systems tend to use low powers and close repeater spacings. In coaxial cable systems and in

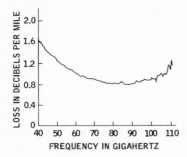

Figure 1.9. Attenuation of millimeter waveguides [from *Bell Lab. Rec.* **53**(10), 406 (1975)].

submarine cables the power for the repeaters is sent over the cable itself and this tends to limit the power that can be supplied and to favor low-power repeaters more closely spaced. The *total* power that has to be supplied over the cable will then be less.

In transmission lines such as wire pairs and coaxial cables the attenuation constant *a* increases as the square root of the frequency. Hence the attenuation is greater for broadband signals such as TV than for narrowband signals such as voice.

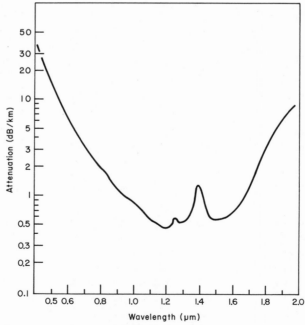

Figure 1.10. Attenuation for best optical fibers [from Olshansky, R., "Propagation in Glass Optical Waveguides," *Rev. Mod. Phys.* **51**(2), 350 (1979)].

In waveguides and optical fibers, a small percentage range of high frequencies (microwave or optical) is used in sending a signal, so in effect, the attenuation constant does not vary with signal bandwidth. This is, of course, the same as in microwave radio systems.

Figures 1.9 and 1.10 give typical data on attenuation of millimeter waveguides and optical fibers. We note that the attenuation of the best optical fibers has been dropping rapidly in recent years, especially for wavelengths around 1 micron.

This first chapter has studied the propagation of electromagnetic signals so that received signal power can be determined. This determines the most important limitation on communication rate. The next chapter studies the second most important limitation on communication rate, noise.

Problems

1. For a path length of 30 miles, how high must two towers of equal height be to just give line-of-sight transmission over a smooth earth? What if one antenna is at zero height?

2. A transmission system using a cable (as, a submarine cable) has a transmitter at one end, a receiver at the other end, and repeaters equally spaced in between, as shown in the diagram. Each repeater consists of a receiver R connected to a

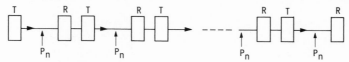

transmitter T. Each receiver adds noise equivalent to a power P_n added to the signal at the receiver input. We assume that the gain in the receiver makes up for the loss in a cable section, so that the signal power is the same at each receiver input. Given an attenuation of A dB for the whole cable and a total power available for all transmitters of P_0 W (watts), what is the ratio of signal power to noise power in dB at the end of the cable if the cable is divided into N sections (transmitter, $N - 1$ repeaters and receiver)? Assume that the noise NP_n is small compared with the signal, P_0/N, so that we can ignore the effect of noise power decreasing the signal power. This is the case in practice.

3. The L5 coaxial cable system can carry 10,800 telephone channels per coaxial pipe for a distance of 4000 miles. The overall attenuation in 4000 miles is about 130,000 dB. Assume a total power P_0 for transmitters of 4000 W (P_0/N per transmitter) and an equivalent noise power at the input of each receiver of 10^{-12} W. What would the signal-to-noise ratio be for repeater spacings of 2, 1, and $\frac{1}{2}$ miles (2000, 4000, 8000 repeaters)?

4. The parameters of this problem are the same as for Problem 3, except that P_0 is not given; but the required signal-to-noise ratio *is* given and is 50 dB. What value of N minimizes P_0? What is this minimum P_0? How does N depend on the specified signal-to-noise ratio?

Sources of Noise

In a microwave transmission system, a *signal* is sent from a transmitting antenna to a receiving antenna. We have seen that only a very small part of the transmitted power is picked up by the receiving antenna. The receiving antenna also picks up as *noise* random electromagnetic radiation from any body that lies within its beamwidth, and some noise from interstellar space, which is filled with electromagnetic radiation. Thus the electromagnetic wave which the receiving antenna feeds to the receiver for amplification and detection is a mixture of signal and noise. The receiver adds further noise, some of it unavoidable, in the process of amplification, so that the amplified signal, prior to detection or demodulation, has some carrier-to-noise power ratio, P_C/P_N, or noise-to-carrier power ratio, P_N/P_C. We will study these phenomena for microwave and optical communication systems in this chapter.

2.1. Johnson or Thermal Noise

The thermal agitation of charges in any conducting or lossy body produces fluctuating currents and voltages. Nyquist evaluated these by considering the average energy in a resonator which is in equilibrium with an environment of temperature T. We will follow in his footsteps. See Chapter 4 of Reference 1 or Chapter 4 of Reference 2 for more.

According to Boltzmann, the probability P that a mode of oscillation will have an energy E is

$$P = Ae^{-E/kT} \tag{2.1-1}$$

We will study probability more formally in Chapter 8. In (2.1-1), A is a constant, and k is Boltzmann's constant:

$$k = 1.380 \times 10^{-23} \, \text{J}/^{\circ}\text{K}$$

(Here we use J/°K to stand for joules/degree kelvin.)

According to quantum mechanics, an electromagnetic mode of frequency f can have $0, 1, \ldots, n, \ldots$ photons in it. Reference 2 of Chapter 1 gives the basics of modes of oscillation. For n photons the energy will be

$$E_n = nhf \qquad (2.1\text{-}2)$$

Planck's constant, h, has a value

$$h = 6.62 \times 10^{-34} \, \text{J sec (joule seconds)}$$

The sum over all n of the probabilities $P(n)$ that a mode will have energy nhf or will have n photons in it must add up to unity. Hence

$$1 = \sum_{n=0}^{\infty} P(n) = A \sum_{n=0}^{\infty} e^{-nhf/kT}$$

We can see by multiplying the left-hand side of the equation below term by term by the denominator of the right-hand side that

$$\sum_{n=0}^{\infty} e^{-nhf/kT} = \frac{1}{1 - e^{-hf/kT}}$$

Hence

$$A = 1 - e^{-hf/kT}$$

The average energy E in a mode will be the sum of the energies weighted by probability of occurrence:

$$E = A \sum_{n=0}^{\infty} nhf e^{-nhf/kT}$$

In Section 8.4 we will study averages or expectations in more detail.

We see that E can be represented as

$$E = -A \frac{d \sum_{n=0}^{\infty} e^{-nhf/kT}}{d(1/kT)}$$

Thus we can use the sum we have already found:

$$E = -A \frac{d[1/(1 - e^{-hf/kT})]}{d(1/kT)}$$

$$= (1 - e^{-hf/kT}) \frac{hf e^{-hf/kT}}{(1 - e^{-hf/kT})^2}$$

$$E = \frac{hf}{e^{hf/kT} - 1} \qquad (2.1\text{-}3)$$

When $hf/kT \ll 1$, the average energy of an electromagnetic mode of frequency f and temperature T is independent of frequency and is

$$E \doteq kT \qquad (2.1\text{-}4)$$

We will now derive Johnson noise. Imagine a very long electrical transmission line of length L on which waves of any frequency travel with a velocity c. Imagine that the line is shorted at each end. The line will have resonant modes having frequencies

$$f = \frac{mc}{2L}$$

where m is an integer. We see that the frequency interval between one mode and the next is $c/2L$. Thus there are $2L/c$ modes per unit frequency.

If these modes are in thermal equilibrium with an environment at a temperature T, each will have an average energy given by (2.1-3). Half of this energy can be regarded as energy traveling to the left. At the end of the line this energy will be reflected and will travel to the right. Half of the energy will travel to the right, and on reflection will travel to the left.

The energy density traveling to the right is $1/L$ times the energy traveling to the right. The power traveling to the right is c times the energy density traveling to the right. Hence the power ΔP_n directed to the right and consisting of waves lying in the frequency range Δf is

$$\Delta P_n = \frac{1}{2} \frac{2L}{c} \frac{1}{L} c \cdot E \cdot \Delta f$$

From (2.1-3), this becomes

$$\Delta P_n = \frac{hf}{e^{hf/kT} - 1} \Delta f \qquad (2.1\text{-}5)$$

When $hf/kT \ll 1$, (2.1-5) is approximately

$$\Delta P_n \doteq kT\Delta f \qquad (2.1\text{-}6)$$

If the transmission line is terminated by resistances equal to its characteristic impedance at each end, (2.1-5) will hold, but the noise power traveling to the right will come from the terminating resistance on the left and will be absorbed by the terminating resistance on the right, and the power traveling to the left will come from the terminating resistance on the right and be absorbed by the terminating resistance on the left. Equation (2.1-5) gives the *available noise power* from any resistor at temperature T.

We get the maximum power from a power source of internal resistance R by attaching a load of resistance R. The voltage falls to half the open circuit voltage V and the available power is thus $V^2/4R$. Thus from (2.1-5) the mean-square open-circuit noise voltage $\overline{\Delta V^2}$ across a resistor of resistance R must be

$$\overline{\Delta V^2} = 4 \frac{hf}{e^{hf/kT} - 1} R\Delta f \tag{2.1-7}$$

When $hf/kT \ll 1$,

$$\overline{\Delta V^2} \doteq 4kTR\Delta f \tag{2.1-8}$$

Equations (2.1-5) and (2.1-7) [or (2.1-6) and (2.1-8)] apply to the radiation resistance of an antenna which sees an environment of uniform temperature T. This tells us, as we already saw in Section 1.4, that a highly directive antenna must have more gain than a less directive antenna, because the highly directive antenna derives from a small part of the environment the same noise power that a less directive antenna derives from a larger part of the environment.

In the next section, we will see how this noise is budgeted for in communication systems.

Problems

1. Compute the temperatures at which $hf = kT$ for the following wavelengths: 10 cm, 1 cm, 1 mm, 6000 Å.

2. For a temperature of 293°K ("room temperature"), at low frequencies, what is the power level of Johnson noise in dB W for bandwidths of 10 Hz, 3000 Hz (telephone), and 4 million Hz (television)?

3. Compute the total power for all frequencies according to (2.1-5). (It may help to know that $\int_0^1 [\ln u/(1 - u)]\, du = -\pi^2/6$.)

4. Give approximate expressions for the fraction of the total power below the frequency f, assuming (a) that $hf/kT \ll 1$, and (b) that $hf/kT \gg 1$.

5. For $T = 293°K$ and $f = 10^7$ Hz, what fraction of the noise power lies below f in frequency?

2.2. Noise Temperature and Noise Factor

In defining noise temperature and noise factor, it is assumed that $hf \ll kT$, so that the noise power output P_n of a receiver whose antenna is pointed at an absorbing body of temperature T is

$$P_n = kTGB \tag{2.2-1}$$

Here GB is the receiver gain-bandwidth product. If the power gain G is a function $G(f)$ of frequency, then (2.1-6) yields a definition of gain-bandwidth product when the gain is not constant:

$$GB = \int_0^\infty G(f)\, df \qquad (2.2-2)$$

Now, imagine two receivers with the same gain-bandwidth product GB. The first receiver is the actual receiver. Its antenna is pointed at an absorbing body of zero temperature. The noise output P_n is entirely a result of noise generated in the receiver.

The second receiver is a noiseless receiver with the same gain-bandwidth product as the first. Its antenna is pointed at an absorbing body of temperature T_n. T_n is adjusted so that the noise power output P_n is exactly the same as the noise power output of the first receiver. T_n is then the noise temperature of the first receiver.

The noise temperature T_n of a receiver is the temperature of the source which would account for the observed noise output if all the noise came from that hot source instead of from the receiver. *It follows that if the receiver were connected to a source at 0°K, T_n is the temperature to which we should have to raise the source in order to double the noise output power.*

Noise figure or noise factor NF is defined as

$$NF = \frac{T_n}{290} + 1 \qquad (2.2-3)$$

This is often expressed in dB.

If we had a receiver of noise factor 1 or 0 dB or a noise temperature of 0°K connected to a source at 290°K (about "room temperature"), the noise would just be amplified room temperature noise. The noise factor of a typical spacecraft receiver for the 2–6 GHz region of the microwave spectrum is about 2–5, or 3–7 dB.

The term "noise factor" is never used in describing the very-low-noise receivers used in radio astronomy and deep space systems. "Noise temperature" is used instead. For sensitive deep space microwave receiving systems, the lowest system noise temperature is around 11°K.

In the next section, we will begin to see how noise affects the "bottom line" of a communication system, which is the rate of transmission of data or information.

Problems

1. The noise temperature of a receiver is 100°K. It is fed by a 100°K source. By what factor will the noise output power be increased if the temperature of the source is increased by 10°K?

2. A receiver is switched between a source at a temperature of 293°K and a source at a temperature of 500°K. The noise output is increased by a factor of 1.5. What is the noise temperature of the receiver?

3. An antenna of 70-dB gain (see Section 1.4) and 20°K system noise temperature when looking into free space looks at the planet Venus when it is 42 million km away. Venus has a diameter of 12,400 km and a temperature of 600°K. What is the system noise temperature now?

4. A satellite receiver has a tunnel-diode preamplifier with a noise figure of 3 dB and a power gain of 10 dB, followed by a FET (field effect transistor) amplifier with a noise figure of 6 dB and a gain of 20 dB. What is the receiver noise temperature? What is the overall receiver noise figure in dB? How does this depend on the gain of the FET stage?

2.3. Noise-Limited Free-Space Transmission System

In a free-space radio transmission system, what we *want* is transmission with some specified signal-to-noise power ratio S/N and some bandwidth B over some distance L. What we have to *pay* for this is some transmitter power P_T, some receiving antenna area A_R, and some transmitting antenna area A_T. The "cost" can be expressed by the product $P_T A_R A_T$. This cost depends not only on S/N and L, but on wavelength λ (or frequency f), and on the temperature T which specifies the unavoidable noise present in space (cosmic microwave background radiation, about 3°K).

The cost will be greater if the receiver is imperfect. Here we assume the receiver to be an ideal amplifier (see Section 2.9) of power gain G.

If the transmitted power is P_T, the received power P_R will be

$$P_R = \frac{P_T A_R A_T}{\lambda^2 L^2} \tag{1.3-1}$$

The carrier power after amplification is

$$P_C = GP_R = \frac{GP_T A_R A_T}{\lambda^2 L^2} \tag{2.3-1}$$

From Section 2.9, the noise power after amplification and frequency changing to a low frequency will be

$$P_N = BGhf \left(\frac{1}{e^{hf/kT} - 1} + 1 \right)$$

The first term is amplified Johnson noise; the second term is of quantum origin. Here we assume a linear amplification. But, for data transmission, photon counting is better (see Sections 12.5 and 12.6). We note that the above relation can be rewritten

$$P_N = BGhf \frac{1}{1 - e^{-hf/kT}} \qquad (2.3\text{-}2)$$

From (2.3-1) and (2.3-2), we see that

$$\frac{P_C}{P_N} = \frac{P_T A_R A_T}{\lambda^2 L^2} \frac{1 - e^{-hf/kT}}{Bhf}$$

Using

$$\lambda = c/f$$

this becomes

$$\frac{P_C}{P_N} = \frac{P_T A_R A_T}{L^2 B} \frac{f}{hc^2} (1 - e^{-hf/kT}) \qquad (2.3\text{-}3)$$

If we are designing a system for a specified carrier-to-noise ratio, it is convenient to rewrite this

$$P_T A_R A_T = (P_C/P_N)(L^2 B)(hc^2/f)(1 - e^{-hf/kT})^{-1} \qquad (2.3\text{-}4)$$

If we are in the microwave region where

$$hf \ll kT$$

we can approximate

$$P_T A_R A_T = (P_C/P_N)BL^2(kTc^2/f^2) = (P_C/P_N)BL^2\lambda^2 kT$$

$$P_T A_R A_T = 1.24 \times 10^{-6}(P_C/P_N)BL^2 T/f^2$$
$$= 1.38 \times 10^{-23}(P_C/P_N)BL^2 T\lambda^2 \qquad (2.3\text{-}5)$$

If we are in the optical region where

$$hf \gg kT$$

the approximation becomes

$$P_T A_R A_T = (P_C/P_N)BL^2(hc^2/f) = (P_C/P_N)BL^2\lambda ch$$

$$P_T A_R A_T = 5.96 \times 10^{-17}(P_C/P_N)BL^2/f = 1.99 \times 10^{-25}(P_C/P_N)BL^2\lambda \qquad (2.3\text{-}6)$$

Equations (2.3-5) and (2.3-6) allow the cost $P_T A_R A_T$ to be rapidly calculated for any desired carrier-to-noise ratio P_C/P_N, if we know the bandwidth B, the communication distance L, the wavelength λ, and, in the lower-frequency case, the noise temperature T.

The next section studies the measurement of noise, not when it is interfering with the signal as here, but rather when the noise itself carries information about a phenomenon we are trying to measure.

Problems

1. Assume $L = 4$ light years and $P_C/P_N = 100$. Assume aperture efficiencies of
   0.8. Assume the temperature of space to be 3°K. What transmitter powers P_T
   are required for bandwidth $B = 10$ Hz and the following combinations of
   antenna diameter D and wavelength λ? (D is the same for transmit and
   receive.)

Case:	a	b	c	d
D (m):	100	50	15	5
λ (cm):	10	1	0.1	6×10^{-7} (6000 Å)

2. Compute the powers also for $B = 3000$ Hz (telephone) and 4,000,000 Hz
   (TV).

3. Compute the antenna beamwidths according to equation (1.5-5). Also compute the width of the beam in kilometers at a distance of 4 light years.

2.4. Detection and Measurement of Faint Isotropic Sources

Radio astronomers and Dr. B. M. Oliver (References 3 and 4) are
interested in receiving and measuring signals from faint sources at great
distances. Presumably these sources radiate more or less isotropically. At
any rate, *we* cannot direct their power into a narrow beam aimed at
earth.

Equation (2.3-6) shows that for antennas of fixed size pointing at one
another, the power required decreases with frequency even at high
frequencies. Let us consider the case of an isotropic source of power P_T
and a receiving antenna of area A_R. The fraction of the power received is

$$\frac{P_R}{P_T} = \frac{A_R}{4\pi L^2}$$

The fraction of the power received is independent of frequency, so that
the carrier-to-noise ratio depends only on the noise power. At low
frequencies this is

$$kTB$$

At high frequencies, if we use an ideal amplifier this noise power is

$$hfB$$

Thus high frequencies are very bad for receiving from isotropic
sources using an amplifier. Photon counting, treated in Section 12.5, is
better, but we shall not consider it here. For a receiver on earth, some
high frequencies are bad because of atmospheric absorption, particularly
absorption by O_2 and H_2O. Low frequencies can be bad, too, because of
noise from the galaxy. The Cyclops report indicates a microwave noise

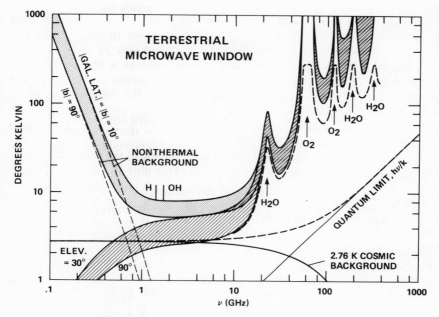

Figure 2.1. Sky noise temperature for coherent receivers.

"window" around 3 GHz or 10 cm. Figure 2.1 is an updated version of Fig. 5-2 from the Cyclops report, Reference 3.

In the next section, we study sources of noise which we regard as a signal. This is the case in radio astronomy.

Problem

1. At what frequency f does the classical thermal noise term kTB become comparable to the quantum noise term hfB, if the receiver system noise temperature T is 5°K? 25°K? 125°K?

2.5. Noise as a Signal

What if we regard noise as a signal? Many signal sources *do* produce mixtures of many frequencies that are incoherent or unrelated in phase. These are therefore noise sources. Most light is such noise, just as the incoherent microwave radiation from a hot body is.

Experience tells us that light from a hot body such as the sun or a hot filament can be a quite steady signal. How can this be? It is because the bandwidth of the detectors with which we observe fluctuations in the power emitted by the noisy light source is very small compared with the bandwidth of the light which we observe with our senses. The wide bandwidth results in an averaging of fluctuations which smooths them out. We shall study averaging formally in Chapter 8.

In noise, the different frequency components of the noise have random and independent phases, so that the phases of the frequency components are of no use to us. In using noise as a signal we use a detector that adds the *powers* of all the frequency components. A square-law detector in which the output voltage is the square of the input voltage is such a detector.

In exploring the use of noise as a signal we must calculate two things. The first is the mean power P_0 of the noise. This is the mean-square noise voltage $\overline{V^2}$. The second is that part of the mean-squared fluctuation in V^2 which lies in some baseband frequency range B, or in the frequency range 0 to B. Reference 5 goes into this quite thoroughly in its Section 7-1d. We shall study frequencies and frequency spectra more formally starting in the next chapter.

Let us use a rather elementary approach to analyzing noise as a signal. Consider a noise voltage V to be made up of a collection of sine waves of powers P_n, frequencies ω_n, and random phases ϕ_n:

$$V = \sum_n (2P_n)^{1/2} \cos(\omega_n t - \phi_n) \qquad (2.5\text{-}1)$$

This is a Fourier series or integral, about which more will be said in Chapter 3. Here we are using power spectrum as an intuitive concept. We will rigorously define it in Chapter 6. In Section 9.4, we will learn more about how to represent noise as a random waveform.

The power P associated with this voltage will be V^2. This will contain terms of the form

$$2P_n \cos^2(\omega_n t - \phi_n) \qquad (2.5\text{-}2)$$

It will also contain terms of the form

$$2(4P_n P_m)^{1/2} \cos(\omega_n t - \phi_n) \cos(\omega_m t - \phi_m) \qquad (2.5\text{-}3)$$

The first sort of term, (2.5-2), can be rewritten

$$P_n[1 + \cos 2(\omega_n t - \phi_n)] \qquad (2.5\text{-}4)$$

Thus each noise component of power P_n gives rise to a constant dc term P_n and a double-frequency term of amplitude P_n.

On summation, the constant term of (2.5-4) gives the total or mean noise power P_0:

$$P_0 = \sum_n P_n \qquad (2.5\text{-}5)$$

The other part of (2.5-4) is the double-frequency term

$$P_n \cos 2(\omega_n t - \phi_n) \tag{2.5-6}$$

The terms of form (2.5-3) can be rewritten

$$2(P_n P_m)^{1/2}\{\cos[(\omega_n - \omega_m)t - (\phi_n - \phi_m)] \\ + \cos[(\omega_n + \omega_m)t - (\phi_n + \phi_m)]\} \tag{2.5-7}$$

Thus every *pair* of noise voltages of power P_n at frequency ω_n and power P_m at frequency ω_m gives rise to two fluctuating terms of amplitudes $2(P_n P_m)^{1/2}$, one at the *difference frequency* $(\omega_n - \omega_m)$ and the other at the *sum frequency* $(\omega_n + \omega_m)$.

Let us now consider a flat or white noise spectrum of power density N_0 that extends from some high frequency $(f_0 \gg B)$ to a frequency $f_0 + B_0$. From (2.1-6), thermal noise is like this if

$$h(f_0 + B_0) \ll kT$$

B_0 is the radio frequency bandwidth of the noise. We will learn more of these spectrum matters in Chapters 6 and 9.

If indeed all the frequencies in the noise waveform which we consider as a signal lie far above the base bandwidth B, the double-frequency fluctuations given by (2.5-6) and the sum-frequency fluctuations given by the second term in (2.5-7) will be at frequencies much higher than the upper baseband frequency B and cannot affect the power in the baseband. They average to zero, and the average does not noticeably fluctuate in bandwidth B. Here B is the base bandwidth over which we shall make use of the noise power as a signal. This is half the reciprocal of the observing or averaging time. The reason for the factor $\frac{1}{2}$ will appear in Chapter 5 when we study sampling.

Thus from (2.5-7) the only terms we need to take into account in determining the fluctuation in the baseband are

$$2(P_n P_m)^{1/2} \cos[(\omega_n - \omega_m)t - (\phi_n - \phi_m)] \tag{2.5-8}$$

The squares or powers of such terms with $\omega_n - \omega_m$ between $-2\pi B$ and $+2\pi B$ will add to give the mean-square fluctuation in the noise power up to the frequency B. The average value of the square of (2.5-8) is

$$2P_n P_m \tag{2.5-9}$$

This fluctuation has a radian frequency $(\omega_n - \omega_m)$ or a frequency $(f_n - f_m)$.

The problem is now to sum terms of the form of (2.5-9) over all pairs of frequencies whose difference is less than B. Here a geometrical diagram will be helpful.

In Fig. 2.2 f_m is plotted vertically and f_n is plotted horizontally. If the

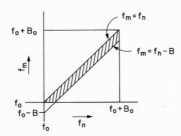

Figure 2.2. Region of integration in frequency space.

noise power density is a constant N_0 for noises of all frequencies f_n and f_m, then for a small frequency range df_n at f_n the power dP_n is

$$dP_n = N_0 \, df_n \tag{2.5-10}$$

Similarly for a small frequency range df_m at f_m the power dP_m is

$$dP_m = N_0 \, df_m \tag{2.5-11}$$

Here

$$2(dP_n)(dP_m) = 2N_0^2 \, df_n \, df_m \tag{2.5-12}$$

The dashed region in Fig. 2.2 represents all pairs of frequencies which (1) lie in the bandwidth B_0 between f_0 and $f_0 + B_0$ and (2) have differences less than the baseband B. This region is bounded by the vertical lines

$$f_n = f_0, \qquad f_n = f_0 + B_0$$

and the horizontal lines

$$f_m = f_0, \qquad f_m = f_0 + B_0$$

and lies between the 45° lines

$$f_m = f_n$$
$$f_m = f_n - B$$

Within this region

$$f_m < f_n$$

Thus within the dashed region we will not be counting a term such as (2.5-9) twice by interchanging f_n and f_m.

From (2.5-9) and (2.5-12) we see that the total fluctuation in the power lying in the baseband B, which we will call P_b^2, will be $2N_0^2$ times the area of the dashed region, or

$$P_b^2 = 2N_0^2[\tfrac{1}{2}B_0^2 - \tfrac{1}{2}(B_0 - B)^2]$$
$$P_b^2 = 2N_0^2 B_0 B(1 - B/2B_0) \tag{2.5-13}$$

We have found the mean-square fluctuation $P_b{}^2$ in the baseband noise power. From the definition of the constant noise power density N_0 it follows that the total average noise power P_0 (which we now regard as our signal) is

$$P_0 = N_0 B_0 \qquad (2.5\text{-}14)$$

Thus the noise-to-signal ratio N/S, the ratio of mean-square baseband power fluctuation to the square of the average power, is

$$N/S = P_b{}^2/P_0{}^2$$
$$N/S = 2(B/B_0)(1 - B/2B_0) \qquad (2.5\text{-}15)$$

Equation (2.5-15) states that for a given observation time ($=1/2B$) the mean-square error decreases as the rf bandwidth is increased, and the use of very large rf bandwidths is highly advantageous in estimating the strength or noise spectral density N_0 of a white noise source.

We see that in order to attain a good signal-to-noise ratio when using an incoherent source as a signal, we must make

$$B_0 \gg B \qquad (2.5\text{-}16)$$

This corresponds to long observations or a wide rf bandwidth B_0. When this is so, very nearly

$$P_b{}^2/P_0{}^2 = 2B/B_0 \qquad (2.5\text{-}17)$$

We will study the performance of microwave radiometry in the next section. The performance depends on the results of this section.

Problems

1. According to the derivation of (2.5-13), what is the mean-square fluctuation $dP_b{}^2$ for a noise measurement in the baseband range df at f? (This gives the *spectral density* of the mean-square fluctuations in the power.)

2. A radio astronomy antenna for measuring the apparent temperature of white noise sources in the sky which completely fill the antenna beam and correspond to noise of spectral density N_1 is replaced by another antenna. The new one has twice the effective area but the same rf bandwidth and the same receiving system noise density N_2. How much is the necessary observing time reduced for the same required measurement accuracy? If the rf bandwidth is doubled, what happens to the observing time, if $B_0 \gg B$?

2.6. Microwave Radiometry

Suppose that we point the antenna of a low-noise microwave receiver at a smooth lake or a tin roof which reflects the sky. The noise temperature of this source will be the sky temperature, which is around 3°K if it is not raining. Suppose that we point the antenna at dense bushes which are quite absorbing, and, by thermodynamic principles, radiating. The noise temperature of the source will be nearly the temperature of the bushes, or around 293°K or 20°C. We can, in effect, measure the noise temperature of a feature of the landscape by measuring the noise output of the receiver. Reference 5 discusses this in Section 7-1.

In making such a measurement, it is very helpful to have a noise reference temperature. If we point the antenna steadily at one object, we can get a noise reference if we periodically push a tapered piece of lossy material into the waveguide which connects the antenna to the receiver. When the lossy material is out, the receiver input sees the effective temperature of the antenna, which depends on what the antenna is pointed at. When the lossy material is in, the receiver sees the temperature of the lossy material. A device in which lossy material is periodically inserted into and withdrawn from a waveguide in order to measure the noise temperature is called a *Dicke radiometer.* Or, we may merely scan the antenna over the scene to be studied. In this case, we get a signal which depends on effective temperature differences in the scene scanned.

Let us assume that the noise temperature of the receiver, including the antenna, is T_n, and that the effective temperature of whatever the antenna is pointed at is T. Then the noise in the output of the receiver will be proportional to

$$T + T_n \tag{2.6-1}$$

and the corresponding noise power P_0 in a radio frequency bandwidth B_0 will be

$$P_0 = k(T + T_n)B_0 \tag{2.6-2}$$

We will note from (2.5-13) that within a baseband B there will be a mean-square fluctuation P_b^2 in P_0 given by

$$P_b^2 = 2k^2(T + T_n)^2 B_0 B(1 - B/2B_0) \tag{2.6-3}$$

Suppose, contrary to fact, that there were no fluctuation in P_0. What fluctuation ΔT about T, the source temperature, would be required to make the mean-square value of the corresponding fluctuation ΔP_0 in P_0 equal to P_b^2? From (2.6-2) we see that if we vary T by an amount ΔT we will have

$$\Delta P_0 = kB_0\Delta T$$

$$\overline{(\Delta P_0)^2} = (kB_0)^2\overline{(\Delta T)^2} \tag{2.6-4}$$

Here the bar above the quantity denotes the average or expectation of the quantity, a concept which we study in more detail in Chapter 8.

From (2.6-3) and (2.6-4), we find that if

$$\overline{(\Delta P_0)^2} = P_b^2$$

then we obtain

$$\overline{(\Delta T)^2} = 2(T + T_n)^2(B/B_0)(1 - B/2B_0) \qquad (2.6-5)$$

We have in $\overline{(\Delta T)^2}$ a measure of the mean-square uncertainty in T attributable to the noisy nature of the receiver noise and of the emissions from the body at temperature T. This is the *variance* of our measurement of T, which we study further in Section 8.4.

Suppose that in scanning our receiving beam over various objects we want to get a very good picture or map of their apparent radio frequency temperatures. To do this we must make $\overline{(\Delta T)^2}$ very small. In order to make this quantity small we should make the noise temperature T_n of the receiver as small as possible and the ratio of the radio frequency bandwidth B_0 to the baseband B over which we observe the noise power as large as possible.

We note from (2.6-5) that for a noiseless receiver ($T_n = 0$) the mean-square uncertainty in temperature is proportional to the square of the source temperature T for a microwave radiometer. The situation is different for optical frequencies. We will study optical radiometry in Section 2.10, after we study optical detection in Section 2.9.

So far, we have studied noise from discrete sources. The next section investigates noise added by the propagation medium. This is very important for communication through the earth's atmosphere.

Problems

1. In an early Dicke radiometer the noise temperature of the receiver was 10,000°K. The rf bandwidth B_0 was 20×10^6 Hz and the base bandwidth was 1 Hz. To about what accuracy in degrees Kelvin could one measure the temperature of an object at room temperature?

2. A modern radiometer scans 200 pixels (picture elements) in 1 sec, which requires a base bandwidth of 100 Hz. The rf bandwidth B_0 is 400×10^6 Hz and the receiver noise temperature T_n is 100°K. Assume that the source temperature $T = 100°K$. What is the mean-square uncertainty in T?

3. What base bandwidth B would be allowable if B_0 were 3×10^8 Hz; $T_n = 30°K$; $\overline{(\Delta T)^2} = 10(°K)^2$ and $T = 100°K$?

4. For cosmological reasons, it is desired to measure the 3°K cosmic background radiation more precisely, to 0.005°K over regions of the sky about 0.1° in angle on a side. This means that a $[\overline{(\Delta T)^2}]^{1/2}$ of 0.005°K is necessary. The receiving system has noise temperature 20°K and an available bandwidth of $B_0 = 100$ MHz at $f_0 = 8500$ MHz. How long should each resolution cell be observed? How long will it take to measure the entire 4π steradian celestial

sphere? What approximate antenna gain should be used to get the desired resolution? What is the antenna diameter if the efficiency is 70%?

2.7. Noise from Partially Absorbing Media

In this section we will assume that $kT \gg hf$, so that we can use expression (2.1-6):

$$\Delta P_n = kT\Delta f \tag{2.1-6}$$

The argument could be easily carried through using (2.1-5) instead.

Suppose that a receiving antenna looks at an absorbing source (trees, for instance) which has a temperature T_a. Suppose that there is an absorbing medium between the receiving antenna and the source, such as a large rainstorm of temperature T_r completely filling the antenna beam. What noise power does the antenna receive?

We argue in the following way: Suppose that the medium, the source, and a resistor which absorbs all the power received by the antenna were all at the same temperature T. The resistor connected to the receiving antenna would radiate a power $kT\Delta f$. Suppose that a fraction a_r of this power is absorbed by the medium and a fraction a_a $(=1 - a_r)$ reaches the source. Then if neither the medium nor the source is to get hot at the expanse of the resistor, the medium must send a power $a_r kT\Delta f$ to the resistor (to the receiving antenna) and the source must send a power $a_a kT\Delta f = (1 - a_r)kT\Delta f$ to the resistor.

There is no reason to believe that these proportions will change just because the temperature of the medium and the source are different. Hence the power received from the medium will be

$$a_r kT_r \Delta f$$

and the power received from the source will be

$$a_a kT_a \Delta f = (1 - a_r)kT_a \Delta f$$

Thus the total noise power received by the receiver will be

$$k(a_r T_r + a_a T_a)\Delta f$$

This corresponds to an *effective temperature* of the antenna, T_e, given by

$$T_e = a_r T_r + a_a T_a$$

More generally, suppose that we feed power to an antenna and a fraction a_n is absorbed by a material at a temperature T_n. If $m - 1$ different absorbing media absorb the total power, the effective temperature T_e which specifies the noise received by the antenna will be

$$T_e = \sum_1^m a_n T_n \tag{2.7-1}$$

Of course,

$$\sum_1^m a_n = 1$$

and T_m is the temperature of the source.

The next section shows how to determine the total noise received from space by an antenna system.

Problems

1. An antenna for receiving microwave signals from a space probe is pointed directly overhead. The temperature of space (the temperature which specifies the noise power received from space) is 3°K. During a rainstorm, rain at a temperature 275°K absorbs half the signal power from the space probe. What is the noise temperature seen by the antenna? What would the absorption have to be to make that temperature equal to 15°K? (The point is that even small absorptions degrade the performance of low-noise systems much more than just by the small drop in signal power.)

2. The same antenna is pointed at a space probe just barely visible above the horizon. The horizon is a wooded hillside which absorbs microwaves. The trees have a temperature of 290°K. What is the noise temperature seen? Explain your reasoning.

3. An antenna 15 ft in diameter with an aperture efficiency of 0.7 is pointed at the sun. At a wavelength of 7.5 cm the effective temperature of the sun is about 6000°K. What effective temperature is seen by the antenna? $L = 1.5 \times 10^{11}$ m, earth to sun; $D_{sun} = 1.4 \times 10^9$ m.

4. Consider Problem 1 of Section 2.3. If, in looking back from Alpha Centauri at the sun, the sun is included in the beam, how much will the effective noise temperature as seen by the antenna be raised from the 3°K for space? (Do all four cases.)

5. A receiver with a noise temperature of 6°K is connected with an antenna by a waveguide with 1-dB loss. The antenna, waveguide, and receiver form a receiving system. The temperature of the waveguide is 293°K. What is the noise temperature of the receiving system? HINT: the noise temperature of a receiving system can be defined as the temperature of the source that would have to be looked at to double the output noise power density. (This problem explains why microwave receivers can have noise temperatures far above the ambient temperature.)

6. If the temperature of space is 3°K and if rain at 275°K absorbs half the power from a satellite, by what factor is the total noise output increased in going from a receiver with a 10°K noise temperature to a receiver with a 100°K noise temperature? What is the carrier-to-noise ratio in each case, if $P_C/P_N = 30$ dB in the absence of rain in both the 10 and 100°K cases?

2.8. Noise Density in Space

The formulas in Section 2.1 give the Johnson or thermal noise power per unit bandwidth for a single mode of propagation. As we have seen, if the noise comes from a waveguide attached to a receiving antenna, the noise temperature is the temperature of the source at which the antenna points.

The antenna receives energy over a solid angle ψ which depends on its dimensions and on the wavelength. In Section 1.5 we found an *effective* solid angle over which an antenna transmits or receives power by comparing the transmission formula

$$\frac{P_R}{P_T} = \frac{A_T A_R}{\lambda^2 L^2} \tag{1.3-1}$$

with the fact that if a transmitting antenna spread power uniformly over a solid angle ψ, the ratio of received power to transmitted power would have to be

$$\frac{P_R}{P_T} = \frac{A_R}{\psi L^2} \tag{2.8-1}$$

This gives us an effective solid angle for an antenna of area A at a wavelength λ:

$$\psi = \lambda^2/A \tag{2.8-2}$$

At low frequencies, the noise power density N_0 (noise power per unit bandwidth) is from (2.1-6)

$$N_0 = kT \tag{2.8-3}$$

This is, as we have noted, the noise per unit bandwidth in one mode of the waveguide coming from the antenna. It must also be the noise radiation of one polarization coming from the solid angle ψ given by (2.8-2) and falling on an area A, the area of the antenna. Hence the power density N of the thermal radiation falling on a surface per unit solid angle per unit area must be

$$N = 2kT/\psi A$$
$$N = 2kT/\lambda^2 \tag{2.8-4}$$

The factor 2 occurs in (2.8-4) because independent noises will arrive polarized in two perpendicular directions, while (2.8-3) applies for noise in one mode only. The λ^2 in the denominator of (2.8-4) is reasonable, because the gain of a directional antenna is inversely proportional to wavelength squared, if we ignore efficiency factors. We saw this in equation (1.4-2).

Expression (2.8-4) gives noise per unit area per unit solid angle. We have assumed that the thermal source is infinitely far away and completely fills the receiving antenna beam for all frequencies that have appreciable power. We also assume that the wavelengths corresponding to these frequencies are small compared with the linear dimensions of the area A. This guarantees that the noise radiation strikes normal to the unit area. We may also want to know the total noise per unit bandwidth which falls on a unit area from all directions to one side of the plane of the area, when the source completely fills space on one side of the area. The power picked up is different for different directions.

Imagine a small area a. If we look at this from an angle ϕ with respect to the normal to the area, the apparent area we see is only $a \cos \phi$. It is this area which will intercept radiation. The solid angle in the angular range $d\phi$ is

$$d\psi = 2\pi \sin \phi \, d\phi \qquad (2.8\text{-}5)$$

Hence, the total power P_0 per unit bandwidth per unit area which comes from radiation lying in a cone of half-angle θ is

$$P_0 = \int_0^\theta (N \cos \phi)(2\pi \sin \phi \, d\phi)$$

$$P_0 = (2\pi kT/\lambda^2) \sin^2 \theta \qquad (2.8\text{-}6)$$

If radiation comes from all directions to one side of a plane,

$$\theta = \pi/2, \qquad \sin \theta = 1$$

$$P_0(f) \, df = (2\pi kT/\lambda^2) \, df \qquad (2.8\text{-}7)$$

Equations (2.8-6) and (2.8-7) are valid for the low-frequency electromagnetic radiation in any isotropic medium. For some peculiar reason, physicists prefer to express thermal noise power per unit wavelength rather than per unit frequency. In free space, if c is the velocity of light,

$$f = \frac{c}{\lambda}$$

$$df = \frac{c}{\lambda^2} \, d\lambda \qquad (2.8\text{-}8)$$

(Here df and $d\lambda$ are taken as positive.) Thus we find that the power $P_1(\lambda)$ per unit area which lies in a wavelength range $d\lambda$ will be

$$P_1(\lambda) \, d\lambda = \frac{2\pi ckT}{\lambda^4} \, d\lambda \qquad (2.8\text{-}9)$$

This is the well-known Rayleigh–Jeans formula for black body radiation. Why *black body*? Because it is the radiation received from a surrounding nonreflecting or black surface at temperature T.

So far we have studied noise which arises from classical or thermodynamical considerations. In the next section, we shall see how quantum mechanical considerations add noise to certain types of receivers.

Problems

1. By using (2.1-5) instead of (2.1-6), find the true (Planck's law) expression corresponding to (2.8-9).

2. Integrate the expression found in Problem 1 with respect to wavelength so as to find the total thermal power striking a unit area as a function of temperature. This gives the total black body radiation. (Use the fact that $\int_0^\infty y^3 \, dy/(e^y - 1) = \pi^4/15$.)

2.9. Quantum Uncertainties in Amplification

The Johnson or thermal noise per unit bandwidth is less at high frequencies than at low frequencies. Does this mean that there will be less noise in a communication system that uses high frequencies than in a communication system that uses low frequencies? Very-high-frequency (optical) amplifiers are indeed noisier than lower-frequency amplifiers, because quantum effects make it impossible to measure a weak signal with perfect accuracy. We shall see a way out of this in Section 12.5.

It is easy to illustrate such quantum effects by considering a simple frequency-changing or superheterodyne receiver. This is sometimes called a double-detection receiver. We will study demodulation more thoroughly in Chapter 4.

Consider two electromagnetic waves which are added together and then impinge on the same photoelectric surface. We will represent these as a signal carrier wave

$$V \cos \omega t$$

of power (into a unit resistance)

$$P = \tfrac{1}{2} V^2$$

and a beating oscillator (local oscillator) wave of frequency ω_0

$$V_0 \cos \omega_0 t$$

which has power (into a unit resistance)

$$P_0 = \tfrac{1}{2} V_0^2$$

The photocurrent I is proportional to the instantaneous power. We will assume a quantum efficiency of unity in using light to produce electrons. Each quantum of energy hf produces one electron of charge e. Hence the current produced, I, is the number of electrons produced times the charge on one electron:

$$I = e[(V \cos \omega t + V_0 \cos \omega_0 t)^2/hf]$$

$$e = 1.602 \times 10^{-19} \, C$$

Here we use C to denote the unit of charge, the coulomb. Also, $\omega = 2\pi f$ and ω_0 is close to ω. The current will have an average value I_0

$$I_0 = \frac{e}{hf}(P + P_0)$$

The current I contains a carrier or intermediate frequency component I_c of radian frequency $(\omega - \omega_0)$, which arises as the low-frequency part of the term

$$2\frac{e}{hf}VV_0 \cos \omega t \cos \omega_0 t$$

in I. This carrier component is the signal we are interested in. The mean-square value of the carrier current is found to be

$$\overline{I_c^2} = \frac{1}{2}\left(\frac{e}{hf}\right)^2 V^2 V_0^2 = 2\left(\frac{e}{hf}\right)^2 PP_0$$

The current will also contain a shot-noise current of mean-square value $\overline{I_n^2}$

$$\overline{I_n^2} = 2eI_0B = 2e\frac{e}{hf}(P + P_0)B$$

This is derived in Section 8.2. Here B is the bandwidth of the amplifier which amplifies the intermediate frequency (IF) signal. This is the signal centered at the frequency $\omega - \omega_0$.

The noise-to-carrier ratio will then be

$$\overline{I_n^2}/\overline{I_c^2} = (hfB/P)(1 + P/P_0) \qquad (2.9\text{-}1)$$

If we make $P_0 \gg P$, as we would expect in any receiver, the noise-to-carrier power ratio N/C will then be

$$N/C = \overline{I_n^2}/\overline{I_c^2} = hfB/P \qquad (2.9\text{-}2)$$

We might think of making the superheterodyne device described above as shown in Fig. 2.3. This will not work. If both P and P_0 are completely absorbed in the photoelectric surface, there can be no component of power (and hence no output current) of frequency $\omega - \omega_0$. Neither the power P nor the power P_0 fluctuates at a frequency $\omega - \omega_0$. The separate electromagnetic fields are separately detected by the photoelectric surface. The two fields are not added coherently before squaring. The integration of the received power over the entire photoelectric surface, which is large compared with a wavelength, adds the fields in random phases.

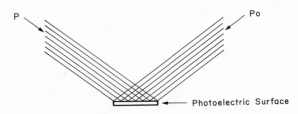

Figure 2.3. A superheterodyne receiver which does not work.

We can use the arrangement shown in Fig. 2.4 instead. The slightly reflecting (mostly transmitting) surface reflects a little of P and transmits most of P_0; it reflects a little of P_0 and transmits most of P. The two electromagnetic fields are first added together coherently before the photoelectric detection takes place. Squaring or power detection then does produce the desired $\omega - \omega_0$ difference frequency term. This is the intermediate-frequency signal we seek in a superheterodyne receiver. There is an equal but opposite-phased power at frequency $\omega - \omega_0$ going off to the left.

Here we have described an almost realizable receiver (almost, because we have assumed unity quantum efficiency and have disregarded IF amplifier noise), a receiver which can, in absence of Johnson noise, attain the noise-to-carrier ratio given by (2.9-2). In this receiver the quantum uncertainty manifests itself as an uncertainty in the time at which a photon will cause the emission of an electron; this gives rise to shot noise.

Let us relate this to the performance of an ideal (least-noise) amplifier. A more general treatment of quantum effects in Chapter 11 or Reference 6 shows that the noise power P_n in the output of an ideal amplifier of power gain $G(\geq 1)$ at frequency f before conversion to baseband is

$$P_n = G \frac{hf}{e^{hf/kT} - 1} B + (G - 1)hfB \qquad (2.9\text{-}3)$$

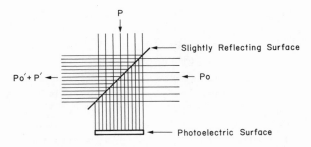

Figure 2.4. A superheterodyne receiver which does work.

The first term of (2.9-3) is simply amplified Johnson noise from a source of temperature T. The second term is *not* amplified noise, because it goes to zero for unity gain ($G = 1$). The second term is rather a fluctuation in a large output signal. We can measure this output signal much more precisely than we can measure the weak input signal, but such measurements cannot tell us more about the weak input signal than the laws of quantum mechanics allow us to know. No matter how much we amplify a high-frequency signal pulse, there will be some small quantum uncertainty in determining its time of arrival. The noise $(G - 1)hfB$ in the output signal, as we will see in Section 11.9, assures us that we will not attain such forbidden knowledge.

It should again be emphasized that the second term in (2.9-3) is not amplified noise. If we do amplify a received signal we are necessarily faced with this noise and its consequences. If we do not amplify a signal—say, if we merely directly detect by letting light fall on a photocell—(2.9-3) is irrelevant.

A convincing argument can be made for the *form* of the second term of (2.9-3). When $G = 1$ (a passive device), $G - 1 = 0$ and we have merely Johnson noise, as must be. Further, it is plausible that two ideal amplifiers of gains G_1 and G_2 should give the same result as one ideal amplifier of gain $G_1 G_2$. Disregarding thermal noise, from (2.9-3) the output of the second amplifier will actually have a noise

$$[G_2(G_1 - 1) + (G_2 - 1)]hfB = (G_2 G_1 - 1)hfB$$

This is as it should be if two ideal amplifiers in tandem are to constitute an ideal amplifier.

Suppose that we amplify a signal of very high frequency f by an ideal amplifier of power gain G. There will be as we have seen a noise power of quantum origin in the output, of power

$$(G - 1)hfB$$

Suppose that we now shift the frequency of the signal down to baseband or to a frequency range so low that quantum effects are negligible. We can do this by means of the sort of double-detection device discussed earlier in this section. If the power gain of this ideal frequency-shifting device is unity, we will find in Section 11.9 that there is an added quantum noise in its output, of power

$$hfB$$

When we add this to the noise power $(G - 1)hfB$ in the high-frequency signal, we find that the total noise power P_n in the signal which has been amplified by a power gain G and shifted to a low frequency is simply

$$P_n = GhfB \tag{2.9-4}$$

This gives a carrier-to-noise ratio in agreement with (2.9-2), the limiting performance for a superheterodyne receiver. Hence in the limit of unity quantum efficiency, high local oscillator power, and no noise added by the following amplifiers, the superheterodyne receiver is an ideal receiver. Reference 7 discusses laser heterodyne receivers and their applications in more detail than we can go into here. Reference 8 is a good reference on laser receivers that will also be some help in Sections 11.9 and 12.6.

The next section considers the quantum limits which apply when trying to measure the strength of sources at optical frequencies.

Problems

1. Suppose that the gain G is very large. Suppose that $T = 293°K$. What is the ratio of the second to the first term in (2.9-3) for wavelengths of 10 cm, 1 cm, 1 mm, 6000 Å?

2. Assume that the photocell has infinite internal impedance and a load resistance R. The gain of the receiver is the ratio of the power at the frequency $\omega - \omega_0$ to the signal power P. What is the gain of the superheterodyne receiver as a function of P_0 and R? What is the gain when $R = 1000$ ohms, $P_0 = 4 \times 10^{-3}$ W, and the signal consists of light of wavelength 6000 Å?

3. Argue from (2.9-2) to show that in shifting the frequency of a high-frequency signal of frequency f down to a low frequency, a noise power hfB appears in the output of the frequency shifter if the power gain of the frequency shifter is unity.

2.10. Optical Radiometry

Sections 2.5 and 2.6 laid the groundwork for microwave radiometry, in which quantum effects are negligible.

In optical radiometry there is a component of noise analogous to that given by (2.6-3). If there were no quantum effects, (2.6-5) would apply and we would be able to detect miniscule temperature differences, since the radio frequency bandwidth or rather the optical bandwidth B_0 is always very large compared with the base bandwidth B over which observations are made. But, quantum effects dominate in optical radiometry (see Reference 7 for more). Actually, (2.6-3) would not exactly apply, since the noise spectral density is not kT in the quantum region. However, this term is dwarfed by the real quantum effects in optical radiometry.

Suppose that we use a photodetector of 100% quantum efficiency as a receiver. The output current I_0 will contain a shot noise current $\overline{I_n^2}$ given by

$$\overline{I_n^2} = 2eI_0B \tag{2.10-1}$$

Here B is the base bandwidth over which radiometric measurements are made.

We may take as a measure of signal-to-noise the ratio of I_0^2 to $\overline{I_n^2}$:

$$S/N = I_0^2/\overline{I_n^2} = I_0/2eB \qquad (2.10\text{-}2)$$

Each quantum releases one electron of charge e. Hence if the power of the incident light is P_0

$$I_0 = P_0 e/hf \qquad (2.10\text{-}3)$$

$$S/N = P_0/2hfB \qquad (2.10\text{-}4)$$

In equation (2.10-4) the optical bandwidth does not appear at all because we have assumed that the noise given by (2.6-3) is small compared with the shot noise in the photodetector current. Nonetheless, optical bandwidth *does* play a part in optical radiometry.

For a source of a given temperature T, P_0 will be proportional to the optical bandwidth B_0 by a factor which we will call $P_0(T)$, if B_0 is not too large compared with the frequency f:

$$S/N = (1/2hf)P_0(T)B_0/B \qquad (2.10\text{-}5)$$

As in microwave radiometry, the signal-to-noise ratio is proportional to the bandwidth of radiation accepted, divided by the bandwidth over which radiometric measurements are made. But the appearance of Planck's constant h testifies to the dominance of quantum effects. The derivation of (2.10-5) was very different from that of (2.6-5).

The preceding chapter derived limitations on received power arising from the physics and geometry of communication antennas. This chapter has initiated our study of the limitations on communication that noise imposes. In the next chapter, we begin our study of the representation of the signals that must be used for communication.

Problems

1. Give a plausible argument that thermal noise can be ignored in radiometry in the quantum region, where $hf \gg kT$.

2. What is S/N as defined by (2.10-5) if we substitute for $P_0(T)$ the black body radiation per unit area at frequency f, given by equation (2.8–7) corrected for the quantum region? How does the area of the collector of radiation come into the expression?

3. Using the parameters of problem 2, how accurately can we measure temperature with an optical radiometer of bandwidth B_0, frequency f, and collector area A? (Assume high signal-to-noise ratios.) Find this uncertainty if $T = 600°K$, $\lambda = 6000$ Å, $A = 1\,\text{cm}^2$, $B = 1$ Hz, and $B_0 = 100$ GHz.

Signals and Frequencies

So far we have considered some aspects of the transmission of sine waves and their admixture with noise. In actual communication systems, we transmit complicated signals. Before we proceed to consider the transmission of such signals, we need to know a little about the representation of complicated waveforms, and even of two-dimensional patterns, by sums of sine waves. Why sines and cosines? Because these functions are the stable or bounded solutions of the differential equations of linear circuit theory and electromagnetic wave propagation.

Everyone is familiar with the fact that frequencies are important in radio communication. They permit tuning to different broadcasting stations. The frequency of a sine wave is easy to define. But, the frequency makeup of the complicated signals that are used in communication is harder to define rigorously. And, we must know the range of frequencies that signals occupy, if we are to allocate different ranges of frequencies to different stations or services, so that they do not interfere with one another.

Figure 3.1 shows the current allocations of the electromagnetic frequency spectrum to different services or uses. We note the logarithmic scale—there is as much bandwidth available to the left of 1 MHz as there is between 1 MHz and 2 MHz. Not all existing uses are indicated, and some uses for radio navigation that we do not study in this book are included. The frequencies used range from 50 Hz for proposed communication to submarines (the 6000-km wavelength penetrates sea water from air) to visible light with a frequency around 600 THz (terahertz) and a wavelength of $5000 \text{ Å} = 0.5 \times 10^{-6} \text{ m}$, or half a micrometer or micron. Light is used for communication via fiber optics and for communication via semaphore flags with the human eye as a receiver. The world-wide navigation services called Omega and Loran-C are shown in the low-frequency (LF) range of around 50 kHz. The standard am (amplitude

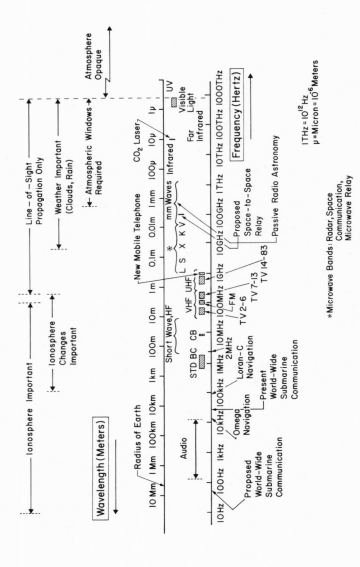

Figure 3.1. Spectrum allocation for electromagnetic waves.

modulation) broadcast band of 540–1600 kHz is in the medium-frequency (MF) range, while short wave and citizen's band (CB) is in the high-frequency (HF) range. Broadcast TV (up to channel 13) and fm (frequency modulation) radio are in the very high frequency (VHF) range, but TV channels 14–83 and the new mobile telephone service are in the ultrahigh-frequency (UHF) range. Beyond this are the microwave (μwave) and millimeter wave (mm wave) bands. Somewhere above 100 GHz (3-mm wavelength), the technology shifts from radio to optical. The CO_2 laser operates at a frequency of 30 THz (a wavelength of 10 microns), and has been used in experimental communication through the atmosphere. Above the visible light region, the atmosphere is opaque. But, such frequencies can be used for communication from space to space.

We speak of the frequencies used for various communication services. But, communication requires more than a single frequency or pure carrier. A range of frequencies must be used. In the standard am broadcast band, the tuning or center or carrier frequency assignments are spaced 20 kHz apart (a maximum or peak power level is assigned as well). Each station is allowed to radiate power in a 10-kHz-wide band 5 kHz on either side of the carrier. This leaves a guard band 10 kHz wide between adjacent station assignments. We must be able to treat these matters analytically because of their great importance to communication.

To define frequency and bandwidth rigorously will require the Fourier transform. Most of this chapter is devoted to this. We also study the two-dimensional Fourier transform, which occurs in several aspects of communication. We will find applications of the Fourier transform to the Friis transmission formula and to the calculation of antenna patterns, as well as to sonograms and their uses in speech analysis.

Table 3.1 gives Fourier transforms of some waveforms that may be used in communication. Most of these transforms will be derived in this chapter, or later in this book. Table 3.1 will be useful throughout the remaining chapters. By using the table in both directions (time functions into frequency functions and frequency functions into time functions), and rescaling if necessary, a large number of transforms can be found. References 2 and 3 are good references on the Fourier transform as used in communications and signal processing.

In Chapter 4 we begin to use the Fourier transform in studying various types of modulation and how they are affected by noise. Chapter 6 provides the basis for a rigorous definition of the power spectrum of a waveform or signal which does not repeat periodically. We note that information-bearing signals are generally not periodic. Chapter 7 shows how the pulses we may use for communication can be shaped to control their spectral occupancy. Chapter 9 provides a rigorous definition of the

Table 3.1. Fourier Transforms

Waveform $x(t)$	Fourier Transform $X(f)$
$ax_1(t) + bx_2(t)$, a, b constant	$aX_1(f) + bX_2(f)$
$x(-t)$ if $x(\pm\infty) = 0$	$X^*(f)$
$x(t - t_0)$	$\exp(-j2\pi ft_0)X(f)$
$(x_1 * x_2)(t)$ convolution	$X_1(f) \cdot X_2(f)$
$x(at)$, a constant	$(1/a)X(f/a)$
$x'(t)$	$j2\pi fX(f)$
$\int_{-\infty}^{t} x(s)\, ds$	$(1/j2\pi f)X(f)$
$\delta(t)$	1
Unit step at time 0	$1/j2\pi f$
Rectangular pulse of height 1 from $-T/2$ to $+T/2$	$T[(\sin \pi fT)/\pi fT]$
Triangular pulse from $-T$ to $+T$ of height T	$T^2[(\sin \pi fT)/\pi fT]^2$
Rectangular pulse of unit amplitude from $-T$ to $+T$ which switches sign from $+$ to $-$ at $T = 0$	$2Tj[(\sin^2 \pi fT)/\pi fT]$
$\exp(-t/t_0)$ for $t \geq 0$; 0 for $t < 0$	$1/(1/t_0 + j2\pi f)$
$\exp(-\pi t^2)$	$\exp(-\pi f^2)$
$x(t) = (1/t^{1/2})$ if $t > 0$; $x(t) = 0$ if $t \leq 0$	$(1 - j)/2f^{1/2}$ for $f > 0$ $(1 + j)/2(-f)^{1/2}$ for $f < 0$

power spectrum, taking into account the probabilistic nature of the signal or noise. This is necessary because the real world is random. Chapter 10 includes a discussion of possible definitions of spectral occupancy or bandwidth, and Chapter 11 shows how the information-carrying capacity of a communication channel increases when the allowed bandwidth increases. So, frequencies, spectra, and bandwidth are very important to communication. To understand them, we need to learn the basic material about Fourier transforms that is covered in this chapter.

3.1. The Peculiar Properties of Sine Waves and Linear Networks

We will deal chiefly with linear systems. Reference 1 can be used to fill in details, but this section contains enough for the purposes of this book.

Suppose that we apply a voltage $V_1(t)$ to the input of a linear system and get an output $V_2(t)$. Suppose we apply a voltage $U_1(t)$ to the input of the same linear system and get an output $U_2(t)$. Then if we apply to the input a voltage

$$AV_1(t) + BU_1(t)$$

we get an output

$$AV_2(t) + BU_2(t)$$

Here A and B are constants.

$V_1(t)$ and $U_1(t)$ can be very complicated functions of time, and $V_2(t)$ and $U_2(t)$ can be very unlike $V_1(t)$ and $U_1(t)$. But, if the system is linear, the output for two time-varying voltages applied together is simply the sum of the outputs for the voltages applied separately.

Ordinarily, $V_2(t)$ will look very different from $V_1(t)$. If the input is a very short spike, we get a long, complicated output. But, the sinusoidal signal

$$V_{01} \cos (\omega t + \phi_1)$$

has a privileged property as an input to linear systems. The output is always

$$V_{02} \cos (\omega t + \phi_2)$$

It has the same shape, though the amplitude and phase may be different.

We see this most easily in the case of lumped linear circuits which are made of resistors, inductors, and capacitors. For such a network, the relation between the input voltage V_1 and the output voltage V_2 has the form

$$A_0 V_1 + A_1 \frac{dV_1}{dt} + A_2 \frac{d^2 V_1}{dt^2} + \cdots = B_0 V_2 + B_1 \frac{dV_2}{dt} + B_2 \frac{d^2 V_2}{dt_2} + \cdots$$

If we assume

$$V_1 = V_{01} e^{pt}$$

we can see that a solution is

$$V_2 = V_{02} e^{pt}$$

$$V_{02} = \frac{A_0 + pA_1 + p^2 A_2 + \cdots}{B_0 + pB_1 + p^2 B_2 + \cdots} V_{01}$$

So, the time variation of the output has the same form as that of the input. If $p = j\omega$, the real part of V_1 can represent a real signal input and the real part of V_2 then represents the real signal output. Both

$$\text{Re}\,(V_{01}e^{j\omega t}) \quad \text{and} \quad \text{Re}\,(V_{02}e^{j\omega t})$$

are sinusoidal.

A sinusoidal wave is the only nongrowing, nondecaying input for which the output has the same kind of time variation as the input. This gives it a privileged position in the mathematical analysis of signals.

Sine waves are also privileged for other reasons. We associate the sensation of pitch with the frequency of sine waves. Some networks or filters pass only sine waves having a limited range of frequencies. And, we can represent any signal, no matter how complicated, as a sum of sine waves. In the next section, we shall summarize the idea of orthogonality which makes such representations easy to work with and understand.

Problem

1. The relation between the input voltage V_1 and the output voltage V_2 of a certain linear circuit is

$$V_1 - \frac{dV_1}{dt} = V_2 + \frac{dV_2}{dt}$$

Find the output $V_2(t)$ if the input is

$$V_1(t) = \cos \omega t$$

Put the answer in the form

$$V_2(t) = V \cos (\omega t + \phi_2)$$

and find the voltage gain V and phase shift ϕ_2 as a function of frequency ω. Are there ω for which the output is in phase with the input? What is the high-frequency behavior of the output?

3.2. Orthogonality

Orthogonality is an important property in connection with sine waves. Reference 2 is a useful compendium on these matters for engineers and scientists, with emphasis on Fourier transforms.

Two functions $f_1(t)$ and $f_2(t)$ of finite energy are orthogonal if

$$\int_{-\infty}^{\infty} f_1(t)f_2(t)\, dt = 0 \qquad\qquad (3.2\text{-}1)$$

Here $f(t)$ has finite energy if

$$\int_{-\infty}^{\infty} f^2(t)\, dt < \infty$$

Two functions are orthogonal over the interval between $t = -T/2$ and $t = +T/2$ if

$$\int_{-T/2}^{T/2} f_1(t) f_2(t)\, dt = 0$$

The two functions

$$f_1(t) = \sin(2n\pi t/T)$$
$$f_2(t) = \cos(2n\pi t/T)$$

are an example of this because

$$\int_{-T/2}^{T/2} \sin(2n\pi t/T)\cos(2n\pi t/T)\, dt = \frac{1}{2(2n\pi/T)} \sin^2(2n\pi t/T)\Big|_{-T/2}^{T/2} = 0$$

Suppose

$$\int_{-T/2}^{T/2} [f_1(t)]^2\, dt \quad \text{and} \quad \int_{-T/2}^{T/2} [f_2(t)]^2\, dt$$

increase approximately linearly with T. We say in this case that $f_i(t)$ has finite average power; Chapter 6 studies this in more detail. This is so for sine and cosine functions of T. It is then proper to say that $f_1(t)$ and $f_2(t)$ are orthogonal if

$$\frac{1}{T} \int_{-T/2}^{T/2} f_1(t) f_2(t)\, dt \qquad (3.2\text{-}2)$$

approaches zero as T approaches infinity. In this sense, $\cos \omega_1 t$ and $\cos \omega_2 t$ are orthogonal if $\omega_2 \neq \omega_1$ because

$$\cos \omega_1 t \cos \omega_2 t = \tfrac{1}{2}[\cos(\omega_1 - \omega_2)t + \cos(\omega_1 + \omega_2)t]$$

This function has no dc component and the integral (3.2-2) approaches zero as T approaches infinity.

The next section makes use of the orthogonality of sines and cosines in representing a waveform which is nonzero over a finite time as a sum of sine and cosine terms with coefficients that depend on the function. This is the Fourier series.

Problems

1. Orthogonality is important in connection with functions other than sine waves. Can two periodic functions which are nowhere negative be orthogonal?

2. Which waveforms $f(t)$ defined for $-\infty < t < \infty$ are orthogonal to them-selves? Consider both the finite energy definition and the finite average power definition. Give an example according to the finite average power definition of a self-orthogonal function which has infinite energy. HINT: start with a waveform somewhat like $|t|^\alpha$ for suitable α.

3.3. Fourier Series

We can represent $f(t)$ over the interval $-T/2$ to $T/2$ as

$$f(t) = \sum_{n=0}^{\infty} [A_n \cos (2\pi nt/T) + B_n \sin (2\pi nt/T)]$$

This representation is periodic with the period T. We have

$$\int_{-T/2}^{T/2} \cos (2\pi nt/T) \sin (2\pi mt/T) \, dt = 0$$

and, if $n \neq m$

$$\int_{-T/2}^{T/2} \cos (2\pi nt/T) \cos (2\pi mt/T) \, dt = 0$$

$$\int_{-T/2}^{T/2} \sin (2\pi nt/T) \sin (2\pi mt/T) \, dt = 0$$

Hence

$$\int_{-T/2}^{T/2} f(t) \cos (2\pi nt/T) \, dt = \int_{-T/2}^{T/2} A_n \cos^2 (2\pi nt/T) \, dt$$

We note that when $n = 0$

$$\int_{-T/2}^{T/2} f(t) \, dt = \int_{-T/2}^{T/2} A_0 \, dt = A_0 T$$

$$A_0 = \frac{1}{T} \int_{-T/2}^{T/2} f(t) \, dt$$

and

$$B_0 = 0$$

When $n \neq 0$

$$\int_{-T/2}^{T/2} f(t) \cos (2\pi nt/T) \, dt = \int_{-T/2}^{T/2} A_n \cos^2 (2\pi nt/T) \, dt = A_n T/2$$

$$A_n = \frac{2}{T} \int_{-T/2}^{T/2} f(t) \cos (2\pi nt/T) \, dt$$

Similarly,

$$B_n = \frac{2}{T} \int_{-T/2}^{T/2} f(t) \sin (2\pi nt/T) \, dt$$

$$B_0 = 0$$

In complex notation, in the interval from $t = -T/2$ to $t = T/2$,

$$f(t) = \sum_{n=-\infty}^{\infty} C_n \exp [j(2\pi nt/T)]$$

$$\int_{-T/2}^{T/2} \exp \{j[2\pi(n - m)t/T]\} \, dt = \begin{cases} T & \text{if } n = m \\ 0 & \text{if } n \neq m \end{cases}$$

Hence

$$C_n = \frac{1}{T} \int_{-T/2}^{T/2} f(t) \exp [-j(2\pi nt/T)] \, dt$$

If $f(t)$ is real then

$$C_{-n} = C_n{}^*$$

and, conversely, if this holds then $f(t)$ will be real. In general, the definition of orthogonality for *complex* $f(t)$ is modified:

$$\int f_1(t) f_2{}^*(t) \, dt = 0$$

We shall not explore this.

Fourier series work for waveforms that are nonzero over a finite time interval only, or that repeat periodically. We can, of course, make a Fourier series representation of an endless function over some finite time interval T, but the series will not give the correct values of the function outside of the interval T unless the endless function is periodic with a period T.

In the next section, we study a representation of functions in terms of sines and cosines or complex exponentials which is more generally valid. This Fourier transform is the basis of much of our later analysis of the behavior of communication systems in the presence of noise.

Problems

1. Find the Fourier components for a sequence of rectangular pulses of amplitudes A and lengths T, which recur at intervals T_0. Assume that one pulse is centered at $t = 0$. Obtain both the real and complex form of the Fourier series.

2. For the pulses of problem 1, assume that A increases and T decreases, keeping $AT = K$ a constant. As T approaches zero, what do the Fourier coefficients

approach? (Use the complex form.) What happens if the pulses are centered at $t = t_0$ instead of $t = 0$?

3. If we have the complex Fourier coefficients C_n for a signal $V(t)$, what are the Fourier coefficients for dV/dt? d^2V/dt^2? $\int_{-T/2}^{t} V(s)\,ds$? In the third case, assume V has no dc. What is the reason for this assumption?

3.4. The Fourier Transform

Let $x(t)$ be a function of time, of finite energy. We define

$$X(f) = \int_{-\infty}^{\infty} x(t) \exp(-j\omega t)\,dt \qquad (3.4\text{-}1)$$

This is the Fourier transform of $x(t)$. Here f is frequency in Hz:

$$\omega = 2\pi f$$

We note that the Fourier transform is a *linear* operation on waveforms or pulses $x(t)$: multiplying pulses by constants and adding them does the same to their Fourier transforms.

We assert that the Fourier transform can be inverted:

$$x(t) = \int_{-\infty}^{\infty} X(f) \exp(j\omega t)\,df \qquad (3.4\text{-}2)$$

We shall derive this. Using (3.4-1) (with t_1 as the variable of integration instead of t), (3.4-2) becomes

$$x(t) = \int_{-\infty}^{\infty} \int_{-\infty}^{\infty} x(t_1) \exp(-j\omega t_1) \exp(j\omega t)\,dt_1\,df \qquad (3.4\text{-}3)$$

In the integration with respect to f, let us replace the ∞ limits with large values $+\omega_0$, $-\omega_0$. Use $df = (1/2\pi)\,d\omega$, and reverse the order of integration:

$$x(t) \doteq \frac{1}{2\pi} \int_{-\infty}^{\infty} x(t_1) \left(\int_{-\omega_0}^{\omega_0} \exp[j\omega(t - t_1)]\,d\omega \right) dt_1$$

$$= \frac{1}{\pi} \int_{-\infty}^{\infty} x(t_1) \frac{\exp[j\omega_0(t - t_1)] - \exp[-j\omega_0(t - t_1)]}{2j(t - t_1)}\,dt$$

$$x(t) \doteq \frac{1}{\pi} \int_{-\infty}^{\infty} x(t_1) \frac{\sin \omega_0(t - t_1)}{t - t_1}\,dt_1 \qquad (3.4\text{-}4)$$

Because t is a constant in the integration with respect to t_1,

$$x(t) \doteq \frac{1}{\pi} \int_{-\infty}^{\infty} x(t_1) \frac{\sin \omega_0(t_1 - t)}{\omega_0(t_1 - t)}\,d(\omega_0(t_1 - t))$$

When ω_0 is very large, the integrand is negligible except very near $t = t_1$. We can take $x(t_1)$ as a constant equal to $x(t)$ over this range, and obtain

$$x(t) = x(t) \cdot \frac{1}{\pi} \int_{-\infty}^{\infty} \frac{\sin x}{x} \, dx \qquad (3.4\text{-}5)$$

The integral of $\sin x / x$ is known to be π (not a trivial result), and the Fourier inversion formula is verified. The integral in (3.4-5) will be found in Section 5.4.

In connection with the Fourier transform of a real function (and signals are real functions), we should note that

$$X(-f) = X^*(f)$$

Thus the values of the transform for negative frequencies tell us nothing that we cannot learn from the values of the transform for positive frequencies.

The Fourier transform $X(f)$ of $x(t)$ is often called the *frequency spectrum* of $x(t)$. It is sometimes called the voltage spectrum, to distinguish it from the energy spectrum and power spectrum of which we shall make much use in this book. The energy spectrum is merely

$$X(f)X^*(f)$$

We shall explain this in Section 3.6. The power spectrum is slightly harder to define rigorously. We have, however, been using it since Chapter 2, and shall make a rigorous definition in Chapter 6.

In the next section, we shall see the relation between the transform of a function and the function shifted in time or frequency.

Problems

1. Find the Fourier transform of a single rectangular pulse of amplitude A and width T centered at $t = 0$. What is the relation between T and the frequency at which the amplitude of the spectrum first goes to zero? Suppose that T goes to zero but A increases so as to make $AT = 1$. This is called a delta function, $\delta(t)$. What is $X(f)$?

2. Find the Fourier transform of the pulse illustrated below:

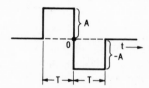

3. What is the integral from $-\infty$ to t of the pulse of Problem 2? What is the Fourier transform of this? What is the relation of the two?

4. Explain why it might be hard to find a Fourier pair $x(t)$, $X(f)$ such that the inversion formula can be easily verified directly by integration, knowing that $\int_{-\infty}^{\infty} (\sin x/x)\, dx$ is hard to evaluate.

3.5. Delay and Phase

We have

$$X(f) = \int_{-\infty}^{\infty} x(t) \exp(-j2\pi ft)\, dt$$

$$x(t) = \int_{-\infty}^{\infty} X(f) \exp(j2\pi ft)\, df$$

The function $x(t - t_0)$ has the same shape as $x(t)$ but is delayed by a time t_0, that is, it reaches a given amplitude a time t_0 later than $x(t)$ does. What about the Fourier transform of $x(t - t_0)$, which we will call $X_1(f)$?

$$X_1(f) = \int_{-\infty}^{\infty} x(t - t_0) \exp(-j2\pi ft)\, dt \qquad (3.5\text{-}1)$$

Let

$$u = t - t_0$$
$$t = t_0 + u$$
$$dt = du$$

Then

$$X_1(f) = \exp(-j2\pi ft_0) \int_{-\infty}^{\infty} x(u) \exp(-j2\pi fu)\, du$$

$$X_1(f) = \exp(-j2\pi ft_0)X(f) \qquad (3.5\text{-}2)$$

The frequency spectrum of $x(t - t_0)$ is $\exp(-j2\pi ft_0)$ times the frequency spectrum of $x(t)$. That is, in the spectrum of $x(t - t_0)$ there is a phase lag of $2\pi ft_0$ with respect to the spectrum of $x(t)$. So, linear phase shift corresponds to time delay.

For completeness we can note that

$$\int_{-\infty}^{\infty} \exp(-j2\pi ft_0)X(f) \exp(j2\pi ft)\, df = \int_{-\infty}^{\infty} X(f) \exp[j2\pi f(t - t_0)]\, df$$

$$= x(t - t_0) \qquad (3.5\text{-}3)$$

We have seen that a phase lag ωt_0, which is proportional to radian frequency, corresponds to a delay of the signal by time t_0, which is the derivative of the phase lag with respect to radian frequency. A constant phase shift which does not change with frequency does not correspond to a delay. When the phase lag ϕ is a function of frequency the quantity

$T(\omega)$,

$$T(\omega) = \frac{d\phi}{d\omega} \tag{3.5-4}$$

is called the *group delay* or simply *delay*. In an approximate sense, at least, the frequency components near a frequency ω are delayed by a time $T(\omega)$ given by (3.5-4).

The next section treats the energy of waveforms, which is insensitive to phase. We shall see that we can compute the energy of a waveform just as easily from the Fourier transform as from the waveform itself.

Problems

1. The Fourier transform of a pulse of amplitude A and duration T centered about $t = 0$ is

$$X(f) = AT\frac{\sin(\pi Tf)}{\pi Tf}$$

Knowing this, write down the Fourier transforms for the pulses sketched below:

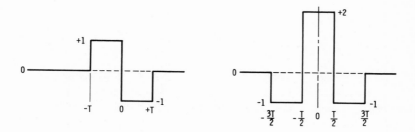

2. If the group delay of a propagation medium or network is proportional to frequency

$$T(\omega) = \omega t_0^2$$

what is the relation of the input and output frequency spectra?

3.6. The Energy Theorem

Any time-varying voltage $v(t)$ is real. Hence

$$v(t) = \int_{-\infty}^{\infty} V(f)e^{j\omega t}\, df = \int_{-\infty}^{\infty} V^*(f)e^{-j\omega t}\, df$$

Here $\omega = 2\pi f$. The energy E is

$$E = \int_{-\infty}^{\infty} [v(t)]^2 \, dt = \int_{-\infty}^{\infty} v(t) \left(\int_{-\infty}^{\infty} V^*(f) e^{-j\omega t} \, df \right) dt$$

If we integrate with respect to time first, we obtain

$$\int_{-\infty}^{\infty} V^*(f) \left(\int_{-\infty}^{\infty} v(t) e^{-j\omega t} \, dt \right) df$$

The integral in the brackets is $V(f)$. Hence

$$E = \int_{-\infty}^{\infty} V(f) V^*(f) \, df$$

$E(f) = V(f)V^*(f)$ is the energy density with respect to frequency, or the *energy spectrum*. This works for complex voltages as well.

We have already noted that $V(-f)$ tells us nothing that $V(f)$ does not tell us. Do we really need to integrate from $-\infty$ to 0 as well as from to 0 to ∞ in order to get the energy? We note that

$$V(-f) = V^*(f) \quad \text{and} \quad V^*(-f) = V(f)$$
$$V(f) V^*(f) = V(-f) V^*(-f)$$

Thus

$$E = \int_{0}^{\infty} 2 V(f) V^*(f) \, df$$

In Section 2.1 we derived the power density of Johnson noise. Power density is similar to energy density. In specifying energy density or power density we must make clear whether we consider both positive and negative frequencies (in this case the total energy is obtained by integrating from $f = -\infty$ to $f = +\infty$) or whether we consider positive frequencies only (in this case the total energy or power is obtained by integrating from 0 to ∞). The power density of Johnson or thermal noise, kT, was so defined that we use positive frequencies only in calculating noise power. The Johnson noise power density appropriate for both positive and negative frequencies would be $kT/2$.

There is a *multiplication theorem* entirely analogous to the energy theorem. If $v(t)$ and $w(t)$ are real voltages, the multiplication theorem states

$$\int_{-\infty}^{\infty} v(t) w(t) \, dt = \int_{-\infty}^{\infty} V(f) W^*(f) \, df$$

Here $V(f)$ is the spectrum of $v(t)$ and $W(f)$ is the spectrum of $w(t)$. The derivation is exactly the same as the derivation of the energy theorem.

In the next section, we study the relation between the Fourier transform of a waveform and the Fourier transform of the waveform after it has been passed through a linear network. The transfer function of the network determines what happens to the transforms.

Problem

1. The pulse

$$x(t) = A \frac{\sin (\pi t/T)}{\pi t/T}$$

has a frequency spectrum $X(f)$ which is constant for $-f_0 < f < f_0$ and zero outside of this range. Find f_0 and the amplitude of $X(f)$ by transforming such a frequency spectrum and comparing the result with the given pulse. Then compute the energy of the pulse from the frequency spectrum. Try computing the energy directly from $x(t)$. What integral does this allow you to evaluate?

3.7. Response of a Linear Network; Transfer Functions

We can measure the output response of a time-invariant linear network to a sinusoidal input of unit amplitude and zero phase. This output, a complex quantity $H(f)$, is a function of frequency which describes the properties of the network. We call the function $H(f)$ a *transfer function* or a voltage transfer function. If we express a time function $x(t)$ as a Fourier transform or spectrum $X(f)$, then the spectrum or Fourier transform of the output waveform of the network when $x(t)$ is applied as an input is

$$H(f)X(f)$$

The time variation of the output, $y(t)$, will be

$$y(t) = \int_{-\infty}^{\infty} H(f)X(f)e^{j2\pi ft} \, df \tag{3.7-1}$$

The output energy E_0 is

$$E_0 = \int_{-\infty}^{\infty} (H(f)X(f))(H(f)X(f))^* \, df$$

$$E_0 = \int_{-\infty}^{\infty} H(f)H(f)^*X(f)X(f)^* \, df \tag{3.7-2}$$

The argument above is made simply on the basis that the Fourier transform of any time function is made up of many (a spectrum of) frequency components. The energies of different frequency components add, and each frequency component is changed in amplitude and phase by the factor $H(f)$, the transfer function of the network, when the signal passes through the network. We shall not go any more deeply into network theory in this book. The transfer function is enough for the purpose of understanding communication systems.

There is a direct way of getting to (3.7-2), without appealing to network theory. Suppose that at $t = 0$ we apply a very short pulse of unit area to the linear network described by $H(f)$. We will study such impulses or delta functions in Section 3.9. The output of the network at a time s later will be the *impulse response* $h(s)$. We will use the impulse response to derive the transfer function.

We can regard the input to the network, $x(t)$, as being made up of a number of successive infinitesimally short rectangular pulses. The strength or area of the pulse of length ds at a time s in the past (at time $t - s$) is

$$x(t - s)\, ds$$

This acts as an input to the network whose response to a pulse of *unit* area at a time s after that pulse is applied is $h(s)$. Because the network is linear and time invariant, at time t the response of the network to the pulse of area $x(t - s)\, ds$ which occurred at time $t - s$ is

$$h(s)x(t - s)\, ds$$

Thus the output of the network to the whole input pulse $x(t)$ will be

$$y(t) = (h * x)(t) = \int_{s=-\infty}^{\infty} h(s)x(t - s)\, ds = \int_{s=-\infty}^{\infty} h(t - s)x(s)\, ds$$

$$(3.7\text{-}3)$$

The quantity $(h * x)(t)$ is called the *convolution* of two functions or waveforms $x(t)$ and $h(t)$. Convolution is a very important concept for network theory. With $t - s$ replaced by $t + s$, the concept of convolution occurs as autocorrelation in Chapter 6. And, in Chapter 8, the convolution of two probability densities occurs when we derive the central limit theorem. This theorem explains the occurrence of gaussian noise in communication systems. The convolution is just a weighted sum of delayed values with weights depending on the delay s. In physical applications to networks, $h(s)$ will usually be 0 for negative values of s.

We can easily show that the Fourier transform $\widehat{h * x}$ of $h * x$ has a special relation to the Fourier transforms of h and x:

$$\widehat{h * x}(f) = \int_{-\infty}^{\infty} (h*x)(t)e^{-j2\pi ft}\, dt = \int_{-\infty}^{\infty}\int_{-\infty}^{\infty} e^{-j2\pi ft}h(s)x(t - s)\, ds\, dt$$

$$\widehat{h * x}(f) = \int_{-\infty}^{\infty}\int_{-\infty}^{\infty} e^{-j2\pi f(t-s)}e^{-j2\pi fs}h(s)x(t - s)\, ds\, dt \qquad (3.7\text{-}4)$$

If we let $t - s = \tau$ in (3.7-4), we can write the double integral (3.7-4) as a product of two single integrals:

$$\widehat{h * x}(f) = \int_{\tau = -\infty}^{\infty} e^{-j2\pi f\tau} x(\tau) \, d\tau \int_{s = -\infty}^{\infty} e^{-j2\pi fs} h(s) \, ds$$

$$\widehat{h * x}(f) = \hat{h} \cdot \hat{x} = \hat{h}(f) X(f) \tag{3.7-5}$$

We see that the Fourier transform $\hat{h}(f)$ is the transfer function $H(f)$ of the network. This is just what we asserted on other grounds at the beginning of this section. And, the Fourier transform of the convolution of two waveforms is the product of the Fourier transforms of the waveforms. We shall meet this result again in Chapters 6 and 8. It is a very important result for understanding linear systems and frequency spectra. But, we have seen that it is not hard to derive.

In the next section, we will determine some important transfer functions. The networks they correspond to occur often in communications.

Problems

1. A pulse

$$x(t) = A \frac{\sin \pi t}{\pi t}$$

   is applied to a network for which $H(f)$ is unity for $-0.01 < f < 0.01$ and zero outside this range. What is the output time function?

2. Derive the Fourier transform of a triangular waveform with apex of height T at time 0 which lasts from $-T$ to T, knowing the Fourier transform of a square pulse of height 1 lasting from $-T/2$ to $+T/2$ and using (3.7-5).

3.8. Some Special Transfer Functions

We have

$$x(t) = \int_{-\infty}^{\infty} X(f) \exp (j\omega t) \, df$$

$$\frac{d(x(t))}{dt} = \int_{-\infty}^{\infty} j\omega X(f) \exp (j\omega t) \, df$$

A network with the transfer function

$$H(f) = j\omega = j2\pi f$$

results in the differentiation of the input time function. Similarly we have Table 3.2. As an example, integrating a sharp spike of unit area gives a

Table 3.2. Transfer Functions Related to Integration and Differentiation

Transfer function of network	Output for input time function $x(t)$
$j\omega$	$dx(t)/dt$
$(j\omega)^2 = -\omega^2$	$d^2x(t)/dt^2$
$1/j\omega$	$\int_{-\infty}^{t} x(s)\, ds$
etc.	etc.

step function of unit height as in Fig. 3.2. The frequency spectrum of this step function is $1/j\omega$ times the frequency spectrum of a sharp spike. The spectrum of a sharp spike or delta function is a constant. We will learn this in the next section.

We saw in Section 3.5 that if $X(f)$ is the Fourier transform of a time function $x(t)$, then

$$\exp(-j2\pi f t_0)X(f)$$

is the Fourier transform of the time function

$$x(t - t_0)$$

We see that

$$\exp(-j2\pi f t_0)$$

is the transfer function of a network which delays a signal applied to it by a time t_0.

Sometimes this transfer function can be advantageously combined with the transfer functions discussed earlier in this section. Thus the output of a network with the transfer function

$$j\omega \exp(-j2\pi f t_0)$$

is the derivative of the input function at a time $t - t_0$. Similarly, the output of a network with the transfer function

$$(1/j\omega) \exp(-j2\pi f t_0)$$

is

$$\int_{-\infty}^{t-t_0} x(t)\, dt$$

Here $x(t)$ is the input function.

Figure 3.2. Step function of unit height.

The next section formally introduces delta functions and finds their Fourier transforms.

Problem

1. What is the transfer function $H(f)$ that integrates the signal over the past T seconds? Do this in two ways, one of which uses the impulse response. (We will later find in Chapter 7 that such integration is useful in detecting the presence or absence of a rectangular pulse.)

3.9. Fourier Transforms and Delta Functions

Before we can discuss Fourier transforms of waveforms which persist forever and do not die out, we have to be more specific or rigorous about the definition of the Fourier transform.

We are computing the Fourier transform at a certain time. We can know the entire past of the signal, but not the future. This implies asymmetric treatment of time. In practice, much use of the Fourier transform is formal or algebraic. Exactly which rigorous definition we use merely changes the formalism but not the conclusions. We shall not explore this any further.

Let us now introduce the concept of delta function, which we have been using implicitly. The delta function of unit strength or area at time t_0, $\delta(t - t_0)$, is an infinitely tall, infinitely thin function which is 0 except at t_0, but has unit area. It is also called a unit impulse at time t_0. The property which defines the delta function is that for every waveform f which is smooth at t_0, we will have

$$\int_{-\infty}^{\infty} f(t)\delta(t - t_0) \, dt = f(t_0) \tag{3.9-1}$$

In other words, $\delta(t - t_0)$ picks out the value of f at time t_0. Or, it *samples* f at t_0. A delta function is much like an antenna of infinite directivity, with the time dimension replacing the space dimension.

What is the frequency spectrum $X(f)$ of $\delta(t - t_0)$? We have

$$X(f) = \int_{-\infty}^{\infty} \delta(t - t_0) \exp(-j2\pi ft) \, dt$$

$$X(f) = \exp(-j2\pi ft_0) \tag{3.9-2}$$

This is because of characteristic property (3.9-1) of delta functions. The spectrum of a delta function at time 0, $\delta(t)$, is constant with frequency, and is equal to 1. The spectrum of a delta function at time t_0, $\exp(-j2\pi ft_0)$, is a pure complex sinusoid of amplitude 1. This phase shift is as it must be according to Section 3.5.

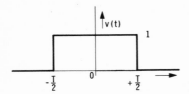

Figure 3.3a. Rectangular pulse of height 1.

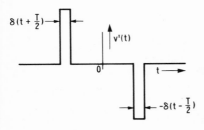

Figure 3.3b. Derivative of rectangular pulse.

We can make use of (3.9-2) in taking Fourier transforms of voltages $v(t)$ if $v(t)$ has only a limited number of derivatives. Consider, for example, $v(t)$ as shown in Fig. 3.3a; $v'(t)$ is two delta functions of unit area, the second with sign reversed, as shown in Fig. 3.3b.

It is now easy to take the Fourier transform of $v'(t)$. We must get

$$\exp(j\pi fT) - \exp(-j\pi fT)$$

From Section 3.8 we have for the transform of the original function $v(t)$

$$V(f) = \frac{\exp(j\pi fT) - \exp(-j\pi fT)}{j2\pi f}$$

$$V(f) = \frac{\sin \pi fT}{\pi f}$$

This, of course, can be obtained directly. $V(f)$ approaches T as f approaches 0, as it clearly must, because $V(0)$ is always the dc component of $v(t)$:

$$V(0) = \int_{-\infty}^{\infty} v(t)\, dt$$

So far we have studied *one*-dimensional Fourier transforms. We are also interested in *two*-dimensional Fourier transforms, especially for antenna patterns, to complete the derivation of Friis' equation of Section 1.3. We also need to consider two-dimensional Fourier transforms for the analysis of the simultaneous relationship of noise occurring at two points in time, which we will study in Chapter 9. The next section discusses the two-dimensional Fourier transform.

Problems

1. Show that

$$\frac{1}{\pi} \frac{\sin \omega_0(t - t_0)}{t - t_0}$$

approaches $\delta(t - t_0)$ as ω_0 approaches ∞, knowing the Fourier inversion formula and using defining property (3.9-1) of the delta function.

2. Using the definition of Fourier transform that we have adopted in this book, one can show that the Fourier transform $Y(f)$ of the *time reversal* $x(-t)$ of the unit step at time 0 is $(-1/j2\pi f) + \delta(f)$. Which rule does this appear to contradict, and which does it support? Be aware of the fact that $x(t) + x(-t) \equiv 1$, whose Fourier transform is $\delta(f)$.

3. Use second derivatives to find the Fourier transform of

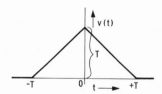

(Compare Problem 2 of Section 3.7.)

4. Find the Fourier transform of the waveform which is 0 for negative t and equal to $(1/t_0) \exp(-t/t_0)$ for positive t, using the relation between the Fourier transform of a function and that of its derivative. (This gives the voltage spectrum of the pulse which results when a unit impulse is presented to an RC filter of time constant $RC = t_0$. We are obtaining the transfer function of the RC filter.)

5. The dc waveform $v(t) \equiv 1$ for $-T/2 \le t \le T/2$ has Fourier series $\equiv 1$. Obtain the complex form Fourier series of

$$f(t) = t + T/2$$

Here

$$f(t) = \int_{-T/2}^{t} v(s)\, ds$$

Explain the relationship of the two Fourier series by referring to the Fourier series for a train of unit impulses, derived in Problem 2 of Section 3,3, and by differentiating an appropriate Fourier series.

6. Use defining property (3.9-1) to show that the delta function $\delta(t)$ is an even function of t.

7. Write the periodic function which is an impulse of strength T repeating every T seconds, with one impulse at time 0, as a sum of delta functions times constants. This is the *sampling function* which we will study in Section 5.1.

8. Find the radar backscatter function $R(\theta)$ of a distant smooth reflecting sphere, given that it is an isotropic scatterer. (Review Section 1.7.) In particular, show

that $R(\theta) = g(\theta)\delta(\theta)$ and find $g(\theta)$ using the radar equation and (1.7-4). Which $g(\theta)$'s will give the same $R(\theta)$?

3.10. Two-Dimensional Fourier Transforms

In Chapter 9, when we study random waveforms, we will encounter two-dimensional Fourier transforms. We also encounter them in the study of two-dimensional sampling in Section 5.7. Suppose that we have a function $v(x, y)$ of two variables, which we may think of as spatial variables x, y. The Fourier transform $V(f, g)$ of this function will involve two spatial frequencies, which we will call f and g. We will have

$$V(f, g) = \int_{-\infty}^{\infty} \int_{-\infty}^{\infty} v(x, y) \exp[-j2\pi(fx + gy)] \, dx \, dy \quad (3.10\text{-}1)$$

This will be a product of two one-dimensional transforms if $v(x, y)$ is the product of a function of x times a function of y. This is because

$$\exp[-j2\pi(fx + gy)] = \exp(-j2\pi fx) \exp(-j2\pi gy)$$

Two-dimensional Fourier transforms are useful in connection with many communications phenomena including those of sonar, radar and passive microwave mapping, picture processing, TV transmission, and antenna patterns. Here we will illustrate the use of the two-dimensional Fourier transform in finding the actual radiation pattern of a microwave reflector antenna.

In the approximate treatment of antenna directivity in Section 1.2 we obtained the contributions from various portions of the antenna aperture, and added them. We took the phase of the arriving contributions into account in a rough way.

There is an exact way to calculate distant electromagnetic fields over a plane. Here we will explore an approximate approach called *scalar diffraction*, which we mentioned in Section 1.2 in discussing antenna directivity. It applies quite accurately to the waves far from a microwave antenna and at moderate angles from the axis.

Imagine that the electromagnetic wave in the x, y plane is everywhere in phase and that the electric field has a magnitude

$$E(x, y)$$

Figure 3.4 illustrates the geometry of the distant field.

We go a long distance L away in the z direction, normal to x and y, and go out normal to the z axis by a distance $L \sin \theta$ in the x, z plane and $L \sin \phi$ in the y, z plane.

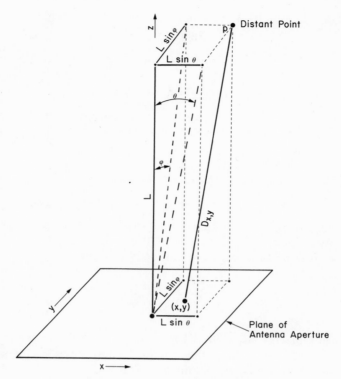

Figure 3.4. Geometry of distant field.

Consider the field $dF_{x,y}(p)$ at the distant point p caused by a little area $dx\,dy$ of the antenna aperture in the z plane. The field at the point x, y is $E(x, y)$. According to scalar diffraction, we have the expression

$$dF_{x,y}(p) = \frac{C}{D_{x,y}} \exp\left[-j(2\pi/\lambda)D_{x,y}\right]E(x, y)\,dx\,dy \qquad (3.10\text{-}2)$$

Here $D_{x,y}$ is the distance from the distant point p to the point x, y in the antenna aperture, and C is a constant which may depend on λ.
  Let us find $D_{x,y}$:

$$D_{x,y} = [(L \sin \theta - x)^2 + (L \sin \phi - y)^2 + L^2]^{1/2} \qquad (3.10\text{-}3)$$

Because L is arbitrarily large while $\sin \theta$ and $\sin \phi$ are small and fixed, and x is fixed, we can ignore the terms x^2 and y^2, and use

$$(1 - \varepsilon)^{1/2} \approx 1 - \frac{\varepsilon}{2}$$

for ε small. Then (3.10-3) becomes

$$D_{x,y} = L'\left(1 - \frac{x\sin\theta + y\sin\phi}{L(1 + \sin^2\theta + \sin^2\phi)}\right) \tag{3.10-4}$$

Here

$$L' = L[1 + \tfrac{1}{2}(\sin^2\theta + \sin^2\phi)] \tag{3.10-5}$$

Because $\sin\theta$ and $\sin\phi$ are small, we can replace the denominator $L(1 + \sin^2\theta + \sin^2\phi)$ by L' in the right-hand side of (3.10-4). Equation (3.10-2) becomes

$$dF_{x,y}(p) = \frac{C}{D_{x,y}}\exp[-j(2\pi/\lambda)L']\exp[+j(2\pi/\lambda)$$
$$\times(x\sin\theta + y\sin\phi)]E(x, y)\,dx\,dy \tag{3.10-6}$$

The full field $F(p)$ at the distant point p is obtained by integrating this over x, y. We can replace the $D_{x,y}$ in (3.10-6) by L because this affects the magnitude little and the phase not at all. We obtain

$$F(p) = \frac{C}{L}\exp[-j(2\pi/\lambda)L']\int_{-\infty}^{\infty}\int_{-\infty}^{\infty}\exp[j(2\pi/\lambda)(x\sin\theta + y\sin\phi)]$$
$$\times E(x, y)\,dx\,dy \tag{3.10-7}$$

We can replace L' by L in (3.10-7) if we interpret $F(p)$ as the field on a sphere at distance L from the origin of coordinates in the aperture plane. When we do this, we obtain the expression

$$F(p) = \frac{C}{L}\exp[-j(2\pi/\lambda)L]\int_{-\infty}^{\infty}\int_{-\infty}^{\infty}\exp[+j(2\pi/\lambda)(x\sin\theta + y\sin\phi)]$$
$$\times E(x, y)\,dx\,dy \tag{3.10-8}$$

This shows that the field at the distant point p at distance L with x and y components $L\sin\theta$ and $L\sin\phi$, respectively, is the two-dimensional inverse (inverse, because of the $+j$ in the double integral) Fourier transform of the field $E(x, y)$ in the plane of the antenna. The spatial frequencies f and g are

$$f = \frac{\sin\theta}{\lambda}\quad\text{and}\quad g = \frac{\sin\phi}{\lambda}$$

Let us now apply this to the case of a constant field of unit strength over a square of side W in the x, y plane centered at $x = y = z = 0$. The distant field strength $F(p)$ at the point p with angular coordinates θ, ϕ is

$$F(p) = \frac{C}{L}\exp[-j2\pi(L/\lambda)]\int_{-W/2}^{W/2}\int_{-W/2}^{W/2}\exp[+j(2\pi/\lambda)(\sin\theta)x]$$
$$\times\exp[+j(2\pi/\lambda)(\sin\phi)y]\,dx\,dy \tag{3.10-9}$$

This integral is a product of two one-dimensional Fourier transforms which we have already found (see Section 3.9). We obtain

$$F(p) = \frac{C}{L} \exp\left[-j2\pi(L/\lambda)\right] W^2 \cdot \frac{\sin(\pi W\theta/\lambda)}{\pi W\theta/\lambda} \cdot \frac{\sin(\pi W\phi/\lambda)}{\pi W\phi/\lambda}$$

(3.10-10)

Here we have used the fact that θ and ϕ are small angles, so that

$$\sin\theta \sim \theta, \qquad \sin\phi \sim \phi$$

For us, the important thing about (3.10-10) is the product of $(\sin x)/x$ terms, which tells us how the field varies with the small angles θ and ϕ from the axis. Physical reasoning tells us that almost all the power is concentrated at small angles θ and ϕ. We can then assume that (3.10-10) is valid for *all* θ and ϕ in calculating the received power. This is so because in integrating $F(p)F^*(p)$ over the entire plane at distance L from the transmitting antenna, almost all the integral is concentrated at small θ and ϕ.

By making use of this fact, we can evaluate the received power by first finding the constant C in (3.10-10). The element of area in the distant plane is

$$d(L\sin\theta)\, d(L\sin\phi) = L^2\cos\theta\cos\phi\, d\theta\, d\phi \sim L^2\, d\theta\, d\phi$$

This permits us to derive the exact form of Friis's transmission formula, improving upon the approximate formula of Section 1.2:

$$\frac{P_R}{P_T} = \frac{A_T A_R}{\lambda^2 L^2}$$

(1.3-1)

It appears that we derived this only for *square* apertures. However, the result is valid for uniformly illuminated apertures of any shape, as we see by thinking back to Sections 1.1–1.3, and Problem 5 of Section 1.3. The argument is this: the received power must be proportional to the area of the receiving antenna. But, the path loss is the same for a pair of antennas if the transmitter and receiver are interchanged, so that the transmitting antenna is used to receive. Hence, the power received must be proportional to the area of the transmitting antenna. Then (1.3-1) must hold for any geometric shape, if it holds for one particular shape. This will be so as long as the illumination is of constant phase and amplitude over the transmitting aperture and as long as all dimensions of the antennas are large compared with wavelength.

There is another kind of two-dimensional representation of spectra. For this, there will only be one frequency dimension. The other dimension will be time. Short-term energy spectra display a spectrum averaged over short time intervals as time progresses, resulting in a two-dimensional plot often used for speech analysis. The next section derives the properties of short-term energy spectra.

Problems

1. Show that the constant C in (3.10-10) is $1/\lambda$ by finding the total power received in the plane at distance L and using conservation of energy. You need to know (see Section 5.4 or Section 3.6, Problem 1) that

$$\int_{-\infty}^{\infty} \frac{(\sin \pi t)^2}{(\pi t)^2}\, dt = 1$$

2. Use Problem 1 to derive Friis's transmission formula, equation (1.3-1).

3. What is the two-dimensional Fourier transform of a function $v(x, y)$ which depends on x only? That is,

$$v(x, y) = h(x)$$

3.11. Short-Term Energy Spectra

The spectrum $V(f)$ of the impulse $\delta(t)$ is

$$V(f) = 1 \tag{3.11-1}$$

That is, all frequency components have the same amplitude and zero phase. The energy spectrum is

$$E(f) = V(f)V^*(f) = 1 \tag{3.11-2}$$

Now consider a function $v(t)$ for which the voltage spectrum is

$$V(f) = \exp[-j(\omega t_0)^2] \tag{3.11-3}$$

The energy spectrum of this signal is clearly given by (3.11-2) and so is the same as that of the delta function $\delta(t)$. Yet, the function $v(t)$ is very different from $\delta(t)$.

What is $v(t)$ like? If we compute the delay $T(\omega)$ for (3.11-3) according to (3.5-4) we find that

$$T = \frac{d(\omega t_0)^2}{d\omega} = 2\omega t_0^2 = 4\pi f t_0^2 \tag{3.11-4}$$

In effect, the signal $v(t)$ is a signal of changing frequency, because higher-frequency components arrive later than lower-frequency components. Locally it looks like a sine wave, but the frequency of this wave is proportional to time (or delay) T after $t = 0$ by the factor $1/4\pi t_0^2$. This is not brought out clearly in the Fourier transform (3.11-3), though, as we have seen, it is implicit in the phase of the transform.

The Fourier transform gives us a complete but implicit description of a time function over all time, but it does not tell us directly what the function is like at a particular time. We can get at least a qualitative feel for the short-term nature of a time function by means of the short-term spectrum and the short-term energy spectrum. See Reference 4 for a description of the use of these concepts in speech research.

To obtain the short-term spectrum $S(f, t)$ we multiply a time function $v(t)$ by a *gating* or *window* function and take the Fourier transform:

$$S(f, t) = \int_{-\infty}^{\infty} g(T)v(T + t) \exp(-j2\pi fT) \, dT \qquad (3.11\text{-}5)$$

Here $g(T)$ is a function which goes to zero for large positive and negative value of T and which in most cases decreases as the departure T from time t increases.

According to the definition given above, we regard the short-term spectrum at time t as the spectrum of the part of v in the vicinity of t. By this, we mean $v(t + T)$ weighted by the function $g(T)$. Thus the short-term spectrum at time t tells about the waveform over a range of time around t.

We can also look at the short-term spectrum in a somewhat different way. Let us use the inverse transform formula to express $g(T)$ in terms of the Fourier transform $G(f)$:

$$g(T) = \int_{-\infty}^{\infty} G(f_0) \exp(j2\pi f_0 T) \, df_0$$

We can now express the short-term spectrum $S(f, t)$ in terms of $G(f)$:

$$S(f, t) = \int_{-\infty}^{\infty} \int_{\infty}^{\infty} G(f_0)v(T + t) \exp[-j2\pi(T + t)(f - f_0)]$$
$$\times \exp[j2\pi t(f - f_0)] \, d(T + t) \, df_0$$

$$S(f, t) = \int_{-\infty}^{\infty} G(f_0) \exp[j2\pi t(f - f_0)] V(f - f_0) \, df_0$$

Here V is the Fourier transform of v. This is also expressible as

$$S(f, t) = \int_{-\infty}^{\infty} G(f - f_0) V(f_0) \exp(j2\pi f_0 t) \, df_0 \qquad (3.11\text{-}6)$$

Equation (3.11-6) shows that the short-term spectrum at frequency f can be regarded as the part of the signal spectrum at frequencies near to f, weighted by $G(f - f_0)$, where $f - f_0$ is the departure from the frequency f. Thus the short-term spectrum at a frequency f is a weighted composite of the spectrum over a range of frequencies near to f.

In order to get the short-term spectrum $S(f, t)$ we must integrate over a range of time [equation (3.11-5)]] or a range of frequency [equation (3.11-6)]. Both integrations must give the same result.

From the short-term spectrum $S(f, t)$ we can obtain the short-term energy spectrum $E(f, t)$:

$$E(f, t) = S(f, t)S^*(f, t) \qquad (3.11-7)$$

Because of the gating function, there is a finite short-term energy spectrum even for functions $v(t)$ which persist forever, or which have infinite bandwidth or infinite energy.

Frequently, the short-term energy $E(f, t)$ is represented as degree of darkness in plot where the horizontal coordinate is time and the vertical coordinate is frequency. Such a plot is called a *spectrogram* or *sonogram*.

Let us consider two special cases. In the first case, we will assume that $g(T)$ is a delta function:

$$g(T) = \delta(T) \qquad (3.11-8)$$

In this case,

$$S(f, t) = v(t) \qquad (3.11-9)$$

The short-term energy spectrum is a function of time only. In a sonogram, the darkness will be equal in the vertical direction for each value of time. There is no frequency information. The darkness plots out $[v(t)]^2$ as a function of time.

As the other special case, we will let the signal, $v(t)$, be

$$v(t) = \cos 2\pi f_0 t = \tfrac{1}{2}[\exp(j2\pi f_0 t) + \exp(-j2\pi f_0 t)] \quad (3.11-10)$$

This may represent a sinusoidal signal, or it may represent one frequency component of a more general signal. We have

$$V(f) = \tfrac{1}{2}[\delta(f - f_0) + \delta(f + f_0)]$$

The short-term spectrum and energy density are

$$S(f, t) = \tfrac{1}{2}\exp(j2\pi f_0 t)G(f - f_0) + \tfrac{1}{2}\exp(-j2\pi f_0 t)G(f + f_0)$$
$$(3.11-11)$$

$$E(f, t) = \tfrac{1}{4}[G(f - f_0)G^*(f - f_0) + G(f + f_0)G^*(f + f_0)]$$
$$+ \tfrac{1}{4}[G(f - f_0)G^*(f + f_0)\exp(j4\pi f_0 t) \qquad (3.11-12)$$
$$+ G^*(f - f_0)G(f + f_0)\exp(-j4\pi f_0 t)]$$

For a $\sin 2\pi f_0 t$ input, the formula is the same, except that the sign of the cross terms is reversed.

Thus in this case there are two terms which are functions of frequency. The first or dc term does not vary with time; the second or cross term varies with a frequency $2f_0$. In spectrograms or sonograms we ordinarily do not see the cross term, but derive our information from the dc term only.

We see because of (3.11-12) that $E(f, t)$ for a given frequency f is affected also by components of other frequencies $f \pm f_0$. Thus the depiction of frequency is blurred.

In general, if $E(f)$ is the total energy of $E(f, t)$ over all time,

$$E(f) = \int_{\infty}^{\infty} S(f, t)S^*(f, t)\, dt \qquad (3.11\text{-}13)$$

then we can write

$$E(f) = \int_{-\infty}^{\infty} \int_{-\infty}^{\infty} \int_{-\infty}^{\infty} G(f - f_0) V(f_0) \exp(j2\pi f_0 t) G^*(f - f_1) V^*(f_1)$$

$$\times \exp(-j2\pi f_1 t)\, dt\, df_0\, df_1$$

Integrate with respect to t first:

$$E(f) = \int_{-\infty}^{\infty} \int_{-\infty}^{\infty} V(f_0) V^*(f_1) G(f - f_0) G^*(f - f_1) \delta(f_1 - f_0)\, df_0\, df_1$$

$$E(f) = \int_{-\infty}^{\infty} V(f_0) V^*(f_0) G(f - f_0) G^*(f - f_0)\, df_0 \qquad (3.11\text{-}14)$$

$E(f)$ as given by (3.11-14) is the convolution of VV^* with GG^*. For fixed f, this can be thought of as the energy output of a tunable band-pass filter when the filter characteristic is centered at frequency f and the voltage transfer function of the filter as a function of the frequency f_0 is $G(f - f_0)$. As in the case of equation (3.11-6) for the short-term spectrum, the energy spectrum $E(f)$ is made up of weighted contributions from frequencies $f - f_0$. Indeed, actual sonograms or spectrograms are constructed by scanning a recorded signal with a tunable receiver, rather than by time shifting and taking a Fourier transform. A new $E(f)$ is computed about every 10 msec, a time comparable to the reciprocal of the bandwidth of the window $g(T)$.

Let us turn our attention to (3.11-5). If $g(T)$ is a narrow time function, $G(f - f_0)$ will be a broad frequency function; time will be sharply depicted and frequency will be blurred. If $g(T)$ is a broad time function, $G(f - f_0)$ will be a narrow frequency function; frequency will be sharp but time will be blurred.

However, how blurred a signal *looks* in time and frequency depends on the nature of the signal. Thus we can regard the signal of (3.11-3) as a signal whose frequency increases with time at a rate that we can find from (3.11-4). For this signal, if we make $g(T)$ very broad we see a broad range of frequencies in $S(f, t)$ because the frequency changes during the time interval encompassed by the gating function $g(T)$. On the other hand, if we make $g(T)$ very narrow, $S(f, t)$ is necessarily blurred in frequency because $G(f - f_0)$ is a broad function of frequency.

About the best we can do if the frequency is changing with the time is to make the blurring due to change of frequency with time and the blurring due to a narrow $g(T)$ and a broad $G(f - f_0)$ about the same.

We may note, for example, that if

$$g(T) = (1/T_0) \exp(-\pi T^2/T_0^2) \tag{3.11-15}$$

then (see Section 8.6)

$$G(f - f_0) = \exp[-\pi T_0^2(f - f_0)^2] \tag{3.11-16}$$

That is, if $g(T)$ falls off by a fraction $e^{-\pi}$ in a time interval T_0, $G(f - f_0)$ will fall off by the same fraction in a frequency interval $1/T_0$.

The particular case given by (3.11-15) and (3.11-16) suggests that the length of the gating function $g(T)$, which we will call ΔT, and the bandwidth of $G(f - f_0)$, which we will call Δf, are related by an uncertainty relation (see Section 11.8)

$$\Delta T \Delta f = \text{constant} \tag{3.11-17}$$

This relation is true (as a lower bound) for general g and G. In this case, if we define ΔT and Δf as T_0 and $1/T_0$ in (3.11-15) and (3.11-16), then

$$\Delta T \Delta f = 1 \tag{3.11-18}$$

If we accept (3.11-18) in general, and if the rate of change of frequency with time is df/dt, then a good compromise is

$$(df/dt)\Delta T = \Delta f$$

In view of (3.11-18), this gives

$$\Delta f = 1/\Delta T = (df/dt)^{1/2} \tag{3.11-19}$$

For the particular example of (3.11-3)

$$\frac{df}{dt} = \frac{1}{4\pi t_0^2}$$

$$\Delta f = \frac{1}{2\pi^{1/2} t_0} \tag{3.11-20}$$

The running energy spectrum, or spectrogram or sonogram, has a wide range of uses. It is particularly important in the depiction of speech. In sonograms of speech, the pitch or periodicity of vibration of the vocal cords appears as a periodic time structure if $g(T)$ is narrow and as a periodic frequency structure (the harmonics of the fundamental pitch frequency) if $g(T)$ is broad. The resonances of the vocal tract appear as

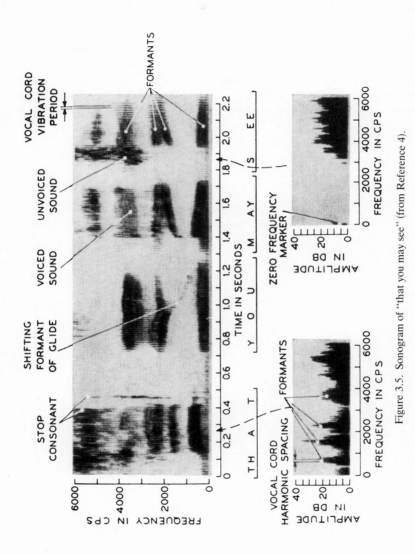

Figure 3.5. Sonogram of "that you may see" (from Reference 4).

dark stripes which are horizontal for a steady vowel (a constant vocal tract configuration) and which bend up or down during transitions between a consonant and a vowel or from one vowel to another, as in Fig. 3.5. The vocoder, which we will study in Section 13.4, transmits the short-term spectrum of speech along with other information to reduce the amount of data that must be transmitted for speech communication.

This chapter has given us some facility with the representation and analysis of waveforms as weighted sums of sinusoidal waveforms. In the next chapter, we use this capability to study ways of generating and receiving signals. We shall also begin to see how to take noise into account quantitatively in signal-to-noise ratios as a limiting factor in communication systems.

Problems

1. Consider the signal involving a high frequency f_0 and a low frequency f_1:
   $\cos (2\pi f_0 t + 10 \sin 2\pi f_1 t)$.
   (a) If we define the rate of change of the argument with time as the instantaneous radian frequency, what is the instantaneous radian frequency? The instantaneous frequency?
   (b) Qualitatively, what range would you expect the energy spectrum to cover for a very broad $g(T)$? (This anticipates Carson's formula of Section 4.7.)
   (c) What is the extreme value of df/dt?
   (d) If we accept (3.11-19), how big should Δf be in order to show the change of frequency with time most clearly?

2. Find the time function $v(t)$ corresponding to

$$V(f) = \exp[-j(\omega t_0)^2]$$

knowing

$$\int_{-\infty}^{\infty} \exp(-jx^2)\, dx = \pi^{1/2} \exp(-\pi j/4)$$

What can we say about the amplitude?

4

Modulation and Noise

In the preceding chapters we have considered two aspects of radio transmission: Path loss between antennas, and the noise which is mixed with a radio frequency carrier signal of constant amplitude. We have specified the noisiness of a receiver in terms of a noise temperature. We have discussed the noise performance of an ideal receiver which is, of course, better than that of any actual receiver. Given transmitter power, antenna sizes including aperture efficiencies and separation, wavelength, the temperature of space, and receiver bandwidth, we can calculate the carrier-to-noise ratio for an electromagnetic transmission system in free space.

Such transmission systems are not used to send an unmodulated carrier. They are used to transmit information.

In this chapter we will consider various analog or continuous modulation systems, as opposed to digital or discrete systems. Reference 3 can be consulted for more details including some of the major circuit concepts. Here we will see how carrier-to-noise ratio affects the signal-to-noise ratio of the demodulated signal. We shall be using what we have learned about the Fourier transform and the representation of complicated signals by sums of sine waves. We shall see how sine waves themselves are directly used in creating the modulated waveforms which get radiated from the antenna. We will study two forms of amplitude modulation, single-sideband, frequency modulation, phase modulation, and some lesser-used techniques.

4.1. Modulation and Detection—Multiplication of a Signal by a Sine Wave

Multiplication of a signal by a sine wave is a powerful process in modulation and detection. It is of both theoretical and practical importance.

Consider a particular frequency component of a *baseband* signal (a signal whose frequencies $f = \omega/2\pi$ lie between 0 and B, the bandwidth of the signal). Let the radian frequency of the component be ω, the amplitude be A, and the phase be ϕ, so that the frequency component is

$$A \cos(\omega t + \phi) \tag{4.1-1}$$

Let us multiply this by

$$2 \cos \omega_0 t$$

This operation takes place at the transmitter. The product is

$$2A \cos(\omega t + \phi) \cos \omega_0 t = A \cos[(\omega_0 + \omega)t + \phi] \\ + A \cos[(\omega_0 - \omega)t - \phi] \tag{4.1-2}$$

We see that on multiplication by $2 \cos \omega_0 t$ a single baseband frequency component of radian frequency ω gives rise to two sideband components of radian frequencies $\omega_0 + \omega$ and $\omega_0 - \omega$, and that the baseband phases appear in these components.

Let us in turn multiply either component by $2 \cos \omega_0 t$. This takes place at the receiver. The product is

$$\begin{aligned}(2 \cos \omega_0 t)A &\cos[(\omega_0 \pm \omega)t \pm \phi] \\ &= A \cos(\mp \omega t \mp \phi) + A \cos[(2\omega_0 \pm \omega)t \pm \phi] \\ &= A \cos(\omega t + \phi) + A \cos[(2\omega_0 \pm \omega)t \pm \phi] \tag{4.1-3}\end{aligned}$$

The second, high-frequency, term can be eliminated by low-pass filtering. We see that the baseband signal component can be recovered in amplitude and phase by multiplying either of the terms in (4.1-2) by $2 \cos \omega_0 t$. This formula is the basis for the study of the modulation schemes to follow. In particular, in the next section we will describe how frequency shifting is used to produce single-sideband modulation.

Problems

1. Graph the waveform of equation (4.1-2) for $A = 1$, $\phi = \pi/2$ radians, $\omega_0 = 3\pi$ rad/sec, $\omega = 4\pi$ rad/sec. What happens if ϕ is π instead?

2. Find the voltage spectrum of the signals given by (4.1-1) and (4.1-2).

4.2. Single-Sideband Modulation (Frequency Shifting)

Consider a band-limited baseband signal. This means its frequencies $f = \omega/2\pi$ lie between 0 and B:

$$0 < \omega < 2\pi B \tag{4.2-1}$$

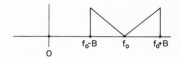

Figure 4.1. Upper and lower sidebands, when $f_0 > B$.

Suppose we multiply the signal by $2 \cos \omega_0 t$. We have seen that if we multiply a frequency component

$$A \cos (\omega t + \phi) \qquad (4.2\text{-}2)$$

by

$$2 \cos \omega_0 t$$

we obtain frequency components

$$A \cos [(\omega_0 + \omega)t + \phi] + A \cos [(\omega_0 - \omega)t - \phi] \qquad (4.2\text{-}3)$$

The two terms in (4.2-3) are called *sidebands*. The term which involves $(\omega_0 + \omega)$ is called the *upper sideband* and the term which involves $(\omega_0 - \omega)$ is called the *lower sideband*.

The range in frequency of the two sidebands [in accord with (4.2-1)] is shown in Fig. 4.1, when $f_0 > B$ and $\omega_0 = 2\pi f_0$. We easily see two things.

If

$$f_0 < B$$

the lower-sideband frequency $f_0 - f$ will be negative for some values of f in the range 0 to B. This means that the same lower-sideband frequency can be produced by two different signal frequencies. The signal cannot be recovered unambiguously from the lower sideband.

If

$$f_0 < B/2$$

the lower sideband will encroach on the upper sideband.

Let us assume that

$$f_0 > B$$

$$\omega_0 > 2\pi B \qquad (4.2\text{-}4)$$

In this case either sideband represents the signal adequately. In principle, we can select either by means of a band-pass filter. We call such a selected sideband a *single-sideband* (*ssb*) signal or a *frequency-shifted* signal.

Let us in turn multiply either sideband by $2 \cos \omega_0 t$. The product is

$$
\begin{aligned}
(2 \cos \omega_0 t)(A \cos [(\omega_0 \pm \omega)t \pm \phi] &= A \cos (\mp \omega t \mp \phi) \\
&\quad + A \cos [(2\omega_0 \pm \omega)t \pm \phi] \\
&= A \cos (\omega t + \phi) \\
&\quad + A \cos [(2\omega_0 \pm \omega)t \pm \phi] \qquad (4.2\text{-}5)
\end{aligned}
$$

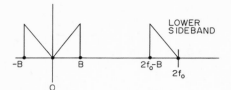

Figure 4.2. Demodulating the lower sideband.

The frequency components in (4.2-5) are represented schematically for the lower sideband in Fig. 4.2. We see that for the − sign in (4.2-5), that is, for recovery from the lower sideband, the frequencies of the second term of (4.2-5) will overlap the frequencies of the first term unless

$$B < f_0 \qquad\qquad (4.2\text{-}6)$$

We should note from Fig. 4.1 that unless (4.2-6) holds, the lower-sideband frequencies obtained by multiplying the baseband signal by $2 \cos \omega_0 t$ will overlap the baseband frequencies. If any baseband frequencies leak past the multiplication, they will be mixed with the sideband frequencies.

Let us turn to another matter concerning frequency shifting. This is oscillator drift at the receiver.

If we multiplied the signal of either sideband by $2 \cos(\omega_0 t + \theta)$ (θ is oscillator drift), we would obtain

$$2 \cos(\omega_0 t + \theta)A \cos[(\omega_0 \pm \omega)t \pm \phi]$$
$$= A \cos(\omega t + \phi \mp \theta) + A \cos[(2\omega_0 \pm \omega)t \pm \phi + \theta] \quad (4.2\text{-}7)$$

We do not recover the baseband frequency component in its proper phase. In fact, all frequency components are shifted by the same phase angle $\mp\theta$. This can seriously distort the wave shape of a recovered baseband signal which contains many frequency components.

At the receiving end, how do we get the waveform $2 \cos \omega_0 t$ which we need in order to recover the baseband signal?

In carrier telephony a crystal-controlled oscillator independent of the oscillator at the transmitter is used. This differs in frequency from the oscillator at the transmitter by as much as a few hertz. In effect, θ in (4.2-7) changes with time, and the received waveform differs from the transmitted waveform by a change in phase.

This failure to reproduce the transmitted waveform is not serious in the case of speech, for hearing is not very sensitive to phase. However, a square data pulse may come out with the wrong amplitude or sign because the phase of the oscillator at the receiving end continually changes with respect to the phase of the oscillator at the transmitting end.

There is a further problem in single-sideband modulation, which is not nearly as critical for amplitude modulation. Filters are imperfect. We cannot remove one sideband completely without removing part of the other. In telephony, not all of the sideband is used; the part corresponding to frequencies near zero baseband frequency is eliminated, and these frequencies (usually below 300 Hz) are not transmitted. See Section 10.2.

Here we have shifted the frequency to obtain a modulated waveform (either upper or lower sideband). The envelope is constant if the baseband signal is a pure sinusoid. In the next section, we use the baseband signal to change the envelope of the carrier. This is called amplitude modulation.

Problems

1. The square wave of unit amplitude, illustrated below, is band limited to

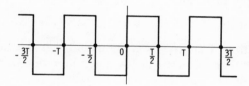

frequency f_0 to produce $v(t)$ which modulates a carrier of frequency $\omega_0 = 2\pi f_0$ by single sideband using the upper sideband. Assume $f_0 \gg 1/T$. The modulated signal is demodulated to $w(t)$ by an oscillator of incorrect phase producing a reference voltage $\cos(\omega_0 t + \theta)$. The oscillator offset θ is defined to between 0 and 2π radians. Using Fourier series, find $w(t)$ as a function of θ. Sketch $v(t)$ and $w(t)$ for $\theta = \pi/2$ (only finitely many harmonics need be considered, because only finitely many will generally pass through the system). What happens for $\theta = \pi$?

2. What happens to the baseband term in (4.2-7) if the reference oscillator phase θ drifts uniformly and very rapidly but never more than $\pm\pi/4$ from its nominal value of 0? ("Very rapidly" means in comparison with the baseband frequency ω.) What happens to the power? HINT: first find the average power near time t, then average that over one cycle of the baseband signal.

3. Design a frequency inverter which produces a baseband spectrum the same as the input but reversed, i.e., components near B appear near 0 and conversely. HINT: demodulate ssb incorrectly, using $2\cos\omega_1 t$ where $\omega_1 \neq \omega_0$.

4. Investigate how constant the envelope of ssb may be by considering as input the sum of two cosines of zero phase and the same amplitudes but of differing frequencies $\omega_1 < \omega_2$. HINT: define a new fictitious modulating frequency ω_0' so that the output looks like *both* sidebands in (4.2-3) with a pure sinusoid $\cos\omega_3 t$ input.

4.3. Amplitude Modulation

In frequency shifting, that is, single-sideband modulation, one sideband is removed. In amplitude modulation (am), both sidebands are retained, and a *carrier* component of the radian frequency ω_0 used in shifting may be a part of the signal as well. Here ω_0 must be chosen greater than the highest radian frequency to be transmitted to avoid distortion in the recovered signal. Or,

$$\omega_0 > 2\pi B$$

where B is the base bandwidth of the signal $x(t)$ to be transmitted.

Let us first write an amplitude-modulated signal in the form

$$x(t)(2P_0)^{1/2} \cos \omega_0 t \qquad (4.3\text{-}1)$$

This signal has zero power when $x(t) = 0$ and a power P_0 when $x(t) = 1$.

If we multiply this signal by $2 \cos \omega_0 t$ we obtain

$$2x(t)(2P_0)^{1/2} \cos^2 \omega_0 t = x(t)(2P_0)^{1/2}(1 + \cos 2\omega_0 t) \qquad (4.3\text{-}2)$$

If (see Section 4.2) $x(t)$ is a baseband signal whose top radian frequency is less than ω_0, we can low-pass filter this signal and recover the modulating signal

$$x(t)(2P_0)^{1/2} \qquad (4.3\text{-}3)$$

Almost universally, am signals are *not* demodulated by multiplying by $2 \cos \omega_0 t$. Rather, they are rectified, or made positive, and low-pass filtered so as to obtain the envelope of the modulated rf wave. This is called *envelope detection*. When this is done, what we recover is not

$$x(t)(2P_0)^{1/2}$$

What we actually recover is

$$|x(t)| \, (2P_0)^{1/2} \qquad (4.3\text{-}4)$$

This is perfectly satisfactory if $x(t)$ is always positive. If $x(t)$ is an audio signal, however, it will sometimes be positive and sometimes negative. Thus it is plausible to express an am signal by

$$[2P_0/(1 + \overline{x^2})]^{1/2}[1 + x(t)] \cos \omega_0 t \qquad (4.3\text{-}5)$$

$$|x(t)| \le 1 \qquad (4.3\text{-}6)$$

$$\overline{x(t)} = 0 \qquad (4.3\text{-}7)$$

The form of (4.3-3) makes the average power equal to P_0. Here the bar denotes a time average.

Linear rectification and low-pass filtering recover

$$[2P_0/(1 + \overline{x^2})]^{1/2}[1 + x(t)]$$

By removing the dc, we can recover $x(t)[2P_0/(1 + \overline{x^2})]^{1/2}$.
Suppose, for instance, that

$$x(t) = a \cos pt \tag{4.3-8}$$

The corresponding am signal will be

$$\left(\frac{2P_0}{1 + a^2/2}\right)^{1/2} (1 + a \cos pt) \cos \omega_0 t$$

$$= \left(\frac{2P_0}{1 + a^2/2}\right)^{1/2} \left(\cos \omega_0 t + \frac{a}{2}\cos(\omega_0 + p)t + \frac{a}{2}\cos(\omega_0 - p)t\right) \tag{4.3-9}$$

The term of radian frequency ω_0 is the carrier and the other two terms are the upper and lower sidebands.

Thus we will consider two different sorts of am signal. If we are allowed to demodulate by multiplying by $2 \cos \omega_0 t$ we write an am signal as

$$x(t)(2P_0)^{1/2} \cos \omega_0 t \tag{4.3-10}$$

Here the transmitted power is $\overline{x^2}P_0$.

If we are allowed only envelope detection, as in broadcast am radio, it is more sensible to write an am signal as

$$[2P_0/(1 + \overline{x^2})]^{1/2}[1 + x(t)] \cos \omega_0 t \tag{4.3-5}$$

Here the transmitted power is P_0.

$$|x(t)| \le 1 \tag{4.3-6}$$

with zero average value of $x(t)$

$$\overline{x(t)} = 0 \tag{4.3-7}$$

In amplitude modulation, as in single-sideband transmission, the power transmitted varies with time. Transmitting tubes are most efficient at a constant high-power level, and their operation near this level may be nonlinear. Also, in waveguides large voltage differences can cause arcing and breakdown. For these and other reasons, it is sometimes desirable to transmit a signal of constant power, such as provided by phase modulation and frequency modulation. Their peak power is equal to their average power. Phase modulation, which we introduce in the next section, is the technique commonly used for transmission of digital voice or data from space to earth or from earth to space.

Problems

1. We wish to contrast the ssb (single-sideband) signal

$$A \cos (\omega_0 + p)t$$

and the am (amplitude-modulated) signal

$$A \cos pt \cos \omega_0 t$$

(a) What baseband signal do we recover by multiplying the ssb signal by $2 \cos \omega_0 t$? What baseband signal do we recover by multiplying the am signal by $2 \cos \omega_0 t$?

(b) What is the power of the ssb signal? What is the power of the am signal?

2. We can recover the same baseband signal by multiplying $(1 + A \cos pt) \cos \omega_0 t$ by $2 \cos \omega_0 t$ as we do by multiplying $A \cos pt \cos \omega_0 t$ by $2 \cos \omega_0 t$. In the first signal (a) what is the power of the carrier and (b) what is the power of the sidebands? What is the power of the second signal? By how much do we have to increase the power in the first case relative to the second so that the sideband power is equal in the two cases? Assume $A = 1$ for this.

4.4. Phase Modulation

The phase-modulated (pm) signal which results from a baseband signal x can be represented as

$$\cos (\omega_0 t + \phi_d x) = \cos \omega_0 t \cos (\phi_d x) - \sin \omega_0 t \sin (\phi_d x) \quad (4.4\text{-}1)$$

Here ϕ_d is a factor which prescribes the amount of phase shift corresponding to the signal x. When ϕ_d is very small we say that the *index of modulation* is small. When

$$\phi_d \, |x| \ll 1$$

a very good approximate result for low-index phase modulation is

$$\cos (\omega_0 t + \phi_d x) \doteq \cos \omega_0 t - \phi_d x \sin \omega_0 t \quad (4.4\text{-}2)$$

It is of interest to consider the case $\phi_d x = A \cos pt$, $A \ll 1$. From (4.4-2) we obtain

$$\cos (\omega_0 t + \phi_d x) \doteq \cos \omega_0 t - A \cos pt \sin \omega_0 t$$
$$= \cos \omega_0 t - (A/2)[\sin (\omega_0 + p)t + \sin (\omega_0 - p)t] \quad (4.4\text{-}3)$$

It is interesting to note that a network with zero phase shift at a radian frequency ω_0 and a phase shift $-\pi/2$ at both $(\omega_0 + p)$ and $(\omega_0 - p)$ will transform signal (4.4-3) into

$$\cos \omega_0 t + A/2[\cos (\omega_0 + p)t + \cos (\omega_0 - p)t]$$

What we have here is an amplitude-modulated signal.

This indicates that if the phase of a transmission system varies nonlinearly with frequency, the transmission system will convert low-index phase modulation partly into amplitude modulation (with a loss of phase modulation) or weak amplitude modulation partly into phase modulation. This is why the filters in communication satellites are carefully phase-equalized so as to give a linear variation of phase with frequency over the bandwidth of the signal. (Linear phase shift corresponds to a time delay, from Section 3.5.)

The power of signal (4.4-1) is simply 1/2, whether $\phi_d x$ is large or small. When $\phi_d x$ is large, the phase-modulated signal contains frequencies over a band much broader than $2B$, where B is the bandwidth of the baseband signal x. This is because a shift of a number of cycles in 1 sec corresponds to a "local" frequency shift of about that many hertz (cycles per second). In this case, $f_0 = \omega_0/2\pi$ must be considerably larger than B if an undistorted baseband signal is to be recovered in this way. It is hard to find the exact spectrum of pm unless $\phi_d |x|$ is small.

The baseband modulating function x of a phase-modulated signal can be recovered in various ways. Here we will merely assume that it can be recovered.

So far we have disregarded noise. We will begin to incorporate it in the next section, for the case of a sine wave plus a small amount of noise.

Problems

1. Suggest a way of recovering the baseband signal from the phase-modulated signal (a) if the phase modulation is small, and (b) if the phase modulation is large.

2. What is the spectrum of pm like when $\phi_d |x| \ll 1$? What fraction of the power is in the sidebands?

3. Find the spectrum $Y(f)$ of phase modulation for $\phi_d x = A \cos pt$ (when it is not necessarily small) in terms of the Fourier transform $G(f)$ of $e^{iA \cos pt}$ (which we do not know explicitly).

4.5. A Sine Wave Plus a Little Noise

In Fig. 4.3, the vector or *phasor* of length C represents a signal voltage of magnitude C. If we add to this another voltage vector or phasor of magnitude n_0, the vector sum of the two voltages will depend on the relative phase of the two voltages. When

$$n_0 \ll C$$

we see from the figure that the maximum difference in phase ϕ between

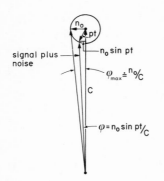

signal plus
noise

Figure 4.3. Signal phasors plus noise phasors.

the signal voltage and the vector sum of the two voltages will be very nearly

$$\phi_{max} = n_0/C \tag{4.5-1}$$

If the noise vector makes an angle pt with respect to C, we have

$$\phi = (n_0 \sin pt)/C \tag{4.5-2}$$

Now imagine a signal voltage

$$C \cos \omega_0 t$$

and an added noise voltage at a nearby frequency

$$n = n_0 \cos (\omega_0 + p)t$$

The relative phases will change with a radian frequency p and the magnitude of the phase fluctuation of the sum will be ϕ as given by (4.5-2).

Now let $\overline{\phi^2}$ be the mean-square value of ϕ, and let

$$P_n = \overline{n^2} = n_0^2/2$$

be the mean-square value of n. This is the noise power. Then

$$\overline{\phi^2} = \frac{\overline{(n_0 \sin pt)^2}}{C^2} = \frac{n_0^2/2}{C^2} = \frac{\overline{n^2}}{C^2} = \frac{P_n}{C^2}$$

The power P_c associated with a sinusoidal voltage of peak magnitude C is

$$P_c = C^2/2$$

Thus

$$\overline{\phi^2} = P_n/2P_c \tag{4.5-3}$$

Here we have assumed that the noise $n(t)$ is all at a single frequency. But (4.5-3) must also represent the differential mean-square phase modulation or phase noise power $d\overline{\phi^2}$ produced by the differential noise power dP_n in the differential frequency range df at the frequency f:

$$d\overline{\phi^2} = \frac{dP_n}{P_c} \qquad (4.5\text{-}4)$$

We may integrate (4.5-4) over all frequencies to obtain the phase noise power $\overline{\phi^2}$. The result is that (4.5-3) holds in general. We conclude that the addition of a small noise power P_n of any distribution in frequency to a carrier of power P_c produces a mean-square phase modulation given by (4.5-3).

We may note that the *amplitude* of the combined voltage will also fluctuate somewhat. An exact treatment of this phenomenon is given in Section 9.6, Representation of Narrowband Gaussian Noise.

The next section uses the simple techniques we have developed to study the effects of noise on four types of modulation, with two more to follow in the section after that.

Problems

1. How large can P_n become relative to P_c before the formula (4.5-3) is off by about 2%? HINT: arcsin $x \doteq x + x^3/6$ for small x, and $\int_0^\pi \sin^4 x\, dx = 3\pi/8$.

2. In a certain phase-tracking system used for radio navigation, the rms phase jitter must be kept below 10°. What is the minimum allowable carrier-to-noise power ratio in dB?

4.6. Effect of White Noise in Single Sideband; Amplitude Modulation and Phase Modulation; Pulse Position Modulation

4.6.1. Baseband

As a standard of comparison, let us consider the addition of noise to a baseband signal $x(t)$ which has a bandwidth B and a power P_s given by

$$P_s = \overline{x^2}P_0$$

Here we transmit the voltage $x(t)(P_0)^{1/2}$. Let the baseband noise power density be N_0. Then the noise power P_n will be

$$P_n = N_0 B$$

The signal-to-noise power ratio S/N will be

$$\frac{S}{N} = \frac{P_s}{P_n} = \frac{\overline{x^2}P_0}{N_0 B} = \frac{\text{transmitted power}}{N_0 B} \qquad (4.6\text{-}1)$$

4.6.2. Single Sideband

In single sideband let the radio frequency signal power be P_r and the noise power density near this radio frequency be N_0. The bandwidth is B, the same as the baseband. Thus the ratio of signal to noise in the single-sideband signal is

$$P_r/N_0B$$

In recovering the baseband signal by multiplying by $2 \cos \omega_0 t$ and low-pass filtering, signal frequency components are treated just like noise frequency components, so the baseband signal-to-noise power ratio S/N is the same as that in a single-sideband signal:

$$S/N = P_r/N_0B$$

And, the rf power P_r is proportional to $\overline{x^2}$. This is obvious for double sideband (am). Since the powers in the upper and lower sideband are equal, it is true for single sideband as well. If we let P_r be $\overline{x^2}P_0$, then

$$\frac{S}{N} = \frac{\overline{x^2}P_0}{N_0B} = \frac{\text{transmitted power}}{N_0B} \qquad (4.6\text{-}2)$$

This is the same as the baseband performance given by (4.6-1).

4.6.3. Amplitude Modulation

Suppose that we allow demodulation by multiplication by $2 \cos \omega_0 t$, and take the am signal as

$$x(t)(2P_0)^{1/2} \cos \omega_0 t$$

The power of this signal is $\overline{x^2}P_0$. If we multiply this by $2 \cos \omega_0 t$ and low-pass filter we obtain a baseband signal

$$x(t)(2P_0)^{1/2}$$

The power of this baseband signal is

$$2\overline{x^2}P_0$$

In receiving both sidebands we must accept noise over a bandwidth $2B$, where B is the bandwidth of the baseband signal $x(t)$. All noise components in this range will be "beaten down" into the baseband by multiplication by $2 \cos \omega_0 t$ and the total baseband noise in the demodulated signal will be

$$2N_0B$$

Thus the baseband signal-to-noise ratio in the demodulated signal will be

$$\frac{S}{N} = \frac{2\overline{x^2}P_0}{2N_0B} = \frac{\overline{x^2}P_0}{N_0B} = \frac{\text{transmitted power}}{N_0B} \tag{4.6-3}$$

This signal-to-noise ratio is the same as in the case of single sideband, despite the fact that the am receiver accepts noise over twice the bandwidth. Let us explain this.

Consider single-sideband or am transmission of a single modulating frequency as an rf signal of power P_r. In single sideband, all the power is in one sideband of amplitude $(2P_r)^{1/2}$. In am there are two sidebands, each of amplitude $(P_r)^{1/2}$. On multiplication by $2\cos\omega_0 t$ these give rise to two components of amplitude $(P_r)^{1/2}$, which, in phase, add to $2(P_r)^{1/2}$. Hence the power of the demodulated sidebands is $2P_r$. The noise power over the bandwidth $2B$ is added to the baseband signal after multiplication by $2\cos\omega_0 t$, the modulated noise power is $2N_0B$. This accounts for the fact that am gives the same noise performance as ssb.

We have considered an am signal of the form $x(t)(2P_0)^{1/2}\cos\omega_0 t$, which has a power $\overline{x^2}P_0$. We have noted, however, that if we are only allowed detection by rectification, it is more realistic to write an am signal as

$$[2P_0/(1 + \overline{x^2})]^{1/2}[1 + x(t)]\cos\omega_0 t \tag{4.3-5}$$

$$|x(t)| \leq 1 \tag{4.3-6}$$

$$\overline{x(t)} = 0 \tag{4.3-7}$$

This signal has a power P_0 rather than $\overline{x^2}P_0$. In considering demodulation of the signal, we will get the same result by rectification and low-pass filtering as we get by multiplying by $2\cos\omega_0 t$ and low-pass filtering. At high carrier-to-noise ratios, the result is correct for noise as well. This is because the composite envelope (signal plus noise) will almost always be of the same sign as the signal envelope. The resulting demodulated signal will be

$$[2P_0/(1 + \overline{x^2})]^{1/2}[1 + x(t)] \tag{4.6-4}$$

Because we have insisted that $x(t)$ have average value zero, this complete demodulated signal has a power $2P_0$. However, the ac part, corresponding to the original signal we are interested in

$$[2P_0/(1 + \overline{x^2})]^{1/2}x(t)$$

has a power

$$2P_0\overline{x^2}/(1 + \overline{x^2})$$

Multiplication of the noise in a bandwidth $2B$ by $2 \cos \omega_0 t$ results in a baseband noise power

$$2N_0 B$$

Thus the signal-to-noise ratio after demodulation is

$$\frac{S}{N} = \frac{[\overline{x^2}/(1 + \overline{x^2})]P_0}{N_0 B} \tag{4.6-5}$$

Here P_0 is transmitted power, so that

$$\frac{S}{N} = \frac{\overline{x^2}}{1 + \overline{x^2}} \cdot \frac{\text{transmitted power}}{N_0 B} \tag{4.6-6}$$

Thus for a given transmitted power, the signal-to-noise ratio in the recovered signal goes to zero as the level of modulation $x(t)$ goes to zero. We should contrast this with (4.6-2) for single-sideband and (4.6-3) for amplitude modulation when we allow the modulating function to go negative; in each of these cases S/N is merely the transmitted power divided by the noise power in the signal bandwidth B. Here there is a multiplier in front of that ratio which goes to zero as the magnitude of the modulating signal x goes to zero.

4.6.4. Phase Modulation

From (4.5-3) we see that a noise power P_n added to a signal of power P_0 produces a mean-square phase modulation

$$\overline{\phi_n^2} = P_n/2P_0$$

Assume a constant one-sided rf noise density N_0. Noise components lying in frequency range df at frequencies at a distance f *above* and *below* the carrier frequency will produce baseband phase noise in the range df at f. Thus the mean-square phase noise $d\overline{\phi_n^2}$ in the *baseband* range df at f will be

$$d\overline{\phi_n^2} = \frac{N_0(2df)}{2P_0} = \frac{N_0 \, df}{P_0} \tag{4.6-7}$$

The total phase noise power is then

$$N_0 B/P_0 \tag{4.6-8}$$

This is because we only retain baseband frequencies 0 to B in (4.6-7), where B is the base bandwidth. This noise will not be much changed by the presence of a modulating signal, if the carrier-to-noise ratio is not too low. The total signal phase modulation ϕ_s is taken as

$$\phi_s = x\phi_d$$

The final signal-to-noise ratio is

$$\frac{S}{N} = \frac{\overline{x^2}\phi_d{}^2 P_0}{N_0 B} = \overline{x^2}\phi_d{}^2 \frac{\text{transmitted power}}{N_0 B} \qquad (4.6\text{-}9)$$

As in the case of amplitude modulation in which no power is transmitted in the absence of modulation [equation (4.6-6)], the signal-to-noise ratio goes to zero as the modulation goes to zero.

4.6.5. Pulse Position Modulation

Pulse position modulation (ppm) is a close relative of phase modulation. In pulse position modulation a pulse which recurs with a radian frequency ω_0 is modulated in *position* or time of occurrence.

We shall consider only a pulse train given by

$$v_p = (2P_0/m)^{1/2} \sum_{n=1}^{m} \cos n(\omega_0 t - x\phi_d/m) \qquad (4.6\text{-}10)$$

This is obtained by taking a train of impulses as in Section 3.3 and low-pass filtering up to maximum frequency $m\omega_0$. The dc term is not included here because it is not possible to modulate its position. In this particular pulse train, all frequency components have the same amplitude $(2P_0/m)^{1/2}$, and the total power of the m components is P_0. The phase modulation is $x\phi_d$ for the highest-frequency component $(n = m)$ and less for lower-frequency components. Here $x = x(t)$ is of bandwidth B. It varies continuously with time. In a sampled data ppm system, we shift the phase of each pulse at certain times only. We shall study a form of sampled data ppm in Section 11.5 in connection with information theory.

We can if we wish phase-detect each frequency component separately, at least if the modulation is small and

$$\omega_0 > 2(2\pi B)$$

This is so because around each frequency $n\omega_0$, the phase modulation extends $2\pi B$ on either side. Consecutive frequencies $n\omega_0$ and $(n + 1)\omega_0$ are ω_0 apart, which is more than twice this. If, however, $m = 1$, $\omega_0 > 2\pi B$ suffices, as in amplitude modulation.

From (4.6-9) the signal-to-noise ratio for the nth component, $(S/N)_n$, will be

$$\left(\frac{S}{N}\right)_n = \frac{(1/m)(n/m)^2 \overline{x^2}\phi_d{}^2 P_0}{N_0 B} \qquad (4.6\text{-}11)$$

The factor $1/m$ enters because each frequency component has $1/m$ of the total power. The factor $(n/m)^2$ enters because the phase modulation of the nth component is n/m times the phase modulation of the component of highest frequency.

Thus we have a number m of demodulated signals, each with a different signal-to-noise ratio. How can we combine these most advantageously? This is an optimization problem. Optimization is important in communications. We shall have a similar optimization problem when we study diversity reception in Chapter 7. If we weight the nth-demodulated voltage by a factor a_n, the total signal power will be proportional to

$$\left(\sum_{n=1}^{m} na_n \right)^2 \tag{4.6-12}$$

The total noise power (which does not depend on the degree of modulation) is the same for each demodulator and will be the sum of the noise power contributions from each frequency. It will be proportional to

$$\sum_{n=1}^{m} a_n^2 \tag{4.6-13}$$

Suppose we constrain the a_n's so as to keep the total noise power constant. Then if we make a small change Δa_n in each, this constraint says that

$$\sum_{n=1}^{m} [d(a_n^2)/da_n]\Delta a_n = 0$$

$$2 \sum_{n=1}^{m} a_n \Delta a_n = 0 \tag{4.6-14}$$

What is the small change in the signal voltage when we make the changes in the a_n's given by (4.6-14)? This is

$$\Delta \sum_{n=1}^{m} na_n = \sum_{n=1}^{m} n\Delta a_n \tag{4.6-15}$$

We will note that if we choose

$$a_n = n \tag{4.6-16}$$

expression (4.6-14) can be rewritten as

$$\sum_{n=1}^{m} n\Delta a_n = 0$$

The consequence is that the change in signal voltage given by (4.6-15) is zero with the noise power constant, for any small changes Δa_n in the a_n's. We see that the signal voltage or power for constant noise is at an extreme value for the choice $a_n = n$. It is clear that this represents the *maximum* signal-to-noise ratio. (The minimum occurs when $a_n = 0$ for $n > 1$.)

We have described an optimum means for detection in which we phase-detect each component of (4.6-10). This results in the same noise for each component and a signal voltage proportional to n. We then multiply both the nth signal voltage and the nth noise voltage by n and add all signal and noise voltages. Thus the contribution to the signal voltage of the nth component is proportional to n^2, but the contribution to the noise voltage of the nth component is proportional to n.

The optimum signal-to-noise ratio for the signal of (4.6-10) is then

$$(S/N)_{max} = (\overline{x^2}\phi_d^2 P_0/N_0 B)\left((1/m^3)\sum_{n=1}^{m} n^2\right) \qquad (4.6\text{-}17)$$

The first factor is just the signal-to-noise ratio for phase modulation on a carrier of the highest-frequency component of the pulse train. The second factor is plainly at most 1, because $m \cdot m^2 = m^3$ and $n^2 \le m^2$. It is known to be equal to a simple expression:

$$(1/m^3)\sum_{n=1}^{m} n^2 = (m + 1)(2m + 1)/6m^2 \qquad (4.6\text{-}18)$$

This decreases from its maximum value of 1 when $m = 1$ to a minimum of $\frac{1}{3}$ when $m = \infty$. Since we get the best result when $m = 1$, this means we should have only one phase-modulated frequency component. We will have ordinary phase modulation (with ϕ_d replaced by $-\phi_d$). The same conclusion holds with a slightly more complicated argument for any kind of pulses, not only for the low-passed impulses of (4.6-10). We should use only one frequency component if we wish to attain the best signal-to-noise ratio.

We demodulate the ppm signal in a sampled data system by noting the position of each pulse with respect to some reference position. Arguments that will be given in Chapter 5 apply here, and we find that the pulse rate used in transmitting sampled data should be greater than twice the highest baseband frequency B to be transmitted. We have seen that this must also be the case for continuous-time ppm if more than one frequency is used ($m > 1$).

Two additional modulation types are discussed in the next section: frequency modulation and pulse rate modulation. We will pay special attention to the output noise spectrum in frequency modulation.

Problems

1. What is the ratio of S/N for an am signal of the form

$$[2P_0/(1 + \overline{x^2})]^{1/2}[1 + x(t)]\cos \omega_0 t$$

to the S/N for a single-sideband signal for the same rf power and the same $\overline{x^2}$? What is the power loss in dB if $x(t) = \pm 1$ (digital data), and if $x(t)$ is a sequence of bursts of sinusoids as in speech?

2. For what phase deviation ϕ_d does phase modulation give the same signal-to-noise ratio as envelope-detected am, assuming a small value $\overline{x^2} \ll 1$ in each case?

3. Show that a single-frequency pulse is optimum for ppm *without* the constraint that the fraction β_n^2 of the power in the nth harmonic of ω_0 is equal to $1/m$ for every n. Here the pulse is

$$v_p = (2P_0)^{1/2} \sum_{n=1}^{m} \beta_n \cos n \, (\omega_0 t - x\phi_d/m)$$

with

$$\sum_{n=1}^{m} \beta_n^2 = 1$$

(The case of low-pass-filtered impulses has all $\beta_n^2 = 1/m$.) In particular, show that all the β_n except one should be taken as 0 to get the overall best $(S/N)_{max}$.

4. What happens in ppm when the modulation x is identically 0 for a long time? Assume that m is very large.

4.7. Frequency Modulation; Pulse Rate Modulation

4.7.1. Frequency Modulation

Phase modulation is seldom used as such in transmitting analog signals; it is reserved for digital signals. We will study this in Sections 10.7–10.9.

We have written the phase-modulated signal produced by a baseband signal x as

$$\cos (\omega_0 t + \phi_d x)$$

The rate of change of the argument is

$$\omega_0 + \phi_d \frac{dx}{dt}$$

This may be taken as the instantaneous radian frequency.

We can also express the instantaneous deviation $f(t)$ from the rest frequency f_0 as $f_d x$. Here f_d is a frequency deviation. The frequency-modulated signal can then be written

$$\cos 2\pi \left(f_0 t + f_d \int^t x(s) \, ds \right)$$

The instantaneous phase $\phi(t)$ of the signal is

$$\phi(t) = 2\pi \int^t f(s)\, ds = 2\pi f_d \int^t x(s)\, ds$$

The phase is differentiated to get the frequency deviation, which is the modulation. The noise in the phase is differentiated as well. Thus we obtain the frequency modulation noise f_n in $f(t)$ by differentiating the phase modulation noise ϕ_n with respect to time and dividing by 2π. From equation (4.6-7) and Section 3.8, we see that for any baseband frequency range df at f the frequency modulation noise $\overline{df_n^2}$ must be

$$\overline{df_n^2} = [1/(2\pi)^2]\omega^2(N_0/P_0)\, df$$
$$\overline{df_n^2} = f^2(N_0/P_0)\, df \qquad (4.7\text{-}1)$$

Here P_0 is transmitted power. We see that the frequency modulation noise density rises as f^2, the square of the baseband frequency. The total mean-square frequency modulation noise in the base bandwidth B is

$$\overline{f_n^2} = \int_0^B (N_0/P_0)f^2\, df = \frac{N_0 B^3}{3P_0}$$

The mean-square frequency modulation due to the signal x is

$$f_d^2 \overline{x^2}$$

Accordingly, after demodulation the signal-to-noise power ratio S/N will be

$$\frac{S}{N} = 3\frac{P_0}{N_0 B}\overline{x^2}\left(\frac{f_d}{B}\right)^2 = 3\frac{P_0}{N_0 B}D^2 \qquad (4.7\text{-}2)$$

In this expression, the quantity

$$(f_d/B)(\overline{x^2})^{1/2} = D \qquad (4.7\text{-}3)$$

is the *rms deviation ratio*. As this deviation ratio increases, the signal-to-noise ratio in the received and demodulated signal increases.

When the deviation ratio is large, the spectrum of the frequency-modulated signal covers a broad band, and B is no longer the rf bandwidth. A reasonable approximation given by Carson (Ref. 1) says that the bandwidth required for transmission of a frequency-modulated signal with approximately equal power at all baseband frequencies is approximately twice the base bandwidth plus the maximum frequency excursion due to frequency modulation. If the largest positive and negative values of x are x_p and $-x_p$, then according to Carson's approximation the required rf bandwidth B_r is

$$B_r = 2(B + x_p f_d) \qquad (4.7\text{-}4)$$

In Section 10.9, we shall define bandwidth of signals more quantitatively so that in principle exact answers could be given to these questions.

Equation (4.7-2) appears to tell us that in using fm, the way to improve the signal-to-noise ratio of the demodulated signal is to increase the frequency deviation. But, according to (4.7-4), this increases the rf bandwidth. Or does it? In transmitting audio signals by fm, we can in a sense have our cake and eat it too.

Part of the fm bandwidth B_r due to an audio signal of baseband B is the $2B$ due to the baseband itself. The added $2x_p f_d$ depends on the peak amplitude x_p of the modulating signal. Conceptually, we can break the audio signal up into different frequency ranges and apply Carson's formula (4.7-4) to each. If a signal component in a particular frequency range is very weak, it adds little to the term $2x_p f_d$. But, in speech and music there is little power at high frequencies (see Fig. 4.4). For example, the power at 4 kHz is about 30 dB less than the power at 400 Hz. There is therefore little power at the upper end of the range given by (4.7-4). So we are not as badly off if we increase the frequency deviation as we might have thought.

There is another way we can take advantage of the low power of speech and music at high frequencies. The above reasoning shows that the high-frequency components will produce little frequency deviation because they represent little voltage. Because of this, we can "preemphasize" the high frequencies before frequency modulation, as in Fig. 4.5. Here we pass the audio signal through a preemphasis network whose gain $H(\omega)$ increases with frequency. This does not increase audio power or voltage much at the output of the network simply because voice and music do not have much high-frequency power.

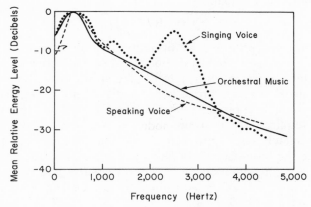

Figure 4.4 Relative spectra for voice, singing, music. [From Johan Sundberg, "The Acoustics of the Singing Voice." *Sci. Am.* (March 1977), page 89.]

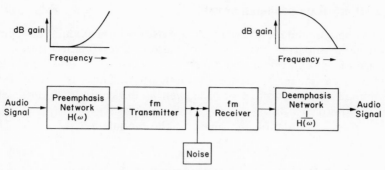

Figure 4.5. Preemphasis for fm.

Because of (4.7-1), the noise density at high frequencies f was increased by the factor f^2. We undo this by preemphasis without a noticeable bandwidth increase. This improves the fidelity at higher audio frequencies.

Because we have preemphasized the high frequencies in the audio signal before using it to frequency modulate the fm transmitter, the fm receiver output will also be preemphasized. That is, its high-frequency components will be too strong. We must filter this preemphasized audio signal with a "deemphasis" network of gain $1/H(\omega)$. The result is that we recover the original audio signal. The deemphasis network reduces the high-frequency noise in the demodulated fm signal, and that is just what we set out to do.

The general form of the preemphasis used in broadcast fm is

$$H(\omega) = 1 + j\omega RC \qquad (4.7\text{-}5)$$

The corresponding deemphasis network is the RC filter with transfer function

$$1/H(\omega) = 1/(1 + j\omega RC) \qquad (4.7\text{-}6)$$

(See Problem 4 of Section 3.9.)

The gains of actual preemphasis networks do not rise indefinitely with frequency, as in equation (4.7-5), but become constant above the top of the audio band. But (4.7-5) and (4.7-6) do illustrate the important features of preemphasis, that is, no preemphasis at low frequencies and a voltage gain proportional to frequency at higher frequencies in the audio band, so that the power gain rises as frequency squared.

We noted from (4.7-1) that in fm the frequency modulation noise power rises as the square of baseband frequency f. Thus deemphasis according to (4.7-6) just undoes this at high audio frequencies, and in the high-frequency range the noise density in the final audio signal is independent of frequency.

4.7.2. Pulse Rate Modulation

As phase modulation has a relative, pulse position modulation (ppm), so frequency modulation has a relative, pulse rate modulation (prm). As in the case of ppm and phase modulation, general prm is inferior in signal-to-noise performance to frequency modulation. There is a time-discrete or sampled data version of prm, as well as the continuous-time version discussed here.

In phase modulation the phase of the modulation is

$$x\phi_d \tag{4.7-7}$$

In ppm, (4.7-7) is the phase modulation of the component of highest frequency in the pulse train (4.6-10). In frequency modulation the phase modulation is

$$f_d \int^t x(s)\, ds \tag{4.7-8}$$

In prm, (4.7-8) is then the phase modulation of the component of highest frequency in the pulse train (4.6-10). Section 4.6.5 shows that $m = 1$ is optimum here as well. This is ordinary fm. In the human body, many nerves communicate by pulse rate modulation with very narrow pulses (large m).

The next section shows what must happen to some of the modulation types we have been studying when the noise becomes strong. This is the phenomenon of breaking.

Problems

1. Frequency modulation braodcasting stations are limited by the FCC to a frequency deviation of ±75 kHz. Modulating frequencies typically cover 30–15,000 Hz, so we must allow the bandwidth B to extend to 15,000 Hz. As the modulation x is more or less sinusoidal, if the peak value of x is 1 the mean-square value will be about $\frac{1}{2}$ at maximum modulation. What are (a) the rms deviation ratio D, (b) the rf bandwidth B_r, and (c) the ratio of the signal-to-noise ratio (for $x^2 = \frac{1}{2}$, corresponding to maximum modulation) to the carrier-to-(baseband)noise ratio $P_0/N_0 B$? What is this "fm gain" in signal-to-noise in dB?

2. An fm detector has to find a carrier of power P_0 in noise in a bandwidth B_r. In Problem 1, what must the carrier-to-(rf)noise ratio $P_0/N_0 B_r$ be in order to attain a signal-to-noise ratio of 40, 50, and 60 dB?

3. For what frequency deviation ratio f_d/B does fm give the same signal-to-noise ratio as an envelope detector for am, given $\overline{x^2} \ll 1$?

4. We see from (4.7-1) that the noise density in a signal which has been transmitted by frequency modulation increases as the square of the frequency.

Suppose that before using it for frequency modulation we put the baseband signal to be transmitted through a network whose power gain varies as f^2. What can you say about the amount of phase modulation for a sinusoidal signal of constant amplitude as the signal frequency is changed?

5. In fm broadcast, it has been common to make the time constant $t_0 = RC$ equal to 75 μsec (microseconds). At what frequency has the preemphasis become 3 dB? How high a frequency makes the noise density in the final audio signal signal 90% of its maximum value?

6. A sinusoid of low frequency is input to a prm system which has a pulse with m large, so that the pulse is a delta function. Draw the output over a complete cycle of the input.

7. We saw in Section 4.3 that peak power is sometimes a constraint in communication. Rank the following systems by their peak power, smallest to largest, for the same signal-to-noise ratio $r = S/N$, or normalized signal-to-noise ratio $rN_0B = \alpha$: (a) ordinary (double sideband suppressed carrier) am; (b) broadcast am; (c) ssb (treat qualitatively); (d) pm with $\phi_d = \pi/2$; (e) fm with $f_d/B = 1$. Assume the modulating signal $x(t)$ of bandwidth B is such that

$$\overline{x^2} = \tfrac{1}{2}x_{max}^2 = \tfrac{1}{2}$$

Here x_{max} is the peak of the modulating signal.

4.8. Breaking

In our analysis of phase modulation and frequency modulation we have assumed that the predemodulation rf noise power is small compared with the signal power, so that the noise somewhat perturbs the phase of the signal. This is illustrated graphically in Fig. 4.6.

With small noise, we represent the transmitted signal by a long vector and the added noise by a short vector. The sum, the received signal, is a vector which differs in phase from the transmitted signal by a small angle θ.

What happens when the noise is larger than the signal? We see that not only is the phase difference between the transmitted and received signals large, it is ambiguous. We really have no way of estimating the phase of the transmitted signal.

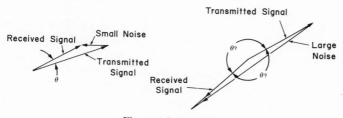

Figure 4.6. Breaking.

When we listen to an fm receiver as the signal fades, we hear bursts of popping or crackling.† During these periods of poor signal-to-noise ratio the fm receiver makes wild errors in trying to estimate the correct instantaneous frequency. This is called *breaking*. If the signal is a video rather than an audio signal, the result of breaking is a rash of dark or bright spots or "speckle." (But, the picture in broadcast television is am, not fm, or rather an am–ssb hybrid called *vestigial sideband*.)

In video, breaking results in a gradual increase of the speckle. In audio, it results in intense + or − pulses which sound very harsh and which obscure the sound. In video, the spatial resolution of the eye helps to avoid obscuration by breaking. But in audio the pulses produced in breaking span the audio frequency range and neither the frequency resolution nor the time resolution of the ear can reject them. The difference is also partly because video signals have more redundancy in them than audio signals do, relative to the capability of the sensor or receiver used, the eye for video and the ear for audio. We shall learn more about redundancy in information sources in Chapter 13.

Breaking is inevitable in systems which gain a signal-to-noise advantage through using a larger rf bandwidth than the bandwidth of the baseband signal to be transmitted. We shall now see how to use information theory to derive this result.

In Chapter 11, we will learn the concept of channel capacity. The capacity of a channel with transmitted power P_0 and noise spectral density N_0 cannot be infinite because the noise prevents this. If (4.7-2) were to hold for large frequency deviations f_d, we could attain arbitrarily large signal-to-noise ratios merely by increasing the frequency deviation f_d. The channel would then approach infinite capacity. What has gone wrong?

The answer is that as f_d increases, so does the total predemodulation noise power $P_n = N_0 B_r$, because B_r must increase with f_d according to (4.7-4). When P_n becomes comparable to the signal power P_0, we experience fm breaking. We cannot continue to increase f_d indefinitely. So, (4.7-2) does not hold when f_d is sufficiently large. We will examine this more rigorously in Section 13.3, where we study rate distortion theory.

Many fm receivers simply amplify the whole broadband spectrum of a large-deviation fm signal, limit it, and feed it to a frequency-sensitive discriminator circuit in order to recover the baseband signal. The signal level at which breaking occurs, or the *fm threshold*, can be reduced by one of two means: (1) By use of feedback we can make a narrower-band receiver track the large-deviation fm signal in frequency. The signal in the feedback path which produces tracking can be used as the received

† If the *muting* is off. Muting turns the output off in the absence of an adequate signal.

baseband signal. This is called *fm with feedback*. An fm with feedback receiver made possible the reception of the weak, large-deviation fm signal reflected by the Echo balloon satellite in 1960. (2) We can track the received signal not only in frequency but in phase. The circuit which does this is called a *phase-locked loop* (see Reference 2). However, no matter what we try to do to decrease the rf bandwidth, information theory assures us that breaking is ultimately inevitable because the predemodulation noise must grow.

This completes our discussion of the basics of modulation and noise. Here we have treated *analog* communication systems. In the next chapter we study the key concept of *sampling*. This will permit us to discuss *digital* communication systems.

Problems

1. Frequency modulation with feedback narrows the rf bandwidth over which the carrier must be searched for from B_r to a little more than B on either side of where the instantaneous frequency is at any particular time. By how many dB does this reduce the fm threshold in broadcast fm? What does this do to the (baseband) signal-to-noise ratio?

2. A phase-locked loop is used in a certain fm receiver to track not only the frequency but also the phase of the fm carrier. In this way, the carrier can be confined to within a bandwidth of 4 kHz on either side of it. How many dB does this reduce the fm threshold below the value it has for the basic limiter-discriminator receiver? Qualitatively, why would we expect phase tracking to narrow the effective predemodulation bandwidth below that for fm with feedback?

3. For reasons of information theory, when P_0/N_0B is large, $\frac{1}{2}\ln(S/N)$ cannot much exceed $\ln(P_0/N_0B)$. Assuming the cleverest receiver possible, how large a frequency deviation f_d could we have in fm if $x^2 = \frac{1}{2}$, $B = 15$ kHz, and the baseband carrier-to-noise ratio P_0/N_0B is 40 dB? What does this say about the maximum possible fm gain?

The Link between Continuous and Pulsed Signals—Sampling and Digitization

We usually think of two distinct sorts of transmission: the transmission of continuous signals such as ssb, am, or fm signals, and the transmission of discrete, successive pulses.

The *sampling theorem* tells us that we can represent a band-limited waveform of bandwidth less than B *exactly* by a sequence of $2B$ samples per second which express the amplitude of the waveform at times $1/2B$ apart. If we could transmit these sample amplitudes exactly, without noise, we could reconstruct the original waveform exactly from the received samples.

But, noise prevents us from ever transmitting the samples exactly. One thing that we can do is to transmit the amplitude of each sample approximately as one of a discrete set of previously chosen amplitudes or *levels*. For instance, we can represent each sample amplitude as an m-digit binary number. There are 2^m such numbers, and hence we can represent each sample by the one of these numbers which corresponds most nearly to the true amplitude.

We can transmit such an encoded or digitized sample amplitude almost without error. We have then replaced noise in the decoded (demodulated) signal by an error of approximation in the encoded (and decoded) signal. Here we study the error of digital representations of analog signals. This is the basis of digital communication.

Why is digital communication becoming so widespread? One reason is that with digital communication we can in theory and in practice reconstruct a signal exactly before amplifying it. We can then transmit it over the next link in a long series of links. But, with analog repeaters, errors will build up. The choice has been digital in common carrier communication, in deep space communication, and in the highest fidelity sound recording and reproduction. A second reason for the trend toward digital communication is that switching of digital signals is easier than switching of analog signals. A third reason is that there is more and more communication of digital data intended for storage or processing by computer. A fourth reason is that the cheapest and most powerful electronic circuits available are very-large-scale integrated circuits (VLSI). And, these circuits are digital. For communication to take advantage of VLSI and its low-cost, high-performance, low-power consumption, small size and weight, and high reliability, it must be made digital. A final reason is that it is inherently easier to encrypt or encipher digital communication so that it is secure and cannot be deciphered except by authorized receivers. This chapter introduces the study of digital communication, which will be pursued more thoroughly in subsequent chapters.

There is another reason for sampling in communication systems, even in systems that do not transmit digits or bits. The revolution in digital components is making it easier to perform digitally many operations in analog communication systems which used to be performed by linear and other analog or continuous-waveform circuits. But, this requires that the incoming waveform be sampled by the receiver and converted to bits by an analog-to-digital converter. Sampling is occurring closer and closer to the first receiver stages or receiver front end as digital logic speeds increase (see Reference 1).

This chapter provides the basics of sampling so that we can understand its effect on communication systems. We need to include both signal-to-noise ratio performance and accuracy of reproduction.

5.1. The Sampling Function and Sampling

The sampling function $s(t)$ consists of a vanishingly narrow pulse (an impulse) of area T every T seconds, that is, $N = 1/T$ impulses per second, with one of the pulses at $t = 0$ (see Fig. 5.1). The average value of $s(t)$ is unity. The sampling function is a train of delta functions of

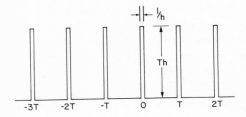

Figure 5.1. A sampling function.

strength T. In a Fourier expansion of this periodic function about $t = 0$, we have, as in Section 3.3,

$$A_0 = \frac{1}{T} \int_{-T/2}^{T/2} s(t)\, dt = \frac{1}{T}\, T = 1$$

$$A_n = \frac{2}{T} \int_{-T/2}^{T/2} s(t) \cos\left(\frac{2\pi nt}{T}\right) dt = 2 \qquad \text{for} \quad n > 0$$

$$B_n = \frac{2}{T} \int_{-T/2}^{T/2} s(t) \sin\left(\frac{2\pi n}{T}\right) dt = 0$$

Hence the sampling function has the series representation

$$s(t) = 1 + \sum_{n=1}^{\infty} 2\cos(2\pi nNt)$$

If we multiply a signal by the sampling function, the result is to take instantaneous samples of the signal amplitude at intervals T apart. To find the spectrum of the sampled signal we consider what happens to a single component $a\cos(\omega t + \phi)$:

$$s(t)a\cos(\omega t + \phi) = a\cos(\omega t + \phi) + \sum_{n=1}^{\infty} 2a\cos(\omega t + \phi)\cos(2\pi nNt)$$

This can be expressed as

$$s(t)a\cos(\omega t + \varphi) = a\cos(\omega t + \varphi)$$

$$+ a\sum_{n=1}^{\infty} \{\cos[(2\pi nN + \omega)t + \varphi] + \cos[(2\pi nN - \omega)t - \varphi]\}$$

This result is expressed graphically in Fig. 5.2 as a voltage spectrum. We see that the baseband spectrum is repeated at higher frequencies. If

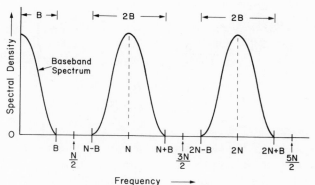

Figure 5.2. Spectrum of sampled signal.

the (one-sided) baseband spectrum lies within the range $f = 0$ to B, the sampled spectrum will lie in ranges

$$0 \text{ to } B, \quad N - B \text{ to } N + B, \quad 2N - B \text{ to } 2N + B, \text{ etc.}$$

We see that if $B < N/2$ these spectral regions do not overlap. When this is so the baseband signal can be recovered by low-pass filtering with a cutoff frequency $N/2$. So, the sequence of N samples per second completely describes a signal of bandwidth less than $N/2$.

Suppose that we filter out the part of the spectrum of the sampled signal that lies between $nN - B$ and $nN + B$. It is clear that by sampling and filtering we have produced an amplitude-modulated rf signal. In this case we could have used $2B$ samples per second and filtering to construct both sidebands of an rf signal of bandwidth $2B$. Had we selected only one sideband, we would have used $2B$ samples per second and filtering to produce an ssb signal of bandwidth B.

The problem of actually reconstructing a function from its samples will be studied in the next section. A remarkable property of the function

$$(\sin x)/x$$

is central to the study of sampled data systems.

Problem

1. (a) What are the samples of $\sin[2\pi(N/2)t]$ and $\cos[2\pi(N/2)t]$? (b) Can we reproduce these functions from the samples? (c) Does one get into trouble if one of the signal components has a frequency *equal to* $N/2$?

5.2. Reconstructing a Signal from Samples: The $(\sin N\pi t)/N\pi t$ or sinc Nt Function

How do we reconstruct a signal from a set of samples? We filter the stream of samples with a low-pass filter of bandwidth $B = N/2$, unity gain, and zero phase shift. The transfer function $H(f)$ of this filter is

$$H(f) = \begin{cases} 1, & -B < f < B \\ 0, & |f| > B \end{cases}$$

What is the time response of this filter to a short pulse which represents a sample? We will assume the short pulse to be centered at $t = 0$, to be very short, and to have an area AT, where A is the sample amplitude and $T = 1/N$ is the interval between samples. We will represent this pulse as $x(t)$. The Fourier transform $X(f)$ of this impulse, or pulse of length zero, is

$$X(f) = \int_{-\infty}^{\infty} x(t)e^{-j\omega t}\, dt$$

$x(t)$ is zero everywhere except at $t = 0$, where, regardless of the value of ω

$$e^{-j\omega t} = 1$$

Thus, regardless of the value of ω, the Fourier transform of the impulse is simply the integral of $x(t)\, dt$, which is AT, where A was the sample amplitude. That is, for all f,

$$X(f) = AT$$

We saw this already in Section 3.9.

We can now find the time response of our ideal low-pass filter to the impulse which represents the sample. The response $y(t)$ is

$$y(t) = \int_{-\infty}^{\infty} X(f)H(f)e^{j2\pi ft}\, df$$

$$= \int_{-N/2}^{N/2} ATe^{j2\pi ft}\, df$$

$$= AT\frac{\sin N\pi t}{\pi t}$$

Because $T = 1/N$

$$y(t) = A\frac{\sin N\pi t}{N\pi t} \tag{5.2-1}$$

The response to a pulse representing a sample of amplitude A_n at $t = nT$ is then

$$y_n(t) = A_n\frac{\sin\,[N\pi(t - nT)]}{N\pi(t - nT)}$$

We should note that the response given in (5.2-1) to a pulse at $t = 0$ is physically unrealizable. Part of the response occurs *before* the pulse is put into the filter. However, if we multiply the transfer function by a phase factor $\exp\,(-j2\pi T_d f)$, the peak of the response will occur at a time T_d after the pulse is applied to the filter. We can approximate the ideal filter with a real filter which has nearly the same response for times $t > 0$ and no response for $t < 0$. As we make T_d greater and greater, we approximate the response of the ideal filter better and better. (See Fig. 5.3.)

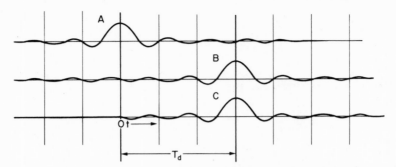

Figure 5.3. Building a realizable low-pass filter. (A) Response of ideal low-pass zero-phase filter; (B) response of ideal low-pass filter, phase $-2\pi f T_d$; (C) response like that of B but with response for $t < 0$ deleted.

The function

$$\frac{\sin N\pi(t - mT)}{N\pi(t - mT)} = \text{sinc } N(t - mT) \tag{5.2-2}$$

is sometimes written in the abbreviated form shown on the right and is called the *sinc* function. This is a convenience of representation and nomenclature. What is important is that (1) the function is band limited to the band $B = N/2$ and (2)

$$\text{sinc } N(nT - mT) = \begin{cases} 0 & \text{if} \quad n \neq m \\ 1 & \text{if} \quad n = m \end{cases} \tag{5.2-3}$$

Let us write out in terms of sinc functions the band-limited function ($f < N/2$) as reconstructed from the samples A_m of an original band-limited function. This recovered signal is

$$\sum_m A_m \text{ sinc } N(t - mT) \tag{5.2-4}$$

This function must be identical with the original function from which the samples were derived. Hence if we sample it at a time

$$t = nT$$

we should obtain the amplitude of the nth sample. We do, because at this time all the other A_m's are multiplied by sinc $N(nT - mT)$, $n \neq m$, which is zero.

We see that a function of the form of (5.2-4) provides an effective way to transmit the values A_m of a succession of analog samples. To recover the amplitude of the nth sample we merely sample the signal [that of (5.2-4)] at time $t = nT$. If we take successive samples at times $t = nT$ we recover all the samples in order. This must be so, because

(5.2-4) is an exact reconstruction of the band-limited function from which the sample amplitudes A_m were derived.

The quantities A_m in (5.2-4) *need not* be derived by sampling a band-limited function. They may be any amplitudes that we wish to transmit by means of a band-limited function. They may be a succession of integers such as 0 or 1; or +1 or −1; or +3, +1, −1, −3.

Thus a band-limited function such as (5.2-4) provides us a means for transmitting a succession of numbers which may be integers. We can recover the succession of numbers by sampling the function at times $t = nT$. As a step in this direction, the next section derives the integral of the sinc function over all time.

Problems

1. Find the total energy of the pulse

$$\frac{A \sin N\pi t}{N\pi t}$$

   This is easier in the frequency domain.

2. One way of estimating the error caused by omitting part of the ideal pulse, as in C of Fig. 5.3, is to estimate the missing power. If one side of the pulse is omitted beyond a time T_d from the center, find an approximate expression for the fraction of energy missing.

3. The difference in the time domain between B and C of Fig. 5.3 is obvious. Show that the difference in the frequency domain can be expressed as the fact that there is a little energy out of the band 0 to $N/2$.

5.3. The Integral of $(\sin \pi t)/\pi t$ or sinc t

Because sinc pulses have become of such central interest to us, we should know the time integral of such a pulse. This integral can be obtained by complex integration, but it can better be obtained by the following very simple argument. We have already used the result in Chapter 3.

Suppose that the band-limited function $f(t)$ that we wish to represent by means of N samples per second is simply a constant voltage of unit amplitude

$$f(t) = 1$$

This is band-limited enough for anyone.

The preceding section tells us that we can represent this constant function by a sequence of samples of amplitude 1 taken at a rate of one sample each second, and that we can reconstruct the constant function of unit amplitude as

$$f(t) = 1 = \sum_{n=-\infty}^{\infty} \frac{\sin \pi(t - n)}{\pi(t - n)}$$

If we integrate $f(t)$ over a time interval M seconds long, the value of this integral must be simply M. But if M is very large, the chief contribution to the integral is made of the M pulses which have peaks in the interval. The sinc pulses with peaks outside the interval contribute little to the integral. Hence the integral over any *one* pulse must be exactly unity if the integral over M pulses is about equal to M when M is large. This gives the following result:

$$\int_{-\infty}^{\infty} \frac{\sin \pi t}{\pi t}\, dt = 1$$

This formula will help us study sampling pulses. But, before we realize the full implications of sinc pulses, the next section summarizes alternatives to sinc pulses for transmitting data samples.

Problem

1. By reasoning similar to the above, derive

$$\int_{-\infty}^{\infty} \text{sinc}^2 t\, dt = 1$$

HINT: use the fact, which will be derived in Section 5.5, that the pulses

$$\text{sinc}\,(t - n)$$

are orthogonal as functions of t for integers n. (We did this problem a different way in Problem 1 of Section 3.6.)

5.4. Signaling Pulses and the Nyquist Criterion

We have seen that we can represent a band-limited function $f(t)$ whose frequencies lie below B by $N = 2B$ samples per second, the samples being spaced $T = 1/N$ apart. Further, we can reconstruct the original function $f(t)$ as a sum of sinc functions from the sample amplitudes $A_n = f(nT)$:

$$f(t) = \sum_{n=-\infty}^{\infty} A_n \frac{\sin N\pi(t - nT)}{N\pi(t - nT)}$$

What options do we have in transmitting the successive sample amplitudes A_n? One way would be to transmit these sample amplitudes as a succession of nonoverlapping pulses of peak amplitudes A_n. But pulses of finite bandwidth *do* overlap, and the sample amplitudes we receive may be affected by the tails of preceding and succeeding pulses. This is called *intersymbol interference*.

We would like to choose our pulses so that at the peak of a given received pulse, the amplitudes of all other received pulses pass through zero. We have seen in Section 5.2 that this is indeed true of properly spaced sinc pulses. The function

$$A_n \frac{\sin N\pi(t - nT)}{N\pi(t - nT)} \equiv A_n \operatorname{sinc} N(t - nT)$$

represents a pulse of amplitude A_n centered at

$$t = nT = n/N = n/2B$$

Such pulses seem an ideal means for transmitting samples of amplitudes A_n or any succession of pulses, including pulses of integer value. If we sample such a train at times $t = nT$, we get the transmitted amplitudes A_n directly, because all the other pulses go to zero at nT.

In using pulses to transmit amplitudes or data, as opposed to using pulses to reconstruct a signal from samples, we see that we need not use sinc pulses. We can use any other pulses which are zero at all other sampling times. Pulse shapes other than sinc might be useful in that they might be more practical to produce than sinc pulses, or have lower energy, or fall off more rapidly away from their peak, so that errors in their tails would be less important.

Let us investigate how this can occur. The spectrum of a pulse

$$\frac{\sin N\pi t}{\pi t}$$

has zero phase and unit amplitude between $f = -N/2$ and $f = N/2$. It is zero outside of these limits. Because the pulse is an even function of frequency (as are all pulses symmetric about time 0), we can think of it as made up of $\cos 2\pi ft$ terms.

At the sampling points t with

$$t = nT = \frac{n}{N}$$

where n is an integer, we cannot distinguish a component of frequency

$$\frac{N}{2} + \Delta f$$

from a component of frequency

$$\frac{N}{2} - \Delta f$$

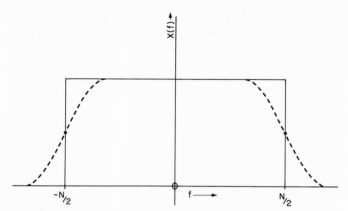

Figure 5.4. The Nyquist criterion for symmetric pulses.

because

$$\cos\left[2\pi\left(\frac{N}{2} + \Delta f\right)\frac{n}{N}\right] = \cos\left(\pi n + 2\pi\frac{n}{N}\Delta f\right)$$

is equal to

$$\cos\left[2\pi\left(\frac{N}{2} - \Delta f\right)\frac{n}{N}\right] = \cos\left(\pi n - 2\pi\frac{n}{N}\Delta f\right)$$

Hence the value of the pulse *at the sampling points* will be unchanged if we reduce the amplitude of the spectrum at one of these frequencies and increase the amplitude by an equal amount at the other. This is illustrated in Fig. 5.4.

If the spectrum is $X(f)$, the pulse will be zero at n/N for any integer n other than zero, if we have

$$X\left(\frac{N}{2} - \Delta f\right) + X\left(\frac{N}{2} + \Delta f\right) = \text{constant} = 1 \qquad (5.4\text{-}1)$$

Pulses which have this property are said to meet the *Nyquist criterion*.† This holds as long as the spectrum of the new pulse is confined to no more than double the original bandwidth. For such symmetric pulses, (5.4-1) is the *exact* condition for no intersymbol interference.

The spectral energy density is $X(f)X^*(f)$. When the spectrum is *rolled off* as above, the pulse height at $t = 0$ is preserved but the pulse energy is reduced. This is because the integral of the square of the spectrum is less.

---

† There is another Nyquist criterion which has to do with the stability of negative feedback amplifiers. Nyquist did a lot for communication.

Keep in mind two distinct problems. One is that of reconstructing a signal from sample amplitudes. If we are to do this perfectly in the absence of noise, sinc pulses are essential. The other problem is transmitting sample amplitudes and receiving them without intersymbol interference. For this, any pulses which meet the Nyquist criterion are satisfactory. In this case, we can recover the amplitudes by sampling at the peaks of the pulses. Rectangular pulses are an instance of this, but these have infinite bandwidth.

Sinc pulses centered at different sampling times are orthogonal, as we see in the next section. This is not true of pulses rolled off as above. Orthogonality is important in studying the accuracy of representing a band-limited function by approximate samples.

Problems

1. Does the following spectrum meet the Nyquist criterion?

$$X(f) = (1 + \cos \pi f / N), \quad -N < f < N$$
$$X(f) = 0, \quad |f| > N$$

2. What is the time function $x(t)$ corresponding to the above $X(f)$? HINT: as far as the $\cos \pi f / N$ term goes, interchange t and f and note that we are am modulating a rectangular pulse. Where is $x(t)$ equal to 0? How fast does it die out for large t? What is the pulse energy as a fraction of the energy of a sinc pulse of the same height at time 0 and band limit $N/2$? HINT: evaluate pulse energies from the spectrum.

3. Show that the Nyquist criterion (5.4-1) is *necessary* if a symmetric pulse of spectrum confined to frequencies below N is to be 0 at all sampling instants other than 0. HINT: use the fact that a pulse of spectrum confined below N which is 0 at all sampling instants other than 0 is a constant times the sinc function. This follows from the sampling theorem.

5.5. The Orthogonality of sinc $(t - n)$ and sinc $(t - m)$

Here we will simplify matters by letting $T = 1$. If we measure time in units of time between samples, sinc $(t - n)$ is zero at all sampling times except $t = n$, and sinc $(t - m)$ is zero at all sampling times except $t = m$.

Beyond this, sinc $(t - n)$ and sinc $(t - m)$ are orthogonal if $m \neq n$.

We can see this most easily by expressing the functions as Fourier transforms; in the first transform we use x for frequency, and use y in the second:

$$\text{sinc} \, (t - n) = \int_{-1/2}^{1/2} e^{-j2\pi nx} e^{j2\pi xt} \, dx$$

Because sinc $(t - m)$ is real, we can also express it as the conjugate of the usual transform

$$\operatorname{sinc} (t - m) = \int_{-1/2}^{1/2} e^{j2\pi my} e^{-j2\pi yt} \, dy$$

Let us integrate the product of these functions over a very long time interval from $t = -T/2$ to $t = +T/2$. The integral I is

$$I = \int_{-T/2}^{T/2} \int_{-1/2}^{1/2} \int_{-1/2}^{1/2} e^{j2\pi(my - nx)} e^{j2\pi(x - y)t} \, dx \, dy \, dt$$

$$I = \int_{-1/2}^{1/2} \int_{-1/2}^{1/2} e^{j2\pi(my - nx)} \frac{\sin \pi(x - y)T}{\pi(x - y)} \, dx \, dy \qquad (5.5\text{-}1)$$

As we let T approach infinity, I approaches the integral of the product of the functions over all time. But, as T becomes very large, the fast oscillations of $[\sin \pi(x - y)T]/\pi(x - y)$ mean that the integral *outside* a thin strip around the line $y = x$ is nearly 0 (see Fig. 5.5). As T becomes large, the oscillating function becomes a delta function of strength 1, as a function of $x - y$. We saw this in Section 3.9, Problem 1.

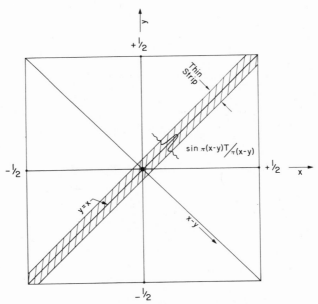

Figure 5.5. Orthogonality of sinc pulses.

In integrating $d(x - y)$, we can now pretend the limits are $-\infty$ to ∞:

$$I = \int_{y=-1/2}^{1/2} \int_{x-y=-\infty}^{\infty} e^{i2\pi(my-nx)}\delta(x - y)\, d(x - y)\, dy$$

$$I = \int_{-1/2}^{1/2} e^{i2\pi(m-n)y}\, dy \tag{5.5-2}$$

This is zero unless $m = n$. If $m = n$,

$$I = 1$$

Thus the functions $\operatorname{sinc}(t - n)$ and $\operatorname{sinc}(t - m)$ are orthogonal unless $m = n$, and the integral of their product over all time is unity when $m = n$. They form an orthonormal set. (We shall see a different way of deriving this in the next chapter.) The representation of a band-limited function by samples can be considered as an orthogonal expansion, much as a Fourier series. A sampler, however, is much simpler than a spectrum analyzer.

Just as we could consider the problem of errors in representing waveforms by Fourier series when the coefficients are inaccurate, so in the next section we study the error in reconstructing a waveform from inaccurate samples.

Problems

1. Based on the orthogonality of $\operatorname{sinc}(t - n)$ and $\operatorname{sinc}(t - m)$, suggest a conceptual way for obtaining the sample values for a band-limited function, other than directly taking samples.

2. If this new way of getting the sample values is applied to a signal which is not band limited, what do the sample values represent? HINT: use the multiplication theorem of Section 3.6.

5.6 Error in Representing Samples Approximately

Figure 5.6 illustrates a signal and some sample levels spaced V apart. The signal can have any of a range of amplitudes at the sampling times t_1, t_2, t_3, t_4, t_5, etc. But, at these times we must choose one of a number of voltages spaced V apart to represent the signal voltage. This will permit us to transmit the signal digitally after we have quantized it in this way. We will study such representations in a more general way in Section 13.3, where we learn about rate distortion theory.

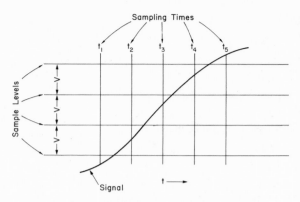

Figure 5.6. Signal and sample levels.

We can estimate the rms error in the representation of the sample voltage if we assume the following:

(1) The signal never exceeds the allowed extreme sample voltage by more than $V/2$. This can be arranged by limiting the signal before sampling.

(2) The signal amplitude is equally likely to fall anywhere between sample levels. This is at least approximately true for any signal as the number of levels becomes large.

If this is so, the maximum voltage error will be $\pm V/2$ and any smaller voltage error is equally likely. Hence $\overline{v^2}$, the mean-square voltage error, will be

$$\overline{v^2} = \frac{1}{V} \int_{-V/2}^{V/2} v^2 \, dv$$
$$\overline{v^2} = V^2/12$$

This can be called the quantization noise power.

We note that m sample levels of spacing V allow us to represent waves up to a peak amplitude of

$$\pm \frac{mV}{2}$$

The average power of this wave is

$$\overline{x^2}\left(\frac{mV}{2}\right)^2 = \frac{\overline{x^2}}{4} m^2 V^2$$

Here we must insist that the peak value of $x(t)$, x_p, be such that $|x_p| \le 1$.

The signal voltage is then

$$x \cdot \frac{mV}{2}$$

Thus the signal-to-noise ratio S/N after quantization is

$$S/N = 3\overline{x^2}m^2$$

Since m is typically a large integer 2^M, where M is the number of bits in the digital word used to represent the levels, the signal-to-noise ratio after quantization is large.

We will show that a given mean-square error in the sample values translates into the same mean-square error in the recovered waveform. Let the true waveform be represented by

$$z(t) = \sum_{n=-\infty}^{\infty} A_n \, \text{sinc} \, N(t - nT) \qquad (5.6\text{-}1)$$

The A_n are the samples of $z(t)$, and the recovered waveform is represented by

$$\hat{z}(t) = \sum_{n=-\infty}^{\infty} \hat{A}_n \, \text{sinc} \, N(t - nT) \qquad (5.6\text{-}2)$$

Here

$$\overline{(A_n - \hat{A}_n)^2} = \overline{v^2}, \qquad \text{all } n \qquad (5.6\text{-}3)$$

Over a time U, the mean-square error E in $z(t)$ (or, more accurately, the mean-square error per unit time), is

$$E = \frac{1}{U} \int_{-U/2}^{U/2} \overline{[z(t) - \hat{z}(t)]^2} \, dt \qquad (5.6\text{-}4)$$

From Section 5.5,

$$\int_{-\infty}^{\infty} \text{sinc} \, N(t - nT) \, \text{sinc} \, N(t - mT) \, dt = \begin{cases} 0 & \text{if } m \neq n \\ T & \text{if } m = n \end{cases} \qquad (5.6\text{-}5)$$

Using (5.6-1) and (5.6-2), (5.6-4) becomes for large U approximately

$$E \doteq \frac{1}{U} \sum_{n=-U/2T}^{U/2T} \overline{(A_n - \hat{A}_n)^2} \cdot T$$

$$E \doteq \overline{v^2} \cdot \left(1 + \frac{T}{U}\right) \doteq \overline{v^2} \qquad (5.6\text{-}6)$$

As U becomes larger and larger, the approximation in (5.6-6) becomes better and better, and

$$E = \overline{v^2} \qquad (5.6\text{-}7)$$

Hence we need not distinguish between the mean-square error in the sample values and the mean-square error per unit time in the recovered analog waveform. The two mean-square errors are equal.

We can say more. E is also the mean-square error in $\hat{z}(t)$ at *any* particular time, not merely at the sampling instants. Thus E is entirely analogous to the noise used in Chapter 4 in signal-to-noise calculations. It can rightly be thought of as quantization noise power.

The reason that the mean-square noise is independent of the time t at which we look is based on the fact that E is both the mean-square noise per unit time and the mean-square noise at the sampling instants. We can sample $\hat{z}(t)$ at the same rate as before, but in any phase or at any times differing by the sampling interval. The value of E as the mean-square error per unit time does not change just because we have changed the phase of sampling. So, E must now be equal to the mean-square error at the new sampling instants.

We close this chapter with a brief section on sampling in two dimensions. Much of the theory is a straightforward extension of what we have learned about sampling in the single dimension of the time domain.

Problems

1. Sample amplitudes are to be represented by M binary digits. Assume $\overline{x^2} = \frac{1}{2}$. How large must M be to give a 30-dB signal-to-noise power ratio? An S/N of 60 dB?

2. We have seen in Section 4.7 that for a given base bandwidth B the fm signal-to-noise ratio S/N varies as $\overline{x^2}(f_d/B)^2$. If $x(t)$ is sinusoidal with a peak value of 1, then $\overline{x^2}$ will be $\frac{1}{2}$. The total rf band B_r will be approximately $2(B + f_d)$. Express S/N as a function of B_r/B.

3. Suppose that a sine wave of frequency less than B is sent by sampling it at a rate $2B$, using M binary digits to represent the amplitude of each sample. (This is called *pulse code modulation* or *pcm*.) Suppose that the rf bandwidth required to transmit the pulses representing the samples is $B_r = 2MB$. What is the expression for the S/N of the recovered signal as a function of B_r/B, assuming errors in the digital system are so rare that they can be ignored? Does this rise faster or slower with increasing B_r/B than in the case of fm? This kind of reasoning was important in the early days of pcm in the late 1940's.

4. For rf noise density N_0 and transmitter power P_0, compute S/N for both fm (Problem 2) and pulse code modulation (Problem 3) for B_r/B equal to 8, 12, and 16. Assume that the rf carrier-to-noise ratio in the band B_r must be 40 dB to allow for fades in microwave transmission of 25 dB. What are the S/N ratios during a 25-dB fade?

5.7. Sampling in Two Dimensions

In Section 3.10, we studied two-dimensional Fourier transforms. Here we will study the use of two-dimensional Fourier transforms in two-dimensional sampling. Such sampling is useful in transmission of video by digital means and in reconstructing radar maps.

Let $l(x, y)$ be a map or brightness pattern in two dimensions. The dimensions here are space dimensions; neither is a timelike quantity. The Fourier transform of l is

$$L(f, g) = \int_{-\infty}^{\infty} \int_{-\infty}^{\infty} l(x, y)e^{-j2\pi(fx+gy)} \, dx \, dy \qquad (5.7\text{-}1)$$

$L(f, g)$ tells how the pattern l varies in each of two spatial frequencies, f in the x direction and g in the y direction. It depends only on the point in two-dimensional frequency space and not on the coordinate system, because the inner or dot product $fx + gy$ is independent of the coordinate system.

The pattern $l(x, y)$ may be band limited in two dimensions. We take this to mean that $L(f, g)$ is 0 outside a square of sides $2B$ parallel to the axes and centered at the origin, as shown in Fig. 5.7. Here B has dimensions reciprocal to the dimensions of x and y. If these dimensions are lengths, B is called *spatial frequency* in units of "per unit length."

If $L(f, g)$ is equal to 1 in this square and 0 outside, $l(x, y)$ is easily found from $L(f, g)$:

$$l(x, y) = \int_{-B}^{B} \int_{-B}^{B} e^{j2\pi(fx+gy)} \, df \, dg$$

$$l(x, y) = 4B^2 \frac{\sin 2\pi Bx}{2\pi Bx} \frac{\sin 2\pi By}{2\pi By} = 4B^2 \operatorname{sinc} 2Bx \operatorname{sinc} 2By \qquad (5.7\text{-}2)$$

Just as in the one-dimensional case, this can be used as the basis of a

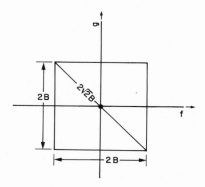

Figure 5.7. Two-dimensional band limiting.

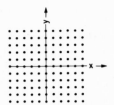

Figure 5.8. Two-dimensional sampling points.

two-dimensional sampling theorem. Any function of baseband bandwidth less than B in the x and y directions can be uniquely recovered from two-dimensional samples on a square grid parallel to the axes spaced $T = 1/2B$ apart in each direction. A few of the sampling points are shown in Fig. 5.8.

There are phenomena in two-dimensional sampling which do not arise in one-dimensional sampling. These arise because the diagonal of a square is longer than its side. Also, human perceptual acuity is greater in the horizontal than in the vertical direction, and this suggests greater horizontal sampling in digital television. We shall not pursue these matters.

This is all we are going to study about general sampling. In the next chapter we go more thoroughly into the structure of signals and noise by studying the autocorrelation function. This is used to rigorously define the power spectrum of a random waveform and, ultimately, to find the error probability in digital communication systems. Sampling will enable us to go from the continuous to the discrete in analyzing random signals and noises.

Problems

1. Write down the formula which tells how to reconstruct a two-dimensional band-limited pattern from its two-dimensional samples.

2. Consider these three functions of two frequency variables f and g: sinc Lf, sinc Lg, and sinc $L(f + g)$. Each represents the two-dimensional spectrum of a bright line. Why? How should the brightness of such a line be defined? What are the comparative brightnesses? What are the comparative lengths of these lines?

3. Is the function

$$l(x, y) = \sin 2\pi B_0(x + y) = \sin 2\pi B_0 x \cos 2\pi B_0 y + \cos 2\pi B_0 x \sin 2\pi B_0 y$$

band limited between $-B$ and B in both dimensions? What is its band limit if the x, y axes are rotated 45°? Assume B_0 is infinitesimally less than B.

4. What happens when we try to reconstruct the function $l(x, y)$ in Problem 3 from regularly spaced two-dimensional samples $1/2B$ apart in each direction? How can we explain this? Switch to coordinates rotated 45°. Can we reconstruct now? Is this disturbing?

6

Autocorrelation and Stationarity

In the foregoing chapters, we have tried to present matters concerning signals and their transmission with as little mathematics as possible. If we are to pursue these matters further and evaluate the potentialities of various transmission schemes, we must consider some properties of signals and noise in a little more detail. Particularly, we must understand autocorrelation and its relation to the power spectrum of stationary signals. Reference 1 is a good introduction to these concepts for communicators. This includes probability, which we study in Chapter 8, and random signals and noises and their power spectra, which we study in Chapter 9.

The chief purpose of this chapter is to learn the necessary details of signal analysis. In subsequent chapters, we shall use the concepts of this chapter to determine the signal-to-noise performance of pulse communication, to find error probabilities in digital communication, to define bandwidth rigorously, to study coded communication, and for many other purposes.

6.1. Autocorrelation

Correlation comes in two varieties: correlation of one signal with another signal, and correlation of a signal with itself. Here we will be concerned only with this second variety of correlation, correlation of a signal with itself. This is called the autocorrelation or covariance or autocovariance of a signal. The other is called the cross-correlation. We will first consider the autocorrelation of signals which have finite energy. Later we will consider the autocorrelation of signals which have infinite energy but finite average power.

We define the autocorrelation $C(\tau)$ of a signal $x(t)$ of finite energy as

$$C(\tau) = \int_{-\infty}^{\infty} x(t)x(t + \tau)\, dt \qquad (6.1\text{-}1)$$

This is the same as the convolution of Section 3.7, but $x(\tau - t)$ there has been replaced by $x(\tau + t)$. Clearly, if $\tau = 0$

$$C(0) = \int_{-\infty}^{\infty} [x(t)]^2\, dt = E \qquad (6.1\text{-}2)$$

Here E is the energy of the signal. Thus the autocorrelation for $\tau = 0$ is the signal energy.

It is clear from (6.1-1) that the autocorrelation of $Ax(t)$ is A^2 times the autocorrelation of $x(t)$.

We will find the autocorrrelation function of a delta function in the next section.

Problems

1. What is the autocorrelation of the following function?

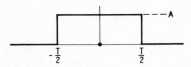

2. What is the autocorrelation of the function shown below? What general result do Problems 1 and 2 suggest?

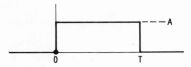

3. What is the autocorrelation of the following function?

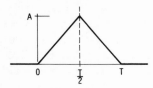

4. What is the general relation between $C(\tau)$ and $C(-\tau)$?

6.2. Autocorrelation of a Delta Function

Consider a pulse $x(t)$ which is a rectangular approximation to a delta function by a pulse of height $1/d$. The width d is very small, the height $1/d$ is very large, but the area $d \cdot 1/d$ is 1. The pulse $x(t + \tau)$ is shifted to the left by the amount τ, which is considered positive in Fig. 6.1. We want to find

$$C(\tau) = \int_{-\infty}^{\infty} x(t)x(t + \tau)\, dt$$

The pulses overlap from $\tau = -d$ to $\tau = +d$. The width of overlap varies linearly between $\tau = -d$ to 0 and $\tau = 0$ to $+d$. The height of the product of the pulses in the overlapping region is always $1/d^2$. When $\tau = 0$ the integral with respect to t is $(1/d^2)d = 1/d$. Here $C(\tau)$ is the triangular autocorrelation function of height $1/d$ and width $2d$ shown in Fig. 6.2.

The area under this function is

$$\tfrac{1}{2}2d\frac{1}{d} = 1$$

As we let d go to zero, we have

$$x(t) = \delta(t)$$

We have a unit impulse at $t = 0$. Or, the autocorrelation function $C(\tau)$ of $\delta(t)$ is

$$c(\tau) = \delta(\tau)$$

The autocorrelation function of a delta function of unit strength is a delta function of unit strength.

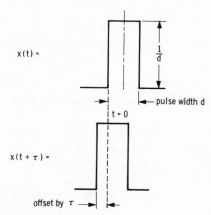

$x(t) =$

$\frac{1}{d}$

pulse width d

$t = 0$

$x(t + \tau) =$

Figure 6.1. Delta function and time-shifted delta function.

offset by τ

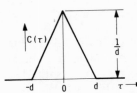

Figure 6.2. Autocorrelation of approximate delta function.

We can do this in another way. If $C(\tau)$ is equal to $\delta(\tau)$, it must be that for every smooth f

$$\int_{-\infty}^{\infty} f(\tau)C(\tau)\, d\tau = f(0) \qquad (6.2\text{-}1)$$

We can show that this is indeed true by writing out $C(\tau)$ as the autocorrelation of $\delta(\tau)$. The integral we obtain is

$$\int_{\tau=-\infty}^{\infty} f(\tau) \int_{t=-\infty}^{\infty} \delta(t)\delta(t+\tau)\, dt\, d\tau = \int_{t=-\infty}^{\infty} \delta(t) \int_{\tau=-\infty}^{\infty} f(\tau)\delta(t+\tau)\, d\tau\, dt$$

The integral in τ on the right side of the above equation is simply $f(-t)$. Then we have

$$\int_{t=-\infty}^{\infty} \delta(t)f(-t)\, dt = f(0)$$

This shows that (6.2-1) must hold, and $C(\tau)$, the autocorrelation of $\delta(t)$, is equal to $\delta(\tau)$.

In the next section we find the autocorrelations of the sinc function.

Problems

1. What is the autocorrelation of
$$x(t) = \delta(t - T) - 2\delta(t) + \delta(t + T)?$$

2. We have so far been considering the autocorrelation only for functions of finite energy. What is the energy in a delta function of strength 1?

6.3. Autocorrelation and Orthogonality of sinc Nt

In Section 5.5, we derived the orthogonality of $\text{sinc}\,(t - n)$ and $\text{sinc}\,(t - m)$ for m and n different integers. To do this, we derived (5.5-2), which we can write as

$$\int_{-\infty}^{\infty} \text{sinc}\,(t - n)\,\text{sinc}\,(t - m)\, dt = \int_{-1/2}^{1/2} e^{j2\pi(m-n)y}\, dy \qquad (6.3\text{-}1)$$

This derivation is valid for arbitrary m and n, not only for integers. When $m - n = \tau$, this is merely the autocorrelation of sinc t evaluated at τ:

$$C(\tau) = \int_{-1/2}^{1/2} e^{j2\pi\tau y} \, dy$$

$$C(\tau) = \text{sinc } \tau \tag{6.3-2}$$

There is another way we can derive the autocorrelation of sinc, knowing that the functions sinc $(t - n)$ form an orthonormal set. Suppose we know only that

$$\int_{-\infty}^{\infty} \text{sinc } (t - n) \, \text{sinc } (t - m) \, dt = \delta(m - n) \tag{6.3-3}$$

Here

$$\delta(m - n) = \begin{cases} 1 & \text{if } m = n \\ 0 & \text{if } m \neq n \end{cases}$$

The autocorrelation of the sinc function will then be equal to 1 if $\tau = 0$ and 0 if τ is a nonzero integer. So, $C(\tau)$ will agree with sinc τ at the sampling instants. If we knew that $C(\tau)$ were itself band limited (to the same band), then the sampling theorem would imply (6.3-2) for all τ. Why is $C(\tau)$ band limited?

We will show that the autocorrelation function of a band-limited function is band limited to the same band, or, more generally, that the cross-correlation of two band-limited functions is band limited.

Suppose $x(t)$ and $y(t)$ are band limited, so that their spectra $X(f)$ and $Y(f)$ are confined to the frequency region $-B$ to B. We will show that the spectrum of the cross-correlation $B(\tau)$

$$B(\tau) = \int_{-\infty}^{\infty} x(t)y(t + \tau) \, dt \tag{6.3-4}$$

is equal to $X^*(f) Y(f)$. Then the spectrum of $B(\tau)$ will also be confined to the region $-B$ to B, and $B(\tau)$ will be band limited.

We can write from (6.3-4)

$$B(\tau) = \int_{t=-\infty}^{\infty} x(t) \left(\int_{f=-\infty}^{\infty} Y(f) e^{j2\pi f(t+\tau)} \, df \right) d\tau$$

$$= \int_{f=-\infty}^{\infty} Y(f) e^{j2\pi f\tau} \left(\int_{t=-\infty}^{\infty} x(t) e^{j2\pi ft} \, dt \right) d\tau$$

$$B(\tau) = \int_{f=-\infty}^{\infty} X^*(f) Y(f) e^{j2\pi f\tau} \, df \tag{6.3-5}$$

We note that when $\tau = 0$, (6.3-4) and (6.3-5) become the multiplication theorem of Section 3.6.

Equation (6.3-5) says that $B(\tau)$ is the inverse Fourier transform of $X^*(f)Y(f)$. Or, $B(\tau)$ has spectrum $X^*(f)Y(f)$. So if $x(t)$ and $y(t)$ are band limited, or even if only one is, so is their cross-correlation $B(\tau)$.

We conclude from this that $C(\tau)$ is band limited to the same $-\frac{1}{2}$ to $+\frac{1}{2}$ band to which sinc t is limited. Since $C(\tau)$ and sinc τ were assumed to agree at the sampling instants, they are the same at all instants of time. The autocorrelation of sinc t is nothing but sinc τ.

Having some examples of autocorrelation functions, we relate them to energy spectra in the next section. The intimate relation between correlation and spectrum is an extremely important one for designing and analyzing communication and signal-processing systems.

Problems

1. What is the autocorrelation of sinc Nt?

2. What is the frequency spectrum of sinc^2 t? HINT: one way is to use (6.3-5) in reverse, going from time to frequency, to express the spectrum of sinc^2 t as an autocorrelation. Is it band limited to any frequency band? Explain why your answer might be guessed in advance.

3. Suppose $x(t)$ has spectrum $X(f)$ which takes only the values 0 and 1, as does the spectrum of sinc t. Show that $x(t)$ is equal to its own autocorrelation. Is this condition necessary?

6.4. Autocorrelation and Energy Spectrum

Consider the autocorrelation $C(\tau)$ of the time function $x(t)$:

$$C(\tau) = \int_{-\infty}^{\infty} x(t)x(t + \tau)\, dt$$

This is a particular case of (6.3-4), and from (6.3-5) we have

$$C(\tau) = \int_{-\infty}^{\infty} X(f)X^*(f)e^{i\omega\tau}\, df$$

We remember that $E(f) = X(f)X^*(f)$ is the energy spectrum or energy spectral density. Hence the autocorrelation and the energy spectrum of a signal are Fourier transforms of one another:

$$C(\tau) = \int_{-\infty}^{\infty} E(f)e^{i2\pi f\tau}\, df \qquad\qquad (6.4\text{-}1)$$

$$E(f) = \int_{-\infty}^{\infty} C(\tau)e^{-i2\pi f\tau}\, d\tau \qquad\qquad (6.4\text{-}2)$$

This forms the basis of many instruments or algorithms for finding the energy density. Such instruments produce the Fourier transforms of the autocorrelation function. We studied a special type of instrument like this, the short-term spectrum analyzer or sonograph, in Section 3.11.

We should note that $X(f)X^*(f)$ is the energy density appropriate to the frequency range from $f = -\infty$ to $f = +\infty$. The energy density appropriate to the frequency range from $f = 0$ to $f = \infty$ is

$$2X(f)X^*(f) = 2\int_{-\infty}^{\infty} C(\tau)e^{-j\omega\tau}\, d\tau$$

In the next section, we define a class of signals of infinite energy but finite average power, called *stationary signals*. In the section following the next, we extend the concept of autocorrelation and energy spectrum to these stationary signals.

Problems

1. What are the autocorrelation and energy spectrum of the following pulse?

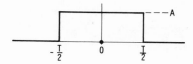

2. Determine the autocorrelation and energy spectrum of

$$A[\delta(t - T/2) + \delta(t + T/2)]$$

Compute the total energy from the energy density.

6.5. Stationary Signals

Many useful signals change with time in such ways that time averages taken over a long time T approach limits, limits that are the same regardless of the location of the time interval. This implies that if the energy is finite for a finite time interval, the total energy over all time must be infinite. However, the energy over a time T divided by T, or the average power P, will approach a limit as T becomes very large.

The function

$$\cos \omega_0 t$$

is a stationary signal. Its average power is $\frac{1}{2}$. There is phase information in this function, but no information on absolute time.

The signal

$$\cos \omega_0^2 t^2$$

also has average power $\frac{1}{2}$. But, it is not a stationary signal because its character (in this case, local frequency) changes with time. We can tell time from this waveform by measuring its instantaneous frequency, or by taking the short-term spectrum.

Often, a sum of sinusoidal terms is a stationary signal. However,

$$\int_{-\infty}^{\infty} e^{j\omega t}\, df$$

is a huge pulse (delta function) at $t = 0$, which is scarcely a stationary signal. We can find time 0 exactly from this delta function.

The basic defining property of stationary signals is very simple in engineering terms. It is that it is not possible to tell anything about absolute time by looking at the signal. So, stationary signals must last forever. There may be *relative* time, or *phase*, information in a stationary signal, but there is no tendency for a stationary signal to decay or to change in any property as time progresses. We shall make this more rigorous in Chapter 9, when we study noise waveforms. We will see in that chapter that individual waveforms of band-limited gaussian noise are stationary signals.

In the next section, we shall find the power spectrum and autocorrelation function of stationary signals, even though they must have infinite energy.

Problem

1. Start with a square wave. Change the signs of the amplitudes randomly by flipping a coin for each interval of constancy (if heads, change sign; if tails, do not). (a) Would you expect such a signal to be stationary? (b) If you change the signs randomly on the positive time axis by the above process, but force the exact *same* change on the negative axis, as in the figure below, will the signal be stationary? Why?

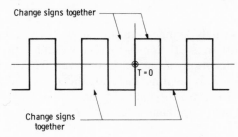

6.6. Power Spectrum and Autocorrelation Function of a Stationary Signal

The energy of a stationary signal is infinite. But, if we divide the energy over any specified time by the length of the time interval, this quotient is finite. The quotient has the dimensions of power. This power approaches a limit, the average power, as the integral becomes infinite, regardless of the position of the time interval.

If we have a stationary function we can truncate it and consider a function $x_T(t)$ which is equal to that function for $|t| < T/2$ and is equal to zero for $|t| > T/2$. This truncated signal of length T is no longer stationary. Its autocorrelation as derived in Section 6.4 can be expressed as

$$C_T(\tau) = \int_{-\infty}^{\infty} X_T(f) X_T^*(f) e^{j\omega\tau} \, df$$

Here $C_T(\tau)$ and $X_T(f)$ are the autocorrelation and the Fourier transform of the truncated signal x_T.

We can see that when $\tau = 0$ the integral on the right is the energy of the truncated function, and so $C_T(0)/T$ must be the average power of the truncated signal. Clearly also, the power spectrum $P_T(f)$ will be

$$P_T(f) = X_T(f) X_T^*(f)/T \qquad (6.6\text{-}1)$$

For a stationary signal, as we let $T \to \infty$, $C_T(\tau)/T$, the correlation per unit time, will approach a limit $C(\tau)$ and $P_T(f)$ will approach a limit $P(f)$. This is a result of the rigorous formal definition of stationary signal.

We see that

$$\frac{1}{T} C_T(\tau) \quad \text{and} \quad P_T(f)$$

form a Fourier pair. Hence their limits

$$C(\tau) \quad \text{and} \quad P(f)$$

form a Fourier pair. We will *define* this $C(\tau)$ to be the autocorrelation of $x(t)$ and $P(f)$ to be the power spectrum or power spectral density of $x(t)$. Because $x(t)$ is stationary, the limits $C(\tau)$ and $P(f)$ are independent of how we define the origin of the time coordinate for truncating $x(t)$.

In summary, for a stationary signal the autocorrelation $C(\tau)$ and the power spectrum $P(f)$ form a Fourier pair. Thus

$$P(f) = \int_{-\infty}^{\infty} C(\tau) e^{-j\omega\tau} \, d\tau \qquad (6.6\text{-}2)$$

$$C(\tau) = \int_{-\infty}^{\infty} P(f) e^{j\omega\tau} \, df \qquad (6.6\text{-}3)$$

We should remember that the total power P_B in the band of frequencies extending from 0 to B is

$$P_B = \int_{-B}^{B} P(f) \, df \qquad (6.6\text{-}4)$$

(The element of integration is f, not ω.)

So, $P(f)$ is a power spectrum appropriate to *negative and positive frequencies*. The power spectrum appropriate to positive frequencies only is

$$2P(f) = 2 \int_{-\infty}^{\infty} C(\tau)e^{-j\omega\tau}\,d\tau$$

The expressions *autocorrelation* or *autocorrelation function* have been used for both $C(\tau)$ as defined in Section 6.1 and $C(\tau)$ as defined in this section. The reader should have no trouble making the appropriate choice in each instance.

In Section 3.7, we defined the transfer function $H(f)$ of a linear network. What happens to the power spectrum $P(f)$ of a stationary signal when it is passed through such a network? It is reasonable to suppose that $P(f)$ is multiplied by $H(f)H^*(f)$, the power gain of the filter, just as we saw that the energy spectrum was. This is because definition (6.6-1) of $P_T(f)$ shows that $P_T(f)$ is multiplied by $H(f)H^*(f)$ when the signal $x(t)$ is passed through the filter *after* truncation. We wish to conclude that this is about the same as passing $x(t)$ through the filter *before* truncation. For filters without infinite memory, at least, truncation only affects the filtered output for a finite time and then has negligible effect. Hence $H(f)H^*(f)$ is the power gain of the filter, when operating on stationary signals as well as when operating on signals of finite duration. This will be used often, starting with the next chapter.

We have seen that the correlation functions of stationary signals always have Fourier transforms. In the next section, we will see that stationary signals themselves can sometimes have Fourier transforms.

Problems

1. What is the autocorrelation of

$$A \cos \omega_0 t?$$

What is the power spectrum? Compute the power spectrum via $C(\tau)$ and via $X_T(f)$. (One of these ways is more complicated than the other in this case.)

2. Find the autocorrelation of a square wave of amplitude $\pm A$ and period T_0. Find the power spectrum by as simple a method as you can, by the use of Fourier series.

6.7. Fourier Transforms of a Stationary Signal

A stationary signal, which has infinite energy, can nevertheless sometimes have a Fourier transform. By the energy theorem of Section

3.6, the transform will also have infinite energy. For example, assume

$$x(t) = \cos \omega_0 t$$

We will first integrate over a finite interval T

$$X_T(f) = \int_{-T/2}^{T/2} \cos \omega_0 t \exp(-j\omega t)\, dt$$

$$= \int_{-T/2}^{T/2} \frac{\exp[j(\omega_0 - \omega)t] + \exp[-j(\omega_0 + \omega)t]}{2}\, dt$$

$$= \frac{1}{2}\left(\frac{\sin(\omega - \omega_0)T/2}{(\omega - \omega_0)/2} + \frac{\sin(\omega + \omega_0)T/2}{(\omega + \omega_0)/2}\right)$$

When T is very large, $X_T(f)$ is essentially zero except very near $\omega = \pm\omega_0$. In these regions there are two spikes with respect to frequency f, as in Section 3.9, Problem 1. The area A of the spikes will be

$$A = \frac{1}{2}\int_{-\infty}^{\infty} \frac{\sin(\omega - \omega_0)T/2}{(\omega - \omega_0)T/2}\, d(fT) = \frac{1}{2}\int_{-\infty}^{\infty} \operatorname{sinc} fT\, d(fT) = \frac{1}{2}$$

So, in the limit as $T \to \infty$, the Fourier transform $X(f)$ of $x(t)$ is

$$X(f) = \tfrac{1}{2}[\delta(\omega - \omega_0) + \delta(\omega + \omega_0)]$$

We saw this already in Chapter 3. The limit does not depend on the choice of the interval of integration of duration T. In taking the inverse transform

$$\int_{-\infty}^{\infty} X(f) \exp(j\omega t)\, d\omega$$

we simply get

$$x(t) = \tfrac{1}{2}\exp(j\omega_0 t) + \tfrac{1}{2}\exp(-j\omega t) = \cos \omega_0 t$$

as we should.

But, stationary functions need *not* have Fourier transforms in any useful sense. We will see this when we study random processes in Chapter 9. The gaussian noise that occurs in communication systems is the most important such example for us, and explains why we use power spectra instead of voltage spectra in analyzing noise.

The next chapter uses the concepts of spectrum to study communication using pulses. We will find pulse shapes that are optimum with respect to properties defined in terms of the energy spectrum of the individual pulses, or in terms of the power spectrum of a stationary source of digital data which is communicated using successive pulses whose amplitudes depend on the data. We will also study related problems of optimum signaling, including in this a study of antenna arraying.

Problems

1. Let $x(t)$ be a periodic signal of period T_0. What is its Fourier transform in terms of its complex Fourier series?

2. Find the Fourier transform of a square wave of period 1 and amplitude 1, which changes sign at time 0. To obtain the Fourier series, differentiate first to obtain a train of impulses.

7

Pulse Shape, Filtering, and Arraying

We now have enough understanding to say something about signals which are optimum in some sense and about signal-to-noise ratios in various types of digital transmission. We will find that variational techniques permit us to match pulses optimally to channel transfer functions. This is useful when we can do something about the pulses but cannot modify the channel, which may be provided by a common carrier. We shall also study optimal signal combining techniques to combat fading. This is one instance of antenna arraying. Reference 1 contains a good summary of variational techniques which will suffice for most applications in engineering and science. Chapter 6 of that book is especially relevant to this chapter.

We will be working with power spectra derived from the communication of a stream of pulses which continues forever. The relation between the resulting power spectrum and the energy spectrum of the individual pulses will be derived in the next chapter.

7.1. An Optimum Filtering or Shaping

In a communication system we desire to have a signal of specified waveform or spectrum at the output of a receiver. In this section we will be concerned with how to attain the best signal-to-noise ratio in the output for a given total transmitted power P_0, by choosing the transmitted power spectrum or transmitter filter. This is useful for analog signaling.

The system we consider is the baseband system shown in Fig. 7.1. The transmitter has a total output power P_0 and an output power spectrum $P_1(f)$. The input spectrum $P(f)$, which is also the signal power spectrum at the output of the receiver, is given.

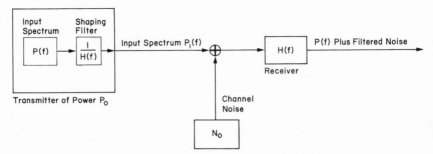

Figure 7.1. Transmission system with shaping filter.

All power spectral densities are chosen suitably for positive frequencies only. Thus

$$P_0 = \int_0^\infty P_1(f) \, df \tag{7.1-1}$$

In the channel a noise of constant power density N_0 is added to the transmitted signal.

At the receiver the sum of the noise and the transmitted signal is passed through a filter of transfer function $H(f)$. This function may if we wish exhibit gain, that is, it may be that $|H(f)| > 1$. The signal output of power spectrum $P(f)$ will be given, according to Section 6.6, by

$$P(f) = P_1(f)H(f)H^*(f) \tag{7.1-2}$$

The total received signal power P_s will be

$$P_s = \int_0^\infty P(f) \, df = \int_0^\infty P_1(f)H(f)H^*(f) \, df \tag{7.1-3}$$

The total noise power P_n will be

$$P_n = N_0 \int_0^\infty H(f)H^*(f) \, df \tag{7.1-4}$$

We assume a constant value of P_0 and a given input (and output) power spectrum $P(f)$. One question to ask is, how can we choose $P_1(f)$ so as to maximize the signal-to-noise power ratio P_s/P_n? Because $P(f)$ is given, this means minimizing the total noise as given by (7.1-4). We do this by choosing $H(f)$ subject to constraint (7.1-1). Physical insight shows that there must be a best shape. What we have to do is find this shape.

We note that

$$P_1(f)H(f)H^*(f) = P(f)$$

$$H(f)H^*(f) = \frac{P(f)}{P_1(f)} \tag{7.1-5}$$

If we vary $P_1(f)$ around its optimum value by an extremely small amount $\delta(f)$ while keeping P_0 constant, we must choose $\delta(f)$ so that

$$\int_0^\infty [P_1(f) + \delta(f)] \, df = P_0 \tag{7.1-6}$$

This technique of infinite-dimensional calculus is studied in Section 7.3. In order to preserve P_0,

$$\int_0^\infty \delta(f) \, df = 0 \tag{7.1-7}$$

Here $\delta(f)$ should be equal to 0 whenever $P(f)$ is equal to 0, and should otherwise be so small that the integrand in (7.1-6) is nowhere negative, and so is a power density. We will also have, from (7.1-4) and (7.1-5),

$$P_n = N_0 \int_0^\infty \frac{P(f) \, df}{P_1(f) + \delta(f)}$$

$$= N_0 \int_0^\infty \frac{P(f) \, df}{[P_1(f)][1 + \delta(f)/P_1(f)]}$$

Except for terms of second or higher order in $\delta(f)$, this is

$$P_n = N_0 \int_0^\infty \frac{P(f)}{P_1(f)} \, df - N_0 \int_0^\infty \frac{P(f)\delta(f)}{[P_1(f)]^2} \, df \tag{7.1-8}$$

From (7.1-7) we see that the integral in (7.1-8) that involves $\delta(f)$ will be zero for every allowed $\delta(f)$ if and only if

$$\frac{P(f)}{[P_1(f)]^2} \equiv K \tag{7.1-9}$$

[This holds where $P(f)$ is greater than 0; $P_1(f)$ is 0 at those f for which $P(f)$ is 0.] Here K is a constant to be determined. This is then the condition that P_n be minimized.

This result is analogous to the one in three-dimensional space which says that any vector perpendicular to everything perpendicular to a given vector must lie along the given vector. Here the given "vector" is the constant function 1, and "perpendicular" is defined as in Section 3.2: the integral of the products equals zero. Examples of this definition of perpendicularity occur in (7.1-7) and in the second integral in (7.1-8). Equation (7.1-9) is then the condition that $P(f)/P_1^2(f)$ lie along the vector defined by the constant function 1.

We have

$$P_1(f) = [P(f)]^{1/2}/K^{1/2} \tag{7.1-10}$$

$$H(f)H^*(f) = P(f)/P_1(f) = [KP(f)]^{1/2} \tag{7.1-11}$$

Here we are at the heart of the matter. The transmitted power spectrum $P_1(f)$ should vary as $[P(f)]^{1/2}$, the square root of the input (or received) power spectrum, and so should the power gain $H(f)H^*(f)$ of the receiving filter. With this spectrum, we get the best received signal-to-noise ratio with the received spectrum $P(f)$. In Fig. 7.1, we have illustrated the fact that we achieve this by generating the power spectrum $P(f)$ and filtering the input by $1/H(f)$ to transmit $P_1(f)$.

We now proceed to find the signal-to-noise ratio P_s/P_n. To do this, we must first find the constant K. The signal power P_s at the output will be

$$P_s = \int_0^\infty P(f)\, df$$

The noise power P_n at the output will be

$$P_n = N_0 K^{1/2} \int_0^\infty [P(f)]^{1/2}\, df$$

We can evaluate K for any specified P_0 and $P(f)$. From (7.1-1) and (7.1-10)

$$P_0 = \int_0^\infty P_1(f)\, df = \frac{1}{K^{1/2}} \int_0^\infty [P(f)]^{1/2}\, df$$

$$K^{1/2} = \frac{\int_0^\infty [P(f)]^{1/2}\, df}{P_0}$$

So

$$P_n = \frac{N_0}{P_0} \left(\int_0^\infty [P(f)]^{1/2}\, df \right)^2$$

$$\frac{P_s}{P_n} = \frac{P_0}{N_0} \frac{\int_0^\infty P(f)\, df}{\left(\int_0^\infty [P(f)]^{1/2}\, df \right)^2} \tag{7.1-12}$$

Relation (7.1-12) gives the optimal signal-to-noise ratio for any *prescribed* signal spectral shape $P(f)$. We may ask, is there any *optimal* spectral shape? Here we will assume that $P(f)$ arises from the communication of digital data.

Let us assume a random and independent data sequence of 0's and 1's, N of them per second. At time

$$T_n = n/N$$

we send a pulse centered at time T_n if the data value is a 0 and invert it if the data value is a 1 at time T_n. In this way, a stationary signal is transmitted. It is not surprising, and we shall assume it, that the *power* spectrum of the resulting signal is N times the *energy* spectrum of the pulse used. We will show this in the next chapter. From this we can conclude that the transmitted power is N times the energy of an isolated pulse, as we may expect. We also want the pulses to have no intersymbol interference, as in Section 5.3.

We want to show that sinc pulses of bandwidth

$$B = N/2$$

maximize (7.1-12), over all pulses which exhibit no intersymbol interference when used at a rate of N pulses per second. The optimum signal-to-noise ratio will be

$$\left(\frac{P_s}{P_n}\right)_{\text{opt}} = \frac{P_0}{N_0 B} \qquad (7.1\text{-}13)$$

Note that in transmitting a flat power spectrum, the optimal receiving filter is flat also, with just the bandwidth of the transmitted signal. This follows from (7.1-11). Thus all of the transmitted power is received, and all of the noise outside of the signal bandwidth is rejected. This makes the optimal property of sinc pulses very plausible. We will now derive this optimal property rigorously.

Here we will make a simplifying assumption. We will treat only symmetric pulses, but this is not essential. We will further assume that the pulse voltage spectrum is nonnegative and confined to a bandwidth of at most $2B$. This assumption is also not necessary for the conclusion. Because of this, $[P(f)]^{1/2}$ is proportional to the *voltage* spectrum of an individual pulse.

Because the pulse contains no frequencies above $2B$, we can use the Nyquist criterion of Section 5.3. This means that if $P(0)$ is normalized to be equal to 1, the function

$$\tfrac{1}{2} - [P(f)]^{1/2}$$

is an *odd* function around the frequency $f = B$, for frequencies up to $2B$. Or,

$$[P(B + f)]^{1/2} + [P(B - f)]^{1/2} = 1 \qquad (7.1\text{-}14)$$

This holds for frequencies f with

$$0 < f < B$$

The pulse voltage spectrum $[P(f)]^{1/2}$ is identically equal to 1 for frequencies between 0 and $B - (B_0 - B) = 2B - B_0$. Here B_0 is the upper frequency limit beyond which the pulse has no energy.

Because of (7.1-14), all the pulses that are in competition for being the optimum will have the same value B for the integral

$$\int_0^\infty [P(f)]^{1/2} \, df$$

So, we maximize (7.1-12) by maximizing the numerator

$$\int_0^\infty P(f)\, df = \int_0^\infty \{[P(f)]^{1/2}\}^2 \, df$$

By squaring (7.1-14), we see that for all frequencies f between 0 and B, we must have

$$P(B + f) + P(B - f) \leq 1$$

We may integrate this inequality over all frequencies f between 0 and $B_0 - B$. The result is that the numerator in (7.1-12) will be less than B unless

$$P(B + f) + P(B - f) \equiv 1$$

This means that $B_0 = B$, or $P(f)$ is identically equal to 1 for f between 0 and B. So, the numerator will be maximized for this spectrum, and its maximum will be equal to B. This shows that sinc pulses are optimum, with signal-to-noise ratio given by (7.1-13).

There is another sense in which sinc pulses are optimum. Here we can waive the requirement that there be no intersymbol interference, but constrain the bandwidth of the pulse to be equal to B. We will use a definition of bandwidth that applies even to pulses with energy at all frequencies.

The definition of bandwidth of a pulse corresponding to voltage spectrum $[P(f)]^{1/2}$ which we shall use here is

$$B = \left(\int_0^\infty [P(f)]^{1/2} \, df \right) \Big/ (P_{max})^{1/2} \qquad (7.1\text{-}15)$$

Here P_{max} is the peak of the spectrum. We shall study various definitions of bandwidth in more detail in Section 10.9. For sinc pulses occupying frequencies 0–B, the bandwidth is B by definition (7.1-15) also. We note that (7.1-14) shows that the bandwidth of pulses of nonnegative voltage spectrum satisfying the Nyquist criterion is at least B. This is because the denominator $(P_{max})^{1/2}$ in (7.1-15) can never be greater than 1 because of (7.1-14), but may be less than 1. Also, the bandwidth of pulses of nonnegative voltage spectrum with no intersymbol interference is at least B by this definition, even if the spectrum extends beyond $2B$. We shall not derive this.

The problem is now to maximize the signal-to-noise ratio subject to constraint (7.1-15). This becomes, from (7.1-12),

$$\frac{P_s}{P_n} = \frac{P_0}{N_0 B^2 P_{max}} \int_0^\infty P(f)\, df$$

We do not need variational techniques here. The situation is simpler than that. We must have

$$P(f) \leq P_{max}$$

Thus

$$P(f) = [P(f)]^{1/2}[P(f)]^{1/2} \leq [P(f)]^{1/2}(P_{max})^{1/2} \qquad (7.1\text{-}16)$$

$$\int_0^\infty P(f)\, df \leq (P_{max})^{1/2}\int_0^\infty [P(f)]^{1/2}\, df = P_{max}B \qquad (7.1\text{-}17)$$

The last equality follows from (7.1-15) which defines B.

If $P(f)$ is P_{max} for $f < B$ and zero for $f > B$, the inequality becomes equality. This is plainly the only way equality can occur in (7.1-17), because of (7.1-16). Hence this choice of $P(f)$ gives the maximum signal-to-noise ratio with bandwidth constraint (7.1-15). Sinc pulses give the optimum signal-to-noise ratio under this bandwidth constraint. So, sinc pulses are good with respect to several different criteria.

The next section considers sinc pulses specifically and interprets the optimum signal-to-noise ratio in terms of detectability of data.

Problems

1. What is the signal-to-noise ratio P_s/P_n attainable for a sinc^2 pulse whose spectrum just meets the Nyquist criterion? This means that the spectrum $[P(f)]^{1/2}$ is equal to $E^{1/2}$ for $f = 0$ and decreases linearly to 0 at $f = N = 2B$. Find $P_1(f)$ and HH^*. Compare with P_s/P_n for the optimum pulse whose signal-to-noise ratio is given by (7.1–13).

2. Find a spectrum of finite energy and finite bandwidth according to definition (7.1-15), but which has some energy at arbitrarily high frequencies. Find the bandwidth of your spectrum.

3. Let $x(t)$ have nonnegative *voltage* spectrum (so that all frequency components are in phase). Let the spectrum have its peak at dc. Show that definition (7.1-15) of bandwidth becomes

$$B = x(0)\Big/2\int_{-\infty}^\infty x(t)\, dt$$

4. A sinc pulse is modulated by a single-sideband transmitter of carrier frequency mB, m a positive integer, and the upper sideband retained to create a new pulse. Show that the signal-to-noise ratio given by (7.1-12) is the same as that for a sinc pulse. Show that there is no intersymbol interference. What is a better way of generating a pulse?

7.2. Signal to Noise for Sinc Pulses

Consider sinc pulses of the form

$$v(t) = (P_s)^{1/2}\frac{\sin(\pi t/T)}{\pi t/T} = (P_s)^{1/2}\,\text{sinc}\,(t/T)$$

The energy of such a pulse is TP_s and the average power of a string of $+$, $-$ pulses spaced T apart is P_s. The power spectrum of such pulses is equal to

$$TP_s$$

in the frequency range

$$-1/2T < f < 1/2T$$

There is no power at other frequencies. Thus the (one-sided) bandwidth B is $1/2T = N/2$. Here there are N pulses per second.

If we have a transfer function $H(f)$ which is unity over this band, the signal power, which is now also the power P_0 transmitted, will still be

$$P_s$$

but the total noise power after filtering will be finite and equal to

$$P_n = N_0 B = N_0/2T$$

The ratio of the squares of the signal voltage to the mean-square noise voltage at $t = 0$ will be

$$\frac{\text{signal voltage squared}}{\text{mean-square noise voltage}} = \frac{2TP_s}{N_0} = \frac{2E_0}{N_0} = \frac{P_s}{N_0 B} \qquad (7.2\text{-}1)$$

Here E_0 is the energy per pulse and N_0 is the noise power density appropriate to positive frequencies only.

Equation (7.1-13) or the arguments of Section 5.6 show that this instantaneous signal-to-noise ratio at time $t = 0$ is also the ratio of the average signal power to the instantaneous or average noise power. (Note that instantaneous noise power must be independent of time, and so is equal to the average of the noise power over a long time.) In using sinc or other pulses for data transmission, we may detect them by sampling at the peak of the pulse after it has been filtered at the receiver. This means that the instantaneous signal-to-noise ratio at the peak of the pulse, i.e., at $t = 0$, is the relevant quantity in measuring the performance of the data transmission system, and not the overall signal-to-noise ratio as defined in Section 7.1. We have just shown that for sinc pulses the two ratios are equal. But, for other pulses which exhibit no intersymbol interference, the two definitions generally give different values. We shall see in Section 7.4 that the ratio defined by (7.2-1) is independent of which pulse shape we use. Because sinc pulses give the optimum value of the signal-to-noise ratio of Section 7.1, that ratio is less than the ratio of this section for pulses other than sinc pulses.

Before we proceed to study optimum pulses for data transmission, we need a brief primer on some simple techniques of the calculus of variations. This is provided by the next section.

Problems

1. How is the signal-to-noise ratio (7.2-1) affected if we make a slight synchronization error at the receiver and sample at time Δt instead of at time 0, where Δt is small? What desirable property of sinc pulses leads to this?

2. In some forms of data transmission using pulses one might be led to find pulses of finite energy which may exhibit intersymbol interference, but instead are orthogonal to time shifts of themselves by integer multiples of the sampling interval. If $P(f)$ is the power spectrum of such a pulse, show that the pulse whose *voltage* spectrum is $P(f)$ exhibits no intersymbol interference, and conversely.

3. Suppose a symmetric pulse is confined to frequencies less than $2B$, and has *both* the properties of Problem 2 above when the sampling rate is $N = 2B$ pulses per second. Find the condition on the voltage spectrum which is equivalent to this. HINT: use equation (5.4-1), the Nyquist criterion.

7.3. A Little Variational Calculus

This section formalizes an argument we have used in Section 7.1. We shall learn a technique of variational calculus which is useful not only in designing pulses and elsewhere in communication and information theory, but indeed in many kinds of technology.

We want to maximize

$$\int_0^B G(f(x))\, dx \qquad (7.3\text{-}1)$$

subject to the constraint

$$\int_0^B f(x)b(x)\, dx = a \qquad (7.3\text{-}2)$$

Here $G(f)$ and $b(x)$ are known functions and f is the unknown function to be determined. Also a is a given constant. We can call $b(x)$ a *weight* function. In many applications it is identically equal to 1. B is an upper limit for x, and can be infinite.

Let us reason as follows. Replace f by $f + \delta f$, where $\delta f = (\delta f)(x)$ is a small function of x, which we can think of as an infinitesimal function. If we make sure that

$$\int_0^B (\delta f)b(x)\, dx = 0 \qquad (7.3\text{-}3)$$

then $f + \delta f$ satisfies the constraint (7.3-2), and conversely. If $f(x)$ maximizes (7.3-1), then for all small δf satisfying (7.3-3), we will have

$$\int_0^B G(f + \delta f)\, dx \le \int_0^B G(f)\, dx \qquad (7.3\text{-}4)$$

Let $G(u)$ be written in the form

$$G(u) = G(u_0) + (u - u_0)G'(u_0) \tag{7.3-5}$$

for real numbers u, u_0 with $(u - u_0)$ sufficiently small in absolute value. Here terms of order $(u - u_0)^2$ or higher are ignored. Using this expansion, with $f(x)$ as u_0, (7.3-4) becomes

$$\int_0^B [G(f) + (\delta f)G'(f)] \, dx \le \int_0^B G(f) \, dx \tag{7.3-6}$$

Canceling common terms, we see

$$\int_0^B (\delta f)G'(f) \, dx \le 0 \tag{7.3-7}$$

This holds for any small δf for which (7.3-3) holds. We can reverse the sign of δf in (7.3-3), and yet (7.3-7) must still hold. Actually, then, we have not merely inequality but equality:

$$\int_0^B (\delta f)G'(f) \, dx = 0 \tag{7.3-8}$$

This is the calculus-of-variations condition that integral (7.3-1) be a maximum. It is the equivalent of the condition for a maximum in ordinary calculus.

If it were not necessary to satisfy the constraint expressed by (7.3-2), (7.3-8) could be satisfied by making

$$G'(f) = 0 \tag{7.3-9}$$

Indeed, sometimes there is no constraint, and then (7.3-9) is the condition for a maximum.

If there is a constraint integral of the form of (7.3-2), then we must maximize subject to satisfying this constraint. We should note that f is a function of x and so $G'(f)$ is a function of x. δf is an arbitrarily chosen small function of x. We now argue that $G'(f)$ as a function of x must be equal to a constant times the weight $b(x)$. If this were not so, we could change the values of δf in two small ranges of x in such a way as not to change integral (7.3-3). If $G'(f)$ were not proportional to $b(x)$, this would change integral (7.3-8). But this integral must be the constant zero for all small functions of δf of x. Figure 7.2 shows this.

The resulting equation is

$$G'(f(x)) = Kb(x) \tag{7.3-10}$$

where K is a constant. We can now solve for $f(x)$ in terms of K. We use constraint (7.3-2) to find K, and thus to determine $f(x)$. This is the method we will use again and again in what follows.

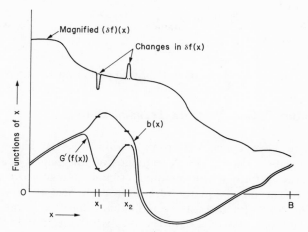

Figure 7.2. Showing $G'(f(x))$ proportional to $b(x)$.

This argument seems to assume that we know that (7.3-1) has a maximum rather than a minimum. The resulting equation (7.3-10) is the same if we know there is a minimum. Which, if either, occurs requires physical or engineering insight, or else some prior calculations.

The method is valid for both finite intervals and infinite ones. Or, we can have a number of such constraints (7.3-2), with functions $b_1(x)$, $b_2(x), \ldots, b_n(x)$, and constants $a_1, a_2, \ldots, a_n$. This situation occurs in finding channel capacities and rate distortion functions in information theory (see Chapters 12 and 13). Finally, we can have inequality constraints as in Section 7.1. There $P(f)$, being a power spectrum, had to be nonnegative. We shall not go into detail on how to handle such inequality constraints. The required modifications are often obvious when these problems arise.

In the very next section, we use the technique for an infinite interval, with

$$G(u) = u^2$$

This will enable us to derive the optimum filter when using pulses for data transmission where we detect by sampling the filtered pulse at its peak.

Problems

1. How should (7.3-10) be modified if we have finitely many, n, constraints like (7.3-2) to satisfy simultaneously? What if we have no constraints ($n = 0$)?

2. Rederive equation (7.1-9) from (7.3-10). First identify the function $G(f)$.

3. What happens when constraint (7.3-2) is replaced by

$$\int_0^B H(f(x))\,dx = a$$

where H is an arbitrary function of f (instead of a linear function)?

7.4. Optimum Pulses for Data Transmission

We might have concluded from Section 7.1 that sinc pulses are optimum for data transmission. A sequence of sinc pulses *does* give the optimum (largest) ratio of mean-square signal voltage to mean-square noise voltage. The mean-square voltages in each case are averages over all times. The time structure of the individual pulses is ignored.

Let us suppose that in a data receiver we detect pulses by filtering and then sampling at the peak of the filtered pulse. This is the best time at which to sample, if we are interested in a high signal-to-noise ratio at the sampling instants. In this section, we shall see that filtering and sampling is the best we can do in data reception. Assuming this, we will want the optimum ratio of signal voltage squared at the sampling time to the mean or expected values of noise voltage squared at the sampling time, rather than the optimum signal-to-noise ratio averaged over all time. We shall study expected values more thoroughly in the next chapter, but we do not need to know very much about them here.

We will for the moment avoid the problem of intersymbol interference by considering the case of a single pulse and using energy spectra instead of power spectra. The system considered is shown in Fig. 7.3. To a pulse of voltage spectrum (Fourier transform) $V_1(f)$ and energy E_0 we add white noise of (one-sided) power density N_0. We then pass the sum of the pulse and the noise through a network of transfer function $H(f)$ to obtain a pulse of *specified* spectrum $V(f)$ plus noise. How do we choose $V_1(f)$ [or $H(f)$],

$$V(f) = V_1(f)H(f) \tag{7.4-1}$$

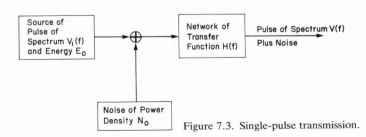

Figure 7.3. Single-pulse transmission.

so as to make the ratio of the square of the peak amplitude of the output pulse to the expected value of the square of the noise voltage greatest, and what is this ratio?

The time function $v(t)$ corresponding to the spectrum $V(f)$ is

$$v(t) = \int_{-\infty}^{\infty} V(f)e^{i\omega t}\, df \qquad (7.4\text{-}2)$$

At $t = 0$, this voltage has a value

$$v(0) = \int_{-\infty}^{\infty} V(f)\, df \qquad (7.4\text{-}3)$$

The function $v(t)$ is real and so $V(0)$ is real. It is obvious that for a given variation of the amplitude of $V(f)$ with frequency, the pulse amplitude at $t = 0$ will be greatest if $V(f)$ is real. This means that all frequency components add in phase at $t = 0$, and accounts for the signal peaking at $t = 0$. Thus, in seeking a high peak pulse amplitude, we will assume that $V(f)$ is real.

We will also assume that $V_1(f)$, and so $H(f)$, are real. This means that we transmit symmetric pulses. This is not essential.

We could also assume phase angles proportional to frequency (linear phase). This merely displaces the pulse in time without altering the conclusions we reach by assuming that $V_1(f)$, $H(f)$, and $V(f)$ are real.

The energy E_0 transmitted from the pulse source is given. It is

$$E_0 = \int_{-\infty}^{\infty} V_1(f)V_1^*(f)\, df \qquad (7.4\text{-}4)$$

Because $V_1(-f) = V_1^*(f)$ and $V_1(f)$ is real, we have the constraint

$$E_0 = 2\int_{0}^{\infty} [V_1(f)]^2\, df \qquad (7.4\text{-}5)$$

The constraint on $[V_1(f)]^2$ corresponds to the constraint on the unknown function $f(x)$ in Section 7.3, and $[V_1(f)]^2$ corresponds to that function. The weight function $b(x)$ is identically 1.

The expected or mean-square noise voltage $\overline{v_n^2}$ (which is equal to the noise power) will be

$$\overline{v_n^2} = N_0 \int_{0}^{\infty} [H(f)]^2\, df = N_0 \int_{0}^{\infty} \left(\frac{V(f)}{V_1(f)}\right)^2 df \qquad (7.4\text{-}6)$$

We wish to minimize this noise voltage while satisfying the transmitted power constraint (7.4-5).

We have seen that the function $[V_1(f)]^2$ corresponds to the unknown function $f(x)$ in Section 7.3. And, $[V(f)]^2/[V_1(f)]^2$ corresponds to the function $G(f(x))$. Its derivative with respect to the unknown function must be a constant for the integral (7.4-6) to be minimized, since $b(x) \equiv 1$. We will call this constant $-K$. This becomes

$$\frac{d}{d[V_1(f)]^2} \frac{[V(f)]^2}{[V_1(f)]^2} = -\frac{[V(f)]^2}{\{[V_1(f)]^2\}^2} = -K \qquad (7.4\text{-}7)$$

$$[V_1(f)]^2 = \frac{1}{K^{1/2}} V(f) \qquad (7.4\text{-}8)$$

From (7.4-8) and (7.4-5)

$$E_0 = \frac{2}{K^{1/2}} \int_0^\infty V(f)\, df$$

$$K^{1/2} = \frac{2}{E_0} \int_0^\infty V(f)\, df \qquad (7.4\text{-}9)$$

By using (7.4-6), (7.4-8), and (7.4-9), we see that

$$\overline{v_n^2} = \frac{2N_0}{E_0} \left(\int_0^\infty V(f)\, df \right)^2$$

Since $V(f)$ is symmetric, (7.4-3) shows that

$$\overline{v_n^2} = \frac{N_0}{2E_0} [v(0)]^2$$

$$\frac{[v(0)]^2}{\overline{v_n^2}} = \frac{2E_0}{N_0} \qquad (7.4\text{-}10)$$

We see that the optimum ratio of peak pulse voltage squared to mean-square noise voltage does not depend on the pulse shape at all. It is the same for all pulses, long or short. It is simply the ratio of the pulse energy E_0 to the two-sided noise power density $N_0/2$. We found this formula to be true for sinc pulses in Section 7.2, if we use an ideal low-pass filter matched to the bandwidth of the sinc pulse. Therefore the low-pass filter must be the one which is optimum for sinc pulses.

A little thought shows that we have also solved the problem of finding the best filter $H(f)$ to use when the transmitted pulse shape $V_1(f)$ is given. From (7.4-1) and (7.4-8), we find

$$H(f) = K^{1/2} V_1(f) \qquad (7.4\text{-}11)$$

The constant $K^{1/2}$ is unimportant here. What *is* important is that $H(f)$ is *matched* to the transmitted pulse spectral shape $V_1(f)$. If the transmitted pulse is not symmetric, (7.4-11) becomes

$$H(f) = K^{1/2} V_1^*(f) \qquad (7.4\text{-}12)$$

This is called a *matched filter*. If we transmit a sinc pulse, this is also an ideal low-pass filter matched to the sinc pulse, as we have already seen.

These concepts arose in the Second World War for the detection of radar pulses. The filters had to be implemented as ordinary analog component filters. With the advent of cheap digital components, and especially large-scale integrated circuits and read-only memories, the matched filter receiver has gradually given way to the *correlation receiver*, which performs the same optimum filtering operation in the time domain. We shall describe this.

Let $v_1(t)$ be the transmitted pulse. To this is added noise, and a time function $r(t)$ is presented to the receiver filter. We shall let the filter act in the time domain, as in Section 3.7. The impulse response of the matched filter whose transfer function is $V_1^*(f)$ is the inverse Fourier transform of $V_1^*(f)$. This means

$$h(t) = v_1(-t) \qquad (7.4\text{-}13)$$

The output $\hat{v}(t)$ of this filter when the input is $r(t)$ will be given by

$$\hat{v}(t) = \int_{-\infty}^{\infty} h(s)r(t-s)\,ds$$

$$= \int_{-\infty}^{\infty} v_1(-s)r(t-s)\,ds$$

$$\hat{v}(t) = \int_{-\infty}^{\infty} v_1(s)r(t+s)\,ds \qquad (7.4\text{-}14)$$

Equation (7.4-14) shows that the output of the matched filter is the *cross-correlation* of the transmitted pulse $v_1(s)$ with the received signal-plus-noise $r(s)$. The cross-correlation was defined in Section 6.3, where we found the spectrum of the cross-correlation in terms of the individual spectra.

We are interested in sampling the filter output $\hat{v}(t)$ at time $t = 0$. This becomes

$$\hat{v}(0) = \int_{-\infty}^{\infty} v_1(s)r(s)\,ds \qquad (7.4\text{-}15)$$

Or, we multiply the received waveform by the transmitted pulse shape and integrate over all time. This is also called the *correlation* of v_1 and r. The block diagram of such a receiver is shown in Fig. 7.4. The same comments about having to delay the output because of the integration from $-\infty$ to ∞ apply here, as they do to any filter whose impulse response lasts forever. We have seen this in Section 5.2.

The problem of intersymbol interference for output pulses $v(t)$ other than sinc pulses must be dealt with separately. In Section 5.4 we found a

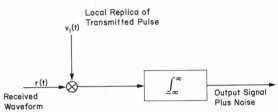

Figure 7.4. Correlation receiver.

means for producing a variety of pulses which meet the Nyquist criterion for having no intersymbol interference.

In the next section we study the case of rectangular pulses in more detail. Our study will confirm that these pulses conform to the findings of this section.

Problems

1. Show that the matched filter for a nonsymmetric pulse has transfer function proportional to $V_1^*(f)$. HINT: rework the optimization without the assumption that $V_1(f)$ and $H(f)$ are real.

2. In a data signaling system it is decided for some strange reason to replace one strong, short transmitted pulse by two weaker nonoverlapping short pulses each of the same shape as the single pulse but each having half the energy of the single pulse. If the voltage outputs of the received short pulses are filtered by the original optimum filter and appropriately delayed and added, what is the ratio of signal voltage squared to noise voltage squared in comparison with that for the single pulse? How can we describe the optimum filter for the split pulse in terms of the optimum filter for the original pulse?

3. When we send pulses over a cable instead of by microwaves the loss is greater for high frequencies than for low frequencies. To recover the transmitted signal we have to amplify the high-frequency components of the received signal more than the low-frequency components, and that means that we will in effect have a greater noise density at high frequencies than at low frequencies. (The noise is added in the receiver rather than in the cables.) Qualitatively, how will this modify the conclusions of this section, if we do not need to take intersymbolic interference into account? What if we *do* have to take it into account?

7.5. Signaling Using Rectangular Pulses—Signal-to-Noise Performance

If we are not concerned with their wide bandwidth, rectangular pulses are very attractive. They are easily and efficiently generated by inverting the sign of a dc source (or the phase of an rf source). Their sharp transitions, which are responsible for their wide bandwidth, make them easy to synchronize when the phase of the pulse is unknown to the receiver. Contiguous rectangular pulses have no intersymbol interference. But, we have seen that pulses must be optimally filtered in order to get the limiting performance of (7.4–10).

Rectangular pulses are used in space communication systems. They are detected by integrating the received positive or negative pulses and sampling at the end of the pulse, at which time the integral of the pulse has the greatest magnitude.

Integration over a fixed time was at one time approximated by a gating operation and a capacitor. At the beginning of the period of integration the capacitor is discharged. During the interval of integration the capacitor is charged by means of a current which is proportional to the received signal plus noise. At the end of the integration period the voltage on the capacitor is sampled and taken as the signal output. With modern digital circuitry, however, integration is performed by high-speed sampling, analog-to-digital conversion to sufficient accuracy, and digitally adding the resulting samples.

We will derive the results of the preceding section independently for rectangular pulses. A rectangular pulse of length T centered at 0 can be detected by integrating from $-T/2$ to $T/2$, the interval over which the pulse is not equal to 0. But, such integration is the same, except for an inessential gain constant, as a multiplication by the same pulse followed by integration over *all* time. Or, integration implements the correlation receiver form, (7.4-15), of the optimum or matched filter. So, we know from Section 7.4 what the signal-to-noise ratio will be using such integration for detection. But here we will derive it directly.

We have seen in Section 3.8 that the transfer function which results in the operation

$$y(t) = \int_{-\infty}^{t-t_0} x(u) \, du \qquad (7.5\text{-}1)$$

is

$$\frac{\exp(-j\omega t_0)}{j\omega} \qquad (7.5\text{-}2)$$

The transfer function $H(\omega)$ that integrates from $t - T/2$ to $t + T/2$ is the difference of two transfer functions as given by (7.5-2), for one of which $t_0 = -T/2$ and for the other of which $t_0 = T/2$. Thus we see that the corresponding filter transfer function $H(\omega)$ is

$$H(\omega) = T\frac{\sin(\omega T/2)}{\omega T/2} \qquad (7.5\text{-}3)$$

This is, of course, the Fourier transform of the pulse itself, and we are using a matched filter.

If noise is added to the pulse stream prior to the integration or filtering, what will the noise power output be? For white noise of noise power density N_0, this total noise power P_n will be

$$P_n = N_0 \int_0^\infty H(f)H^*(f)\, df$$

$$P_n = N_0 T^2 \int_0^\infty \frac{\sin^2(\omega T/2)}{(\omega T/2)^2}\, df \qquad (7.5\text{-}4)$$

The integral is $2/T$; this follows from Section 5.5. So

$$P_n = N_0 T/2$$

Here N_0 has been chosen as appropriate to positive frequencies only. As T is increased, the bandwidth of the transfer function decreases, but the accumulation due to integration over the longer time period dominates this to produce a noise output proportional to pulse width. This calculation can be done more naturally in the time domain using techniques we will learn in Chapters 9 and 11.

For a signal consisting of square $+$ or $-$ pulses of duration T, power P_s, and voltage $(P_s)^{1/2}$, the voltage v due to integration over the period T will be simply

$$v = \pm T(P_s)^{1/2}$$

$$v^2 = T^2 P_s \qquad (7.5\text{-}5)$$

The ratio of the square of the signal voltage to the mean-square value of the noise voltage, or noise power, P_n, will be

$$\frac{v^2}{P_n} = \frac{2TP_s}{N_0} = \frac{2E_0}{N_0} \qquad (7.5\text{-}6)$$

This is the optimal result of Section 7.4 and the result of Section 7.2 for sinc pulses. This must be, for we have seen that, for rectangular pulses, integration is matched filtering.

We can also build a bit synchronizer for a sequence of noisy rectangular pulses of known constant amplitude which arrive initially in an unknown phase. This uses three integrations over consecutive intervals of length $T/2$, half the pulse width. Let the incoming data be alternately a positive pulse and a negative pulse, corresponding to an alternating 1, 0 data sequence. Whatever the phase of the data may be relative to the phase of the integrators, we can learn something of the phase offset by comparing the result of the first two integrations with the result of the last two. For example, if the first two integrators are exactly in phase, their sum will tend to be of large magnitude while the sum of the second two integrals will tend to be zero or very small. This is because there is a pulse

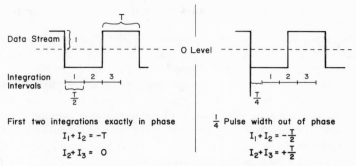

Figure 7.5. Bit synchronizer timing relationships.

transition that reverses the sign of the third integral relative to the second. If the two sums of integrals are closer to being of equal magnitude but opposite sign, the integrators will be $\pm T/4$ from being in phase. Figure 7.5 shows these relationships. We may build a feedback system for bit synchronization. This shifts the phase of the integrators based on the last few values of integrator outputs, so as to drive the sum of the right (or left) two integrals to zero. We shall do no more with this here.

In the next section we consider filtering, not to maximize signal-to-noise, but to minimize mean-square errors in analog communication.

Problems

1. We have seen that the optimal filtering for rectangular pulses corresponds to an integral over a time interval equal to the pulse length. Draw a succession of noisy rectangular pulses of random amplitudes $+1$ or -1. Draw the *running* integrals over a past time equal to the pulse length. This gives the optimally filtered pulses ready for detection. Where should the integral be sampled? Will there be any intersymbol interference?

2. In space communication systems the "rectangular" pulses are constant-amplitude bursts of radio frequency of 0 or 180° phase. How do we optimally detect these according the means outlined in this section? How do we know this is optimum?

3. Such "square" rf bursts are not used in ground microwave systems? Why?

4. A bit synchronizer is used on a square-wave input of amplitude 1 and pulse width 1. The sum of the outputs of the first two integrators is $+0.7$. What values must the sum of the outputs of the last two integrators take on? How far should the intervals of integration be shifted so that the first two integrators are exactly in phase, and so that the sum of their outputs can be used for pulse detection? What if the pulse amplitude is not known in advance?

5. How should the conceptual design of a bit synchronizer be modified for pulses which are *not* rectangular (but still exhibit no intersymbol interference)?

7.6. Another Optimal Filtering Problem

In Sections 7.1–7.5 we put ourselves in the place of engineers who are faced with a flat noise spectrum added to a fixed transmitted power but are free to choose a signal shape or spectrum.

In Section 7.1 we found an optimum filter shape and transmitter power spectrum for any specified recovered signal power spectrum, an optimum in the sense that the ratio of total signal power to total noise power is a maximum. We also found that the greatest ratio of total signal power to total noise power can be obtained with a rectangular spectrum as the recovered signal power spectrum.

In Section 7.4 we found that if the criterion is the ratio of the square of the peak signal voltage to mean-square noise, one pulse shape is as good as another.

In Sections 7.2 and 7.5 we explored two pulsed systems with no intersymbol interference—sinc pulses and rectangular pulses.

There is another way of posing an optimal filtering problem. Suppose that a transmitted spectrum or signal shape is imposed on us, noise is added, and we are asked to filter the mixture of signal and noise so as to obtain an output which differs as little as possible in a mean-square sense from the transmitted signal. This is an analog communication problem, as in Section 7.1. In Section 7.4 we treated a digital communication problem.

Let $X(f)$ be the spectrum of the transmitted signal, $H(f)$ be the filter transfer function, and N_0 be power density of the noise added to the transmitted signal. We do not need to assume that the transmitted signal has a voltage spectrum. We need only use the power spectrum. But, the notation is simplified by using a voltage spectrum.

The noise power output P_n will be

$$P_n = \int_{-\infty}^{\infty} N_0 H(f) H^*(f) \, df \qquad (7.6\text{-}1)$$

Here N_0 is the two-sided noise density.

The mean-square difference, P_d, between the signal component in the output of the filter and the original signal can be evaluated in the frequency domain by the energy theorem of Section 3.6. The result is

$$P_d = \int_{-\infty}^{\infty} X(f) X^*(f) [1 - H(f)][1 - H^*(f)] \, df \qquad (7.6\text{-}2)$$

The total mean-square error is the sum $P_n + P_d$. P_n is the noise component and P_d can be thought of as the distortion component.

It helps to represent $H(f)$ as

$$H(f) = A(f) e^{i\phi(f)}$$

Here $A(f)$ is real and nonnegative. Then the square error $P_n + P_d$ becomes

$$P_n + P_d = \int_{-\infty}^{\infty} (X(f)X^*(f)\{1 + [A(f)]^2 - 2A(f)\cos\phi(f)\} + N_0[A(f)]^2)\, df$$

(7.6-3)

It is immediately apparent that the integral will be *smallest* for given $A(f)$, which is nonnegative, if

$$\cos\phi(f) = 1, \qquad \text{or } \phi(f) = 0 \text{ (positive transfer function)}$$

What happens is that making $\phi(f) = 0$ preserves the value of P_n from (7.6-1), but lowers P_d from (7.6-2). If we make $H(f)$ equal to $A(f)$, then (7.6-3) becomes

$$P_n + P_d = \int_{-\infty}^{\infty} \{X(f)X^*(f)[1 - A(f)]^2 + N_0(A(f))^2\}\, df \quad (7.6\text{-}4)$$

Here the unknown function is $A(f)$, and the integrand in (7.6-4) is the function $G(f(x))$ of Section 7.3. As there is no equality constraint on $A(f)$, the derivative of the integrand with respect to $A(f)$ must be zero in order to minimize $P_n + P_d$. This gives

$$-2X(f)X^*(f)[1 - A(f)] + 2N_0A(f) = 0$$

$$A(f) = H(f) = \frac{1}{1 + N_0/X(f)X^*(f)}$$

(7.6-5)

This expression is never negative. It is zero whenever $X(f)$ is zero, as we would expect by physical reasoning. It is never more than 1, so there is no amplification in $H(f)$.

For this value of $A(f)$ the minimum square error (noise plus distortion) can be found. The result is

$$(P_n + P_d)_{\min} = \int_{-\infty}^{\infty} \frac{N_0\, df}{1 + N_0/X(f)X^*(f)} = \int_{-\infty}^{\infty} N_0 H(f)\, df \quad (7.6\text{-}6)$$

We can reach some conclusions from (7.6-5). If N_0 is very *small*, the transfer function of the filter should be approximately unity for frequencies strongly present in the signal $X(f)$. But when the noise density N_0 is *large* compared to the signal power density, we should have $H(f)$ small:

$$H(f) \doteq \frac{X(f)X^*(f)}{N_0}$$

(7.6-7)

This is not unreasonable.

In the above derivations we did not need to assume N_0 to be constant with frequency, and (7.6-5)–(7.6-7) hold for noise densities N_0 that are functions of frequency.

In the next section, we use variational techniques to derive the optimum signal-combining method when the channel is not the simple channel of microwave space communication. We will consider channels which experience phenomena of multipath and ray bending which cause signal fades. These fades may vary slowly with time. We may use several or many receivers in different locations, and we shall find the optimum way to process the outputs of the receivers.

Problems

1. Give a heuristic explanation for the behavior of the optimum transfer function $H(f)$ for weak and strong noise (or for strong and weak signal).

2. For flat (white) noise (N_0 constant) and a flat signal power spectrum from $-B$ to B of given height M_0, what is the minimum mean-square error? Interpret as N_0 becomes infinite. What happens in the general case as N_0 becomes infinite?

3. Suppose $X(f)X^*(f) = P(f)$ is given by

$$P(f) = \frac{M_0}{1 + (f/f_0)^2}$$

(This is the spectrum of white noise passed through an RC filter.) White noise of two-sided spectral density N_0 is added by the receiver before filtering. What is the minimum mean-square error attainable, in terms of M_0, N_0, and f_0? What is the optimum filter $H(f)$? How much of the minimum mean-square error is a result of noise, and how much a result of distortion? HINTS:

$$\int \frac{dv}{1 + v^2} = \arctan v$$

$$\int \frac{dv}{(1 + v^2)^2} = \int \cos^2 \theta \, d\theta \qquad \text{if } v = \tan \theta$$

4. In the above problem, instead of using the optimum filter, an ideal low-pass filter of one-sided bandwidth $B = bf_0$ and height 1 is for some reason used instead. What is the mean-square error as a function of b? Which value of b minimizes the mean-square error over all ideal low-pass filters? What is the loss, in dB of receiver noise spectral density required, compared to the overall optimum filter of Problem 3 when the above optimum low-pass filter is used and $M_0 = 10N_0$? What is the minimum mean-square error as a fraction of total signal power for the best low-pass filter and for the Problem 3 optimum filter in this case?

7.7. Diversity Reception

We noted in Section 1.8 that in microwave reception on earth, both a direct and a reflected wave may reach the receiving antenna. If these two waves have equal amplitudes and opposite phases, their sum will be zero and the antenna will receive no signal. Thus two received waves can interfere with one another and can even cancel one another.

Such interference or cancellation also occurs when variations in the density of the atmosphere cause microwaves to arrive at an antenna from two slightly different directions. It occurs when short radio waves reflected from the ionosphere arrive at an antenna over two paths, reflected different numbers of times between ionosphere and earth. It occurs when signals reach the receiving antenna of a mobile radio in a car as reflections from buildings.

The phenomenon occurs also in the reflection of laser light from a diffuse surface. We see such a reflection as "speckle" with dark areas because waves reach a given point on the retina of the eye from slightly different angles. Over some small areas of the retina the waves interfere and nearly cancel one another.

This speckle phenomenon occurs in radar mapping as well. Speckle is really a form of "surface noise." It limits the quality of the radar maps that can be made, even in the absence of receiver noise. It is *noise* because the mapmaker is not interested in the small random fluctuations of the order of a wavelength in distance to the surface. But, these fluctuations are the cause of speckle. The mapper is really interested in differences in radar reflectivity due to differences in the topography of the reflecting surface. He can attempt to overcome speckle and get at the underlying surface reflectivity by taking several or many looks at the same area. In different looks the returns from the same point on the surface differ randomly in phase by an amount on the order of a wavelength, or by a larger amount. This phase difference is caused by small random changes in the relative position of the target and the radar between looks. These changes are caused by relative motions of the transmitter and the radar target. Averaging several looks then increases the signal-to-noise ratio. Here the signal is the average brightness in the absence of speckle. We shall say no more about this.

When there are two or more incoming waves, we have seen that there will be a strong signal at some locations, where the waves add, and a weak signal at other locations, where the waves cancel. If we use a number of receiving antennas at different fixed locations, at several heights on a microwave tower, for example, or at several positions on the roof of a car, separated by substantial fractions of a wavelength, it is likely that we will get a strong signal from at least one antenna. But, there will usually be *some* signal from every other antenna. How can we use all of these signals to get the best possible received signal?

This is the problem of *diversity reception*. It is illustrated in Fig. 7.6. Signals from several antennas go into a diversity receiver. Out comes one signal.

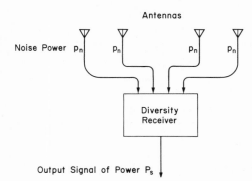

Figure 7.6. Diversity reception.

For diversity to be useful, the output signal must merely be better than the worst signal from any one antenna. We can accomplish this merely by switching to the antenna which gives the best signal-to-noise ratio. This is called *switched diversity*. It is used also to combat fades due to rain in microwave communications above 10 GHz. For this use the antennas had best be miles apart, so that it is not raining heavily on all antennas at once. Switched diversity is simple and effective. But, in principle, we can do better.

If we seek to combine the signals from all antennas, we can either combine the rf signals or we can demodulate the signals and combine the baseband signals. Combining baseband signals is easier because we need not worry about rf phase, since we have discarded that phase information anyway. However, baseband diversity is hard to analyze because the baseband signal-to-noise ratio may be a nonlinear function of rf signal-to-noise ratio. Breaking in fm, which we studied in Section 4.8, is nonlinear. So is the detection of am signals by rectification when the signal-to-noise ratio is low. And so are other detection procedures that are used in practice.

Not only is rf diversity simpler to analyze than diversity after demodulation; rf diversity must always be as good as or better than diversity after demodulation. Here we will boldly assume that we can solve the problem of finding and then shifting the phases of the signals from the several antennas, at rf or after conversion to an intermediate frequency, so that all the signals will be in the same phase when they are added. By what factors should we multiply the amplified signal (and noise) from each antenna in order to get a combined signal with the best signal-to-noise ratio? We solved a problem very much like this in Section 4.6, in connection with pulse position modulation.

We will assume (this is really just a normalization) that the noise power output p_n is the same in the amplified output from each antenna. This means that we have identical receivers. Hence if we multiply both the signal and the noise powers from the xth antenna by a factor a_x before adding, the total noise power P_n in the combined output will be

$$P_n = p_n \sum_{x=1}^{N} a_x \qquad (7.7\text{-}1)$$

Here N is the number of antennas.

Suppose that the signal power from the xth antenna is p_{sx}; then the rms signal voltage will be $(p_{sx})^{1/2}$. We multiply this by $(a_x)^{1/2}$ to obtain the adjusted output voltage from the xth antenna. The signal voltages have been phase shifted so as to add in phase, so the total rms signal voltage V_s will be

$$V_s = \sum_{x=1}^{N} (a_x)^{1/2}(p_{sx})^{1/2} \qquad (7.7\text{-}2)$$

The signal-to-noise ratio S/N in the combined signal will be

$$S/N = V_s^2/P_n \qquad (7.7\text{-}3)$$

The problem is now to maximize S/N through a proper choice of the weighting coefficients a_x. Here, we have a discrete optimization problem. The coefficient a_x is a function of the index x, but x is not a continuous variable; it has certain values only. We may, if we wish, solve this discrete problem by traditional techniques of advanced calculus. This was done to optimize the reception of pulse position modulation in Section 4.6-5.

But it turns out that it is perfectly permissible to use exactly the same procedure that we would use if a_x were a continuous but unknown function of the continuous variable x, as we took $f(x)$ to be in Section 7.3. The technique of Section 7.3 is a good one even for discrete problems.

If we wish, we can without loss of generality let P_n be a constant, thus setting a constraint on the total amount of noise. We then want to maximize V_s, the total output signal voltage, by choice of the a_x function. The known function G of the unknown function a_x is simply

$$(a_x)^{1/2}(p_{sx})^{1/2} \qquad (7.7\text{-}4)$$

By applying the methods of Section 7.3, we can show that the optimum rf signal-to-noise power ratio S/N is

$$S/N = P_s/p_n \qquad (7.7\text{-}5)$$

Here P_s is the sum of the signal powers p_{sx} output from each of the antennas, and p_n is the noise from one antenna, taken as the same for each antenna. This optimum is attained by making

$$a_x = (\text{constant})p_{sx} \qquad (7.7\text{-}6)$$

The weights should be proportional to received signal power. The performance given by (7.7-5) is as if the N antennas were combined into one antenna with the same noise temperature or p_n. This is a physically reasonable and satisfying result.

Equation (7.7-6) represents the solution to the ideal diversity problem that we have posed. In practice, when the signal-to-noise ratio is low, it may be hard to find the phase of the rf signal. Yet, for an optimum result we must combine all rf signals in phase.

An alternative approach is to vary the phases of the signals from the various antennas in some orderly manner in searching for a good signal-to-noise ratio in the combined signal. What is usually done is to systematically vary the relative phases of the outputs of various antennas in an array so as to steer the beam of the array. We shall discuss this in the next section where we study the arraying of antennas.

Problems

1. Justify the fact that the formulas of the calculus of variations that were derived for continuous functions of a continuous variable in Section 7.3 also apply to the sort of discrete problem of this section.

2. Derive (7.7-6) and from it (7.7-5).

3. In a switched diversity system, the signal-to-noise ratios in two identical but separated receiving systems are equal in dry weather. The output of antenna A is used. During a 7-dB fade caused by wet weather over antenna A, the output is instead taken from receiver B, where the weather is dry. How many dB does optimum diversity gain over switched diversity when the weather is dry over both stations, and when it is wet over A and dry over B?

7.8. Antenna Arrays

In the preceding section we have discussed a receiving system made up of several antennas placed at different locations. If we wish to combine the signals from several antennas in order to obtain the best signal-to-noise ratio, we must shift the phases of the signals from the antennas so that all signals are in phase, and then weight the signals powers according to (7.7-6) before adding them.

Let us consider a special case in which the received signal comes from a single source lying in some particular direction. Then, the rf field strength at each receiving antenna will be the same, and equation (7.7-6) tells us that we should combine the antenna outputs with equal weights. This means simply that we should shift the phases of the signals from the various antennas so that all signals are in phase, and then add the signals. If we know a little basic electromagnetic theory, we see that this amounts to steering the receiving (or transmitting) beam of the array of antennas in the direction of the source.

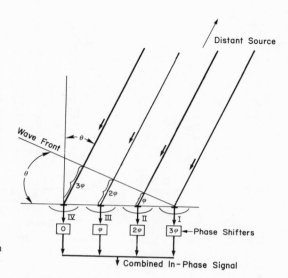

Figure 7.7. Four antennas in equally spaced linear array.

Consider Fig. 7.7, which shows four antennas (I–IV) arranged in an equally spaced linear array. The slanting lines from the upper right represent waves reaching the antenna along the indicated direction from a single far-distant point source. The wave front is perpendicular to these lines or rays. We see that the waves reaching antenna II will lag those reaching I by some phase angle ϕ. With respect to the waves reaching I, those reaching III and IV will lag by 2ϕ and 3ϕ, respectively.

If we put in additional phase shifts or delay as shown in the figure before combining the antenna outputs, all outputs will add in phase. What we have done is to steer the beam of the antenna array so as to receive a wave whose wave front makes some angle θ with respect to the linear array. If we change ϕ we change the direction θ in which the array is steered.

If the phase shifts are controlled electronically, we have an electronically steerable array. The first electronically steerable arrays were the MUSA (multiple unit steerable antenna) arrays devised as early as 1928 (Reference 2) and later used for short-wave reception (Reference 3).

In short-wave reception, signals arrive not from one but from several directions, after bouncing different numbers of times between the ionosphere and the surface of the earth. Beam steering is not the optimum means for combining the signals that reach the antenna, though beam steering is pretty good. But in many cases, beam steering is ideal. It is ideal for sending a signal to a satellite, or for receiving a signal from a satellite.

How far can we steer a signal by means of an array? Only within the

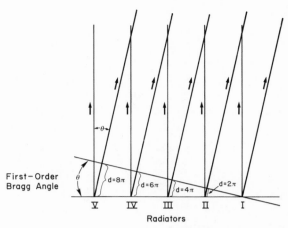

Figure 7.8. Array of five isotropic radiators (d is the phase delay relative to antenna I).

directional pattern of the array elements. Let there be N array elements, each of effective area A. Let the power picked up by each antenna vary with angle θ as $F(\theta)$, where $F(0) = 1$. Then the effective area of the array when the phases are shifted properly before signal addition will be

$$NAF(\theta) \qquad (7.8\text{-}1)$$

When $\theta = 0$, the effective area is the number of array elements N times the effective area A of each, as it must be from the same arguments that led to the Friis transmission formula of equation (1.3-1).

Figure 7.8 shows an array consisting of five isotropic radiators I–V. If these are all driven in phase we will get a strong signal directly upwards, with a wave front parallel to the array. But, if the array elements are widely enough spaced, greater than one wavelength, we will also get radiation at an angle up and to the right, at an angle θ such that a path from antenna II is 2π (one whole cycle of phase) longer than a path from antenna I, and paths from antennas III, IV, and V are successively 4π, 6π, and 8π longer than the path from antenna I. The angle θ for which this is true is the first-order *Bragg angle* familiar from diffraction theory. In fact, a linear array is very much like a diffraction grating. For a *second-order* Bragg angle the phase shifts would be 4π, 8π, 12π, 16π. There can be higher-order Bragg angles as well.

If we are to avoid having several strong beams from an array, the radiation from the array elements must be low at the Bragg angles. That is, $F(\theta)$ in (7.8-1) must be low at the Bragg angles. We note that $F(\theta) \equiv 1$ for an isotropic radiator.

Figure 7.9 shows two antenna patterns (or two-dimensional slices of

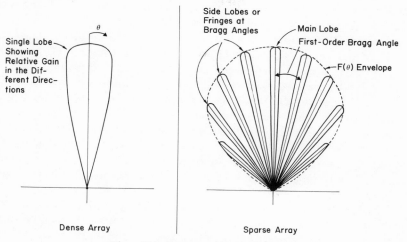

Figure 7.9. Patterns of antenna arrays.

patterns), one for a dense array and one for a sparse array. The length of the radius vector from the origin is the relative gain. This is proportional to the power per unit solid angle that would be received from a test radiator of fixed strength and distance in the given direction, when the phasing is set to steer the beam vertically. We see that the gain in the vertical direction is the same in either case, as must be from (7.8-1). But, for the sparse array, pointing is more critical because the main lobe is narrower. It has been split into several narrow sidelobes at the Bragg angles. Here the sidelobes are reduced from the main lobe by the envelope $F(\theta)$, which is the pattern of a single array element normalized to have peak value 1. The sidelobes are also called fringes or interference fringes by radio astronomers.

Figure 7.10 shows a Van Atta array of antennas equally spaced along a line. A signal is picked up by an antenna on the left of center, amplified, and fed to an antenna symmetrically placed to the right of center. All

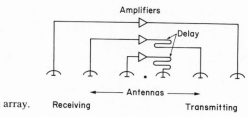

Figure 7.10. Van Atta array.

paths between the receiving and transmitting antennas have the same electrical length or phase shift because of the delay that has been added by inserting the proper elements in the connecting cables. The basic nature of the phase of waves (see Reference 2 of Chapter 1) shows that the Van Atta array transmits a signal back in the very same direction from which the received signal came. It is a *retrodirective* array.

If we insert additional phase shifters in the paths between the antennas of the Van Atta array of Fig. 7.10, we can send the received signal off in some different direction. Indeed, through ingenious phase control electronics, the array can be made self-steering, so that it sends any signal that reaches it from any direction in the direction of a guide beam or pilot signal that is radiated from a moving receiver location (Reference 4).

The pattern of interconnection of receiving and transmitting elements in the Van Atta array of Fig. 7.10 is important, for it can be shown to assure that small changes in the orientation of the linear array structure will not much change the direction of the retransmitted beam in a self-steering array. If the receiving and transmitting frequencies are different so that the amplifier is also a frequency shifter, as it will be in communication satellites, interconnected receiving and transmitting elements must be spaced *different* distances from the center of the array in order to preserve this property. When the right condition is observed, the required accuracy of orientation of the array depends on $F(\theta)$, the directivity of a single array element, but not on the narrow beamwidth of the array. This reduces the sensitivity of the array performance to small changes in orientation.

Today, phase shifters can be controlled digitally by microprocessors, on the basis of data stored in a digital memory. Thus an array of receivers and transmitters aboard a satellite can be steered to successively interconnect earth transmitter–receiver pairs, so that one satellite and one array can serve many widely separated pairs of earth stations in turn, but with high directivity toward each (Reference 5).

We should note, however, that many or most arrays are two-dimensional rather than linear. We shall not explore this more complicated theory.

Radio astronomers use sparse arrays in which the spacing of the antennas is much larger than the antenna diameter. This gives a beam with a very sharp center lobe, but with many Bragg angles. High resolution is obtained for mapping the radio brightness of small regions of the sky. Communicators, on the other hand, usually want dense arrays in order to pick up as much power as possible from the transmitter direction, even when there may be a slight pointing error.

Array elements can be so phased as to give *nulls*, instead of peaks, in certain directions—for instance, in the directions of interfering transmitters. We shall discuss such problems in Section 12.7.

We were drawn into the discussion of steerable arrays by contrasting them with ideal diversity combining to maximize received signal power. We see, however, that arrays have other imaginative uses. They are becoming increasingly important in microwave communication on the earth and to, from, and within space. This is especially true as physical limits on the sizes of single apertures are being approached by larger and larger antennas.

The chapter has taken noise into account, but only in a gross or mean-square sense. The next chapter studies the actual probability distributions of noise. We will learn how the frequency of occurrence of various amounts of noise can be determined. The performance of both analog voice and digital data communication is ultimately determined by the relative frequency of occurrence of various amounts of noise. In this, the gaussian probability distribution plays a central role.

Problems

1. What would a Van Atta array do if the equal delays were provided not between the pairs symmetrical about the center but rather between pairs that are a constant distance apart? Give a simple physical explanation.

2. Suppose that rf plane waves of equal power come from two different directions. Suppose that we receive these by means of four antennas that lie equally spaced along a line parallel to one of the wave fronts. Suppose that the waves exactly cancel at one end antenna, have relative phases of $2\pi/3$ and $\pi/3$ at the two intermediate antennas, and are exactly in phase at the other end antenna. What is the ratio of the best S/N that can be attained with ideal diversity to the best S/N that can be attained by taking the strongest output as the signal (switched diversity) and to the S/N obtained by electronic steering? (In such steering, the signals from all antennas are weighted equally and the phases are shifted by amounts 0, ϕ, 2ϕ, and 3ϕ, where the parameter ϕ, which steers the antennas, is chosen for maximum signal output.) Show that $\phi = \pi/6 = 30°$. HINT: observe from a phasor diagram that the phase of a sum of equal-strength signals is the average of the two individual phases.

3. In a Van Atta array such as shown in Fig. 7.10, how should the transmitting and receiving antennas be spaced with respect to the center of the array if the array is to pick up a 30-GHz signal and send back an 18-GHz signal in the same direction?

4. Suppose that we put phase shifts in the amplifying paths of the horizontal array shown in Fig. 7.10, such that the array receives signals arriving at an angle θ_1 with respect to the vertical and retransmits signals at an angle θ_2 with respect to the vertical. If the array is tipped by an angle ψ, how does the angle θ_1 change? How does the new transmit angle θ_4 with respect to the old vertical vary with ψ? What happens when ψ is small and θ_1 is about equal to θ_2, as it

would be in a self-steering array? (You may find it convenient to define the angle $\theta_3 = \theta_4 + \psi$.)

5. A certain domestic communication satellite has a beam shaped approximately to the area of the continental United States, which has an area of 3.6×10^6 sq. mi. It services the principal metropolitan areas by time-division multiple access (TDMA). This means that different regions receive their transmissions in different time slots. The next generation satellite has a phased array antenna system. When the time slot for handling traffic to and from a metropolitan area comes up, the phasing is arranged so that a spot beam 200 miles in diameter is created on that metropolitan area. If the satellite and ground station power, receiver noise temperatures, antenna efficiencies, ground antenna diameters, and communications frequency are the same for the new generation satellite, by how much has the traffic-handling capacity of the satellite been increased? If each element of the array is the same size as the old satellite antenna, how many array elements would be needed?

6. Consider antenna patterns such as in Fig. 7.9. If we replace the length of radius vector in the pattern by its cube root, we get another figure. Show that the volume of this figure is one-third the total noise power received from space (as in Section 2.8 and the end of Section 2.1). Conclude that the volume does not depend on the array spacing. (This gives some insight as to how thin the beams of a sparse array are compared to those of a dense array.)

Random Pulses and the Gaussian Distribution

So far we have said nothing about the statistical properties of noise, except to assign to noise a mean-square voltage or mean power and a spectral density distribution.

Errors in digital transmission and breaking in fm occur when the noise voltage is large enough to cause an ambiguity in the received signal. What is the probability that this will happen? This depends on more than the noise power alone. It depends on the probability distribution of the noise amplitude.

All modern books on communication rely on probability to describe performance in a statistical sense. And, there are books on probability intended for the electrical engineer or communicator, such as Reference 1 of Chapter 6. Here we will not assume a prior course on probability or any formal experience with it. We develop what is necessary here as we find that we need it.

In this chapter we will consider noise produced by random or chance events. When very many of the independent random events that cause a noise occur in a time equal to the reciprocal of the bandwidth of the circuit or network or system that transmits the events, the noise is always of a kind called *gaussian* noise. Shot noise is like this, and many other noises too. We shall explain why gaussian noise arises so much in communication.

8.1. Regularly Spaced Impulses of Random Sign

Consider equally spaced impulses, spaced $T = 1/N$ apart, whose strengths or areas are randomly and with equal probability $+1$ or -1.

We shall find the autocorrelation of such a random waveform, using the definition of autocorrelation of stationary signals given in Section 6.6. For $\tau = 0$ contributions to the autocorrelation will be of the form $(+1)$ $(+1)\delta(0) = +\delta(0)$ or $(-1)(-1)\delta(0) = +\delta(0)$. There are N of these delta functions per second. For τ not equal to an integer times the pulse spacing T the autocorrelation will obviously be zero. For τ an integer (other than zero) times T, we will with equal likelihood have $(+1)(+1)$, $(+1)(-1)$, $(-1)(+1)$, or $(-1)(-1)$. These will tend to cancel over a long stretch of time T_0. At least, it is highly likely that the net contribution is very small when divided by T_0. So, in the limit when T_0 becomes very large, the autocorrelation of Section 6.6 will be zero in this case. We have completely determined the autocorrelation function of this stationary random signal, and, by taking its Fourier transform, we can find its power spectrum:

$$C(\tau) = N\,\delta(\tau)$$

$$P(f) = N$$

$$(8.1\text{-}1)$$

The power spectrum $P(f)$ is independent of frequency. Because the spectrum is flat, we call it *white* noise. We shall later meet white *gausian* noise. (We have already met it without a name, in Section 2.5.) The spectrum has constant magnitude N, the number of pulses per second. Had the areas of each pulse been A, the autocorrelation and the power spectrum would have been A^2 times as great.

As given above, $P(f)$ is the noise power spectrum appropriate to both positive and negative frequencies. The spectrum appropriate to positive frequencies only is

$$P(f) = 2N \qquad\qquad (8.1\text{-}2)$$

The next section looks at impulses that occur at *random* times, but with known sign. This occurs in shot noise.

Problems

1. A function of time $x(t)$ is $x(t) = \delta(t) - \delta(t - T)$. What is the autocorrelation? What is the *energy* spectrum?

2. A signal is obtained by choosing randomly and with equal probabilities $x(t)$ followed by $-x(t)$, or $-x(t)$ followed by $x(t)$, at intervals $2T$ apart. What is the autocorrelation of this signal? What is the *power* spectrum? What is the relation of this to the energy spectrum of Problem 1?

3. Consider four signals:
    (a) $x_1(t) = \delta(t) - \delta(t - T)$
    (b) $x_2(t) = -\delta(t) + \delta(t - T)$
    (c) $z_1(t) = \delta(t) + \delta(t - T)$
    (d) $z_2(t) = -\delta(t) - \delta(t - T)$

What are the autocorrelations and energy spectra of these four signals? What is the sum of the four autocorrelation functions? What is the sum of the energy spectra?

4. A signal is made up by choosing among $x_1(t)$, $x_2(t)$, $z_1(t)$, $z_2(t)$ randomly and with equal probabilities (1/4) at intervals of $2T$ as in Problem 2. What is the autocorrelation function of this signal? What is the power spectrum? What is the relation to the sum of the energy spectra of Problem 3? (These pulses are used in hearing experiments to determine whether the human ear can distinguish different signals of the same spectrum.)

5. Consider the short-term energy spectrum $E(f, t)$ (as defined in Section 3.11) of regularly spaced unit impulses of spacing T_0 and random sign. Let the window or gating function be $g(T)$. Show that if $g(T)$ is sufficiently broad so as to accommodate many impulses, and slowly varying compared with T_0, then

$$E(f, t) = \sum_{n=-\infty}^{\infty} g^2(nT_0 - t)$$

Conclude that $E(f, t)$ is independent of frequency, and has period T_0. Let $g(T)$ be an even function of T. Show that $E(f, t)$ has symmetry around $T_0/2$.

8.2. Derivation of Shot Noise

We consider the case in which the probability that an impulse of unit area will occur in any time interval dt is simply $N\,dt$. This gives an average of N impulses per second, but we do not know when these will occur. We will arrive at a derivation of shot noise, which we first discussed in Section 2.9 in connection with quantum uncertainties in amplification.

What is the autocorrelation function? All impulses will coincide when $\tau = 0$. Hence one component of the autocorrelation is simply N times the correlation for a single impulse. We see this by referring to the definition of Section 6.6:

$$C(\tau) = \lim_{T \to \infty} \frac{C_T(\tau)}{T} = \frac{NT}{T} \delta(\tau) = N\,\delta(\tau) \tag{8.2-1}$$

There is another component of the autocorrelation, however. The following argument is not fully rigorous, but it can be made so and leads to the correct result.

Let us move a train of randomly spaced impulses past itself. Individual impulses will coincide at various values of τ, giving rise to a series of spikes in the autocorrelation $C(\tau)$. Here the train of impulses has length T, and τ is fixed and small compared with T. In the interval of length $2T$, each of the approximately NT pulses will coincide with every other for exactly one shift, so there will be $(NT)^2$ coincidences spread

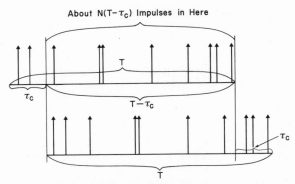

Figure 8.1. Autocorrelation of truncated impulse sequence.

over the range $2T$. The NT of these that occur at $\tau = 0$ have already been taken into account, but this number is negligible in comparison with $(NT)^2$ when T becomes very large.

Figure 8.1 shows the sequence of impulses during a time T, shifted with respect to itself by the delay τ_C. There will be about $N(T - \tau_C)$ impulses eligible to have another impulse occur at delay τ_C and so contribute a spike to the autocorrelation. Consider delays from τ_C to $\tau_C + \Delta\tau_C$. The probability of getting a hit for one pulse with delay τ_C in this range is $N\,\Delta\tau_C$. We multiply this by the number of pulses eligible, which is $N(T - \tau_C)$. This gives a contribution of

$$N^2(T - \tau_C)\,\Delta\tau_C \tag{8.2-2}$$

delta functions to $C_T(\tau_C)$.

We do not expect to have infinitely fine time resolution when we look at a pulse train or a correlation function or at any other function. Thus we are led to average the delta functions $\delta(\tau - \tau')$, the number of which is given by (8.2-2). Here the location τ' is in the range τ_C to $\tau_C + \Delta\tau_C$. The average of this delta function over the interval of length $\Delta\tau_C$ is $1/\Delta\tau_C$. So the contribution to $C_T(\tau_C)$ is obtained by dividing (8.2-2) by $\Delta\tau_C$. The result is

$$C_T(\tau) = N^2(T - \tau) \tag{8.2-3}$$

Here we have replaced τ_C by the more familiar τ.

We divide (8.2-3) by T with τ fixed, and then let T become large. The result is a contribution to the autocorrelation from the train of impulses. This contribution is

$$C(\tau) = N^2 \tag{8.2-4}$$

This is a constant independent of τ. To this we add the component given by (8.2-1). We find that the total autocorrelation function of the sequence of positive impulses occurring at random times is

$$C(\tau) = N\delta(\tau) + N^2 \tag{8.2-5}$$

The power spectrum $P(f)$ is the Fourier transform of this:

$$P(f) = N + N^2\delta(f) \tag{8.2-6}$$

The second term corresponds to "direct current" (zero frequency). There is a physical explanation of this. The average current corresponding to N pulses per second of unit area is N, and so the corresponding zero-frequency power (really, current squared) is N^2.

$P(f)$ in (8.2-6) is the spectrum appropriate to both positive and negative frequencies. The noise component of the spectrum appropriate to positive frequencies only is $2N$.

We can give a more concrete example of this sort of noise. A current I_0 carried by electrons consists of

$$N = \frac{I_0}{e} \tag{8.2-7}$$

electrons per second, where e is the charge of the electron. Each current pulse corresponding to the passage of an electron has an area A given by

$$A = \int I\,dt = e \tag{8.2-8}$$

Sometimes such a current of electrons can flow through a vacuum tube or a semiconductor device in such a way that passage of one electron does not affect the passage of any other electron at any other time no matter how close. In this case, the above result applies, and the power spectrum $P(f)$ (really, the spectrum of the noise *current* squared) is

$$P(f) = A^2 N + A^2 N^2\,\delta(f)$$
$$P(f) = eI_0 + I_0^2\delta(f) \tag{8.2-9}$$

It is common to treat the second part, the dc power or steady signal, separately from the first or ac part eI_0, which is the shot-noise spectrum. We will see in Section 8.6 that this must correspond to a gaussian noise. This is because it was obtained as a result of an averaging process.

Suppose a low-pass filter passes all frequencies lying between 0 and B. We are considering negative as well as positive frequencies, and such a filter will also pass frequencies lying between $-B$ and 0. Thus the total portion of the power spectrum it will pass is $2B$ wide. Hence the mean-square noise current $\overline{i_n^2}$ passed by a filter of bandwidth B is

$$\overline{i_n^2} = 2eI_0B \tag{8.2-10}$$

This is the expression for *shot noise*. We used this in Section 2.9 to study the performance of an optical superheterodyne receiver.

If the current I_0 (that is, the number of electrons per second) varies with time, there will be an ac component owing to this variation as well as to the shot noise. In space-charge-limited vacuum tubes the fluctuation of the electric potential minimum causes a variation of I_0 which, at low enough frequencies, partially cancels the shot-noise current.

The next section studies trains of pulses of arbitrary shape. We will find that the results for impulses can be applied directly.

Problems

1. How small should the bandwidth B of (8.2-10) be taken in order that the mean-square noise current be at most 10^{-8} times the dc current squared, if there are 10^{12} electrons per second in the current flow?

2. What does (8.2-6) become if the impulses have a random sign as well as a random time of occurrence? How does this compare with the result of the previous section?

8.3. Trains of Pulses of Arbitrary Shape

The spectrum of an impulse of unit area is simply

$$X(f) = 1 \qquad (8.3\text{-}1)$$

If the impulse is passed through a linear network whose transfer function is $H(f)$ the output will be a pulse whose time variation $x(t)$ is

$$x(t) = \int_{-\infty}^{\infty} H(f)e^{j\omega t}\,df \qquad (8.3\text{-}2)$$

This is because the voltage spectrum of the output pulse is $H(f)$. The energy of the output due to the impulse will be

$$E_0 = \int_{-\infty}^{\infty} H(f)H^*(f)\,df \qquad (8.3\text{-}3)$$

When we deal with a train of impulses, the output must be the same whether we pass them individually through a linear network and then combine the filtered pulses that result, or whether we pass the whole train of impulses through the same linear network. Hence we can use results for a train of impulses to obtain results for a train of pulses of finite duration. For N equally spaced impulses per second whose areas are randomly and with equal probability +1 or −1, the power spectrum $P(f)$ is

$$P(f) = N \qquad (8.3\text{-}4)$$

This is also true of the ac part of the spectrum of N randomly spaced unit impulses.

If the train of impulses is applied to a network of transfer function $H(f)$ the power spectrum of the output must then be

$$P(f) = NH(f)H^*(f) \qquad (8.3\text{-}5)$$

Integrating this implies that for any pulse shape the average power is simply the energy per pulse times the number of pulses per second. Equation (8.3-5) says that the power spectrum of pulses of random sign occurring regularly N times per second is simply N times the energy spectrum of an individual pulse. We used this result in Section 7.1.

We could also argue to this end by saying that each frequency component of a pulse is in phase with the same frequency component of that pulse but on the average is as often out of phase as in phase with the same frequency component of any other pulse in the train. Hence in calculating the energy per pulse of the train we merely integrate $H(f)H^*(f)$ for each pulse separately and add these energies.

The power spectrum $NH(f)H^*(f)$ is appropriate if we get the power by integrating with respect to frequency from $f = -\infty$ to $f = \infty$. If we integrate from $f = 0$ to $f = \infty$ only, we should use a power spectrum

$$P(f) = 2NH(f)H^*(f) \qquad (8.3\text{-}6)$$

We begin the study of gaussian noise in the next several sections. We start by defining the variance and characteristic function (or Fourier transform) of a random variable. The goal is to show how the rapid occurrence of pulses leads to a gaussian distribution when there is some averaging or filtering.

Problem

1. Equation (8.3-5) gives the power spectrum when pulses of voltage spectrum $H(f)$ are used in a pulse amplitude modulation scheme with random ± data occurring N times a second. What is the power spectrum if the data has amplitude +1 ($\frac{2}{3}$ the time), +3 ($\frac{1}{3}$ the time), and random sign?

8.4. Random Variables, Variance, and the Characteristic Function

We are interested in random variables because noise is a random variable. Sometimes signals are, as well. A random variable is unpredictable. The numbers turning up on successive throws of an honest die are random and we say the succession of numbers constitutes a random variable.

A random variable X might be a particular noise generator. Then x represents its output. X is in a way the ensemble of the x's, which are the particular values that X may have. Questions such as the probability that X be equal to x or *take on* the actual value x:

$$\Pr(X = x)$$

are meaningful. In other words, x represents a *particular* constant. Or, in some cases, x may represent an entire waveform, signal, or function.

If the variable which we will call X assumes a number of discrete values x_n, we can list the probabilities $p(x_n)$ that it will assume the various values x_n. If the variable has a continuous range of values, there may be a *density function* $P(x)$ that characterizes the variable completely by specifying the probability that a particular value of X lies in the range dx at x as

$$P(x)\, dx$$

If X is certain to take on the value x_0, we sometimes say that the density of X is $\delta(x - x_0)$.

Two random variables may have the same distribution or probabilities of occurrence or density but may be entirely independent of one another. Receiver noise powers in the same receiver, measured hours apart, are independent. But, if the receiver is not drifting, the random variables which represent the noise powers will have the same probability distribution. If the voltage control on a continuously variable dc power supply is randomly turned to establish a random voltage, the voltages measured several hours apart have the same distribution. More than this, they have the same values. The difference between having the same distribution and having the same values should be kept in mind.

We can also characterize a random variable completely by its *characteristic function*. The characteristic function always exists, even if there is no density function. The characteristic function will be the expectation of

$$e^{-j2\pi fX}$$

We will define the concept of expectation. But, before we do so, we note that if the random variable has a density function, the characteristic function is the Fourier transform of the density function and the density function is the inverse transform of the characteristic function.

Besides the characteristic function, there are certain quantities which are very useful despite the fact that they do not describe a random variable completely. The *expectations* or *averages* of X and X^2 are two such quantities. The expectation of any random variable is the average of its values weighted by probability of occurrence. We usually write

$$E(X) = \bar{X}$$
$$E(X^2) = \overline{X^2}$$

If X is a random variable such as a random voltage whose expected energy $E(X^2) = \overline{X^2}$ is finite, we will see that X also has an expectation $E(X) = \bar{X}$. Then

$$\bar{X}^2 \leq \overline{X^2} \tag{8.4-1}$$

Equality holds if and only if the random variable does not vary randomly, but is certain to have a fixed or constant value. This follows from the fact that

$$\overline{(X - \bar{X})^2} = E(X - \bar{X})^2 \geq 0 \tag{8.4-2}$$

By writing out the square of $X - \bar{X}$ we see that

$$E(X^2 - 2X\bar{X} + \bar{X}^2) \geq 0 \tag{8.4-3}$$

Because $\bar{X}$ is a constant, this is

$$\overline{X^2} - 2\bar{X}^2 + \bar{X}^2 = \overline{X^2} - \bar{X}^2 \geq 0 \tag{8.4-4}$$

This is an alternate form of (8.4-1).

Furthermore, (8.4-4) is zero if and only if $E(X - \bar{X})^2 = 0$. This is the same as the condition $(X - \bar{X})^2 = 0$, because $(X - \bar{X})^2$ is never negative. So, equality holds if and only if X is always equal to its average value $\bar{X}$, that is, X must be a constant.

The nonnegative quantity

$$\sigma_X{}^2 = \overline{X^2} - \bar{X}^2 = \overline{(X - \bar{X})^2} \tag{8.4-5}$$

is called the *variance* of X.

Let X and Y be two *independent* random variables with average values $\bar{X}$ and $\bar{Y}$. In this case, we can easily find the variance of the sum $X + Y$ of the variables:

$$\sigma_{X+Y}^2 = E[X + Y - \overline{(X + Y)}]^2 \tag{8.4-6}$$

$$\sigma_{X+Y}^2 = E[(X - \bar{X}) + (Y - \bar{Y})]^2 \tag{8.4-7}$$

This is because $\overline{X + Y} = \bar{X} + \bar{Y}$. Expand (8.4-7):

$$\sigma_{X+Y}^2 = E[(X - \bar{X})^2] + E[(Y - \bar{Y})^2] + 2E[(X - \bar{X})(Y - \bar{Y})] \tag{8.4-8}$$

Because X and Y are independent, the value of Y in no way depends on the value of X. This is the definition of independence. So, in taking the expectation of $(X - \bar{X})(Y - \bar{Y})$ we can first average with respect to Y, giving us $(X - \bar{X})E(Y - \bar{Y})$, and then average with respect to X, giving us $E(X - \bar{X})E(Y - \bar{Y})$, which is the product of the expectations of $Y - \bar{Y}$ and $X - \bar{X}$. Since $E(X - \bar{X}) = E(X) - \bar{X} = \bar{X} - \bar{X} = 0$, we conclude that $E[(X - \bar{X})(Y - \bar{Y})] = 0$, when X and Y are independent. This gives us the following conclusion. For any independent random variables X and Y

$$\sigma_{X+Y}^2 = \sigma_X{}^2 + \sigma_Y{}^2 \tag{8.4-9}$$

Let $\bar{X}$ and $\bar{Y}$ be zero, where X and Y are voltages. Since variance is power, the powers add. This tells us that the power of the sum of two independent mean zero random noise voltages is the sum of the powers of the noise voltages taken separately.

The above result is very important in communication. It permits us to add two independent noises as power while we add signals as voltage. We have already used this result many times in this book without calling attention to it. In Section 7.6, we used the result to add the noise power due to quantization error to the mean-square error due to distortion. This is correct, because the error due to quantizing a voltage in no way influences the distortion in the signal caused by receiver filtering. In Section 7.7, we used the result to determine the total noise power of an array of antennas. The total is the sum of the noise powers from each individual antenna, because the different receivers add their own noise which is independent of the noise of the other receivers. But, if each element of an array is looking through the same cloud, the noises from the different array elements will not be independent. Part of the noise from each antenna will be the common thermal noise radiated by the cloud, as in Section 2.2. This common part will add in the output as voltage, not power. So, power noise in the summed signal will be *greater* than the sum of the individual noise powers.

We have noted that a random variable has a *characteristic function*, which we will call $G_X(f)$. This is the expectation of $e^{-j2\pi Xf}$:

$$G_X(f) = E(e^{-j2\pi Xf}) \tag{8.4-10}$$

This exists for all "frequencies" f, no matter what X may be. If X has a density function $P(x)$, $G_X(f)$ is the Fourier transform of that density function:

$$G_X(f) = \int_{-\infty}^{\infty} P(x)e^{-j2\pi xf}\,dx \tag{8.4-11}$$

We can use what we know about Fourier transforms from Chapter 3 to find characteristic functions from densities, or densities from characteristic functions. In this way, we can determine the distribution of a random variable uniquely from its characteristic function, just as we can determine a function from its Fourier transform.

If X and Y are independent, let us find $G_{X+Y}(f)$ in terms of $G_X(f)$ and $G_Y(f)$:

$$G_{X+Y}(f) = E(e^{-j2\pi(X+Y)f}) \tag{8.4-12}$$

$$G_{X+Y}(f) = E(e^{-j2\pi Xf} \cdot e^{-j2\pi Yf}) \tag{8.4-13}$$

Since X and Y are independent and the values of X in no way depend on the values of Y, the expectation of the product in (8.4-13) is the product of the expectations. We saw this in our discussion of variance. This gives

$$G_{X+Y}(f) = E(e^{-j2\pi Xf}) \cdot E(e^{-j2\pi Yf}) \qquad (8.4\text{-}14)$$

This shows that the characteristic function of the sum of two independent random variables is the product of their characteristic functions:

$$G_{X+Y}(f) = G_X(f) \cdot G_Y(f) \qquad (8.4\text{-}15)$$

We will use this equation in a subsequent section to explain why gaussian noise arises so frequently in communication systems. We will define the gaussian distribution in the next section.

Problems

1. Find the characteristic function of the random variable which is $+1$, and -1, each with probability 1/2. Find the characteristic function of the sum of two such independent random variables, in two different ways.

2. Take the inverse Fourier transforms of the above two characteristic functions and show that the density functions are recovered. (It may help to review the Fourier transform table of Chapter 3.)

3. Find the characteristic function of the random variable X whose density function is 1 between $x = -\frac{1}{2}$ and $x = \frac{1}{2}$, and is 0 elsewhere. What does this imply for the characteristic function of the sum of two independent copies of X? Using the Fourier transform table, find the density function of this sum random variable. Check by integration that the result *is* a density function.

4. If X and Y are independent with density functions $P(x)$ and $Q(y)$ respectively, explain why the random variable $Z = X + Y$ has density $R(z)$ given by

$$R(z) = \int_{-\infty}^{\infty} P(x)Q(z - x)\, dx \qquad \text{or} \qquad R = P * Q \quad \text{(convolution)}$$

Deduce (8.4-15) for these random variables using properties of Fourier transforms and convolutions from Section 3.7.

5. Suppose we have N adjacent and nonoverlapping rectangular pulses of width $1/N$ occurring in 1 sec. Let the height of each be randomly ± 1. Suppose we filter or average these by integrating over the 1 sec and then dividing by the number N of pulses. We call the integral over the ith time interval X_i, where i is between 1 and N.

   (a) What is the power of this signal before filtering?
   (b) What values can X_i take, and with what probabilities?
   (c) Are the X_i independent? What happens if the intervals of integration are not exactly synchronous with the times at which the signal can change sign?
   (d) What is the variance of X_i?
   (e) What is the average power after the integration but before dividing by N?
   (f) What is the power after dividing by N?

(g) What would the original heights have had to be to get *unit* power after the division by N? (The resulting random variable approaches the *unit gaussian* as N becomes large.)

8.5. Gaussian Noise

Thermal or Johnson noise, and the shot noise discussed in Sections 2.9, 2.10, and 8.2 are examples of gaussian noise. We will see that any noise which represents a filtered response to many random events is gaussian noise. If we are to understand errors in digital transmission and other matters concerning the effects of noise in communication we must understand gaussian noise.

The probability that a noise voltage with zero mean and variance (or power) σ^2 which has a gaussian distribution will lie in the range from v to $v + dv$ is

$$P(v)\, dv = \frac{1}{(2\pi)^{1/2}\sigma} \exp\left(\frac{-v^2}{2\sigma^2}\right) dv \qquad (8.5\text{-}1)$$

The function $P(v)$ is called the *gaussian density*.

We will show that $P(v)$ integrates to 1 and that the random voltage has mean 0 and variance σ^2. In order to show this, let U and V be two independent gaussian variables with density given by (8.5-1). Because the variables are independent, the *joint* density of U and V, $P(u, v)$, is the product of the separate densities (this can be taken as a definition of independence):

$$P(u, v) = \frac{1}{2\pi\sigma^2} \exp\left(-\frac{u^2 + v^2}{2\sigma^2}\right) \qquad (8.5\text{-}2)$$

We will study joint distributions in Section 9.1, but the meaning of (8.5-2) is the natural extension of the meaning of density for a single random variable. The value A of the integral of $P(u, v)\, du\, dv$ over the entire (u, v) plane must be the square of the integral of $P(v)\, dv$ over the v line, because U and V are independent, or because $\exp[-(u^2 + v^2)/2\sigma^2] = \exp(-u^2/2\sigma^2) \cdot \exp(-v^2/2\sigma^2)$.

Let us note, however, that the double integral can be transformed to polar coordinates (r, θ):

$$A = \int_{-\infty}^{\infty} \int_{-\infty}^{\infty} P(u, v)\, du\, dv = \int_{0}^{\infty} \int_{0}^{2\pi} \frac{1}{2\pi\sigma^2} \exp\left(\frac{-r^2}{2\sigma^2}\right) r\, dr\, d\theta$$

$$(8.5\text{-}3)$$

We can integrate with respect to θ first, giving

$$A = \int_0^\infty \frac{1}{\sigma^2} \exp\left(\frac{-r^2}{2\sigma^2}\right) r \, dr$$

$$A = \int_0^\infty \exp\left(\frac{-r^2}{2\sigma^2}\right) d\left(\frac{r^2}{2\sigma^2}\right) = -\exp\left(\frac{-r^2}{2\sigma^2}\right) \Big|_0^\infty = 1 \quad (8.5\text{-}4)$$

This says that square A of the integral of $P(v)\,dv$ is equal to 1, and so the integral $A^{1/2}$ of $P(v)\,dv$ must be 1 also. In this way, we have defined a true density function. This is the density function of a gaussian random variable.

A mean voltage exists, because $P(v)$ drops off very rapidly for large values of v. This mean is zero, because $P(v)$ is symmetric about 0. So, the variance of V is just the expected value of V^2:

$$\sigma_V^2 = \int_{-\infty}^\infty v^2 P(v) \, dv \quad (8.5\text{-}5)$$

$$\sigma_V^2 = \frac{1}{(2\pi)^{1/2}\sigma} \int_{-\infty}^\infty v^2 \exp\left(\frac{-v^2}{2\sigma^2}\right) dv$$

$$= \frac{1}{(2\pi)^{1/2}\sigma} \int_{-\infty}^\infty v \cdot v \exp\left(\frac{-v^2}{2\sigma^2}\right) dv \quad (8.5\text{-}6)$$

We can do a little algebra. Integrate (8.5-6) by parts:

$$\sigma_V^2 = \frac{(-1) \cdot \sigma^2}{(2\pi)^{1/2} \cdot \sigma} \int_{-\infty}^\infty v \, d\exp\left(\frac{-v^2}{2\sigma^2}\right)$$

$$= \frac{-\sigma^2}{(2\pi)^{1/2}\sigma}\left[v \exp\left(\frac{-v^2}{2\sigma^2}\right)\right]\Big|_{-\infty}^\infty + \frac{\sigma^2}{(2\pi)^{1/2}\sigma} \int_{-\infty}^\infty \exp\left(\frac{-v^2}{2\sigma^2}\right) dv$$

The term $v \exp(-v^2/2\sigma^2)$ is zero at both limits. The second term is $\sigma^2 \cdot 1$ or σ^2, because $P(v)$ is a density function. We have the result that the variance of the gaussian density (8.5-1) is σ^2:

$$\sigma_V^2 = \sigma^2 \quad (8.5\text{-}7)$$

The random variable V given by (8.5-1) is a zero-mean random variable. V may represent some noise voltage. The total voltage will then be $v_0 + V$, where v_0 is the signal voltage. The resulting random variable, signal plus noise, is called a *gaussian random variable of mean v_0*. The variance is still σ^2.

In digital transmission of binary or two-level data, we are interested in the probability P_e that V will be greater than some value v_0, half the spacing between levels. Here e denotes error. This is

$$P_e = \frac{1}{(2\pi)^{1/2}\sigma} \int_{v_0}^{\infty} \exp\left(\frac{-v^2}{2\sigma^2}\right) dv = \frac{1}{(2\pi)^{1/2}} \int_{x=v_0/\sigma}^{\infty} \exp\left(\frac{-x^2}{2}\right) dx$$

$$(8.5\text{-}8)$$

Since the integral

$$\frac{1}{(2\pi)^{1/2}} \int_{-\infty}^{\infty} \exp\left(\frac{-x^2}{2}\right) dx = 1$$

we may write (8.5-8) as

$$P_e = 1 - \Phi(v_0/\sigma) \tag{8.5-9}$$

Here the function $\Phi(x)$ is the *gaussian distribution*, also called the normal distribution, or the unit or standard normal or gaussian distribution. Sometimes the word "cumulative" is added, to indicate that $\Phi(x)$, the cumulative gaussian distribution, is the probability that a standard (mean 0, variance 1) gaussian random variable X be less than or equal to x. So, the definition of $\Phi(x)$ is

$$\Phi(x) = \frac{1}{(2\pi)^{1/2}} \int_{-\infty}^{x} \exp\left(\frac{-x^2}{2}\right) dx \tag{8.5\text{-}10}$$

This is graphed in Fig. 8.2.

   In earlier days, it was more usual to have tables of the *error function* or *probability integral* erf (x), as in Reference 1. Nowadays tables of $\Phi(x)$ are more usual, although some hand calculators have an "ERF" key. The

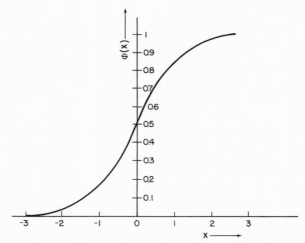

Figure 8.2. The gaussian distribution.

Table 8.1. The Gaussian Distribution[a]

$$\Phi(x) = \frac{1}{(2\pi)^{1/2}} \int_{-\infty}^{x} \exp(-y^2/2)\, dy$$

x	$\Phi(x)$	$1 - \Phi(x)$	$\Phi(x)$	$1 - \Phi(x)$	x
0.0	0.5000	0.5000	0.90	0.10	1.282
0.5	0.6915	0.3085	0.95	0.05	1.645
1.0	0.8413	0.1587	0.975	0.025	1.960
1.5	0.9332	0.0668	0.99	0.01	2.326
2.0	0.9773	0.0227	0.995	0.005	2.576
2.5	0.99379	0.00621	0.999	0.001	3.090
3.0	0.99865	0.00135	0.9999	0.0001	3.719
3.5	0.99977	0.00023	0.99999	0.00001	4.265
4.0	0.999968	0.000032	0.999999	10^{-6}	4.753

[a] For $x < 0$, use $\Phi(-x) = 1 - \Phi(x)$.

definition of erf is

$$\mathrm{erf}\, x = 1 - \left(\frac{2}{\pi}\right)^{1/2} \int_{0}^{x} \exp(-y^2)\, dy$$

Then

$$P_e = \tfrac{1}{2}[1 - \mathrm{erf}\,(v_0\sqrt{2}/\sigma)]$$

Table 8.1 is an abbreviated table of the gaussian distribution. Sometimes communicators speak of the probability of events in terms related to the gaussian distribution. A "three sigma" event is an event whose probability of occurrence is the same as the probability that a gaussian random variable exceed its mean by more than three times σ. Here σ is the standard deviation, or the square root of the variance. This can also be expressed as the probability

$$1 - \Phi(3) = 0.00135 \qquad \text{(three sigma event probability)}$$

In practical cases we are usually interested in small probabilities of error. For large arguments, Chapter 7, Section 6 of Reference 2 gives an approximation

$$1 - \Phi(x) \sim \frac{\exp(-x^2/2)}{(2\pi)^{1/2}x} \tag{8.5-11}$$

This is $1/x$ times the gaussian *density* at x. Thus for large enough values of v_0, the probability P_e that $V > v_0$ (which is equal to the probability that $V < -v_0$) is approximately

$$P_e \doteq \frac{1}{(2\pi)^{1/2}(v_0/\sigma)} \exp\left(\frac{-v_0^2}{2\sigma^2}\right) \tag{8.5-12}$$

In two-level transmission using sinc pulses of $+v_0$ and $-v_0$ peak pulse amplitude, there will be an error in the transmission of a positive pulse

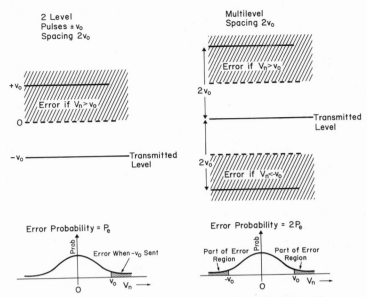

Figure 8.3. Probability of error in two- and multilevel signaling.

$(v_s = v_0)$ if the noise voltage $V_n < -v_0$. For, in this case we receive a negative value and think it more likely that $-v_0$ was transmitted. There will be an error in the transmission of a negative pulse $(v_s = -v_0)$ if $V_n > v_0$. These two probabilities are equal. Hence, P_e is the probability of error in two-level transmission. In multilevel transmission where $2v_0$ is the spacing between levels, the probability of error is approximately $2P_e$. This is explained in Fig. 8.3. We see that $2P_e$ will be the probability of error given that any of the levels except the two extreme levels were sent. So, if there are many levels, $2P_e$ will approximately be the overall average error probability.

The signal power P_s in two-level transmission will be

$$P_s = v_0^2 \qquad (8.5\text{-}13)$$

The signal-to-noise ratio will be

$$\frac{P_s}{P_n} = \frac{v_0^2}{\sigma^2} \qquad (8.5\text{-}14)$$

$\log_{10} P_e$ is plotted vs. v_0^2/σ^2 in Fig. 8.4.

In the next section we show that we get a gaussian distribution by averaging many independent random variables, each of which has the same probability distribution which has a variance. We will note that physical systems have some memory and so perform an average over past

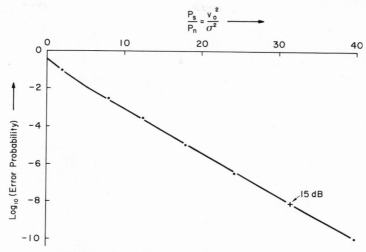

Figure 8.4. Error probability vs. signal-to-noise ratio.

inputs. For this reason, it is not surprising that gaussian noise occurs so widely. The mathematical result which expresses this is called the *central limit theorem*.

Problems

1. When the signal-to-noise ratio v_0^2/σ^2 in a binary signaling scheme is large and the resulting error probability P_e is small, what is the increase in the required signal-to-noise ratio needed to cut the error probability in half? What is this in dB when $v_0^2/\sigma^2 = 30$ (corresponding to about a 15-dB signal-to-noise ratio)? What is the error probability before the increase?

2. When m-level pulses are transmitted, $\log_2 m$ binary digits per pulse, or bits, can be sent. Assume that we keep the error probability small and constant as we change the number m of levels. What is the average signal power per bit for 2, 3, 4, 5 levels? In compouting the average power, assume that the levels are $2v_0$ apart and are equally likely to be sent. (The previous problem indicates that for low error probabilities and m not too large, we do not need to distinguish the *baud* or level or symbol error probability and the *bit* error probability.)

3. Use (8.5-12) to find the ultimate slope in Fig. 8.4.

8.6. Obtaining the Gaussian Distribution

Here we propose to show how a sequence of randomly positive and negative pulses can lead to a gaussian distribution. We will find that the distribution of a suitably normalized *sum* of m successive pulses approaches a gaussian distribution as m becomes very large. This is an example of the central limit theorem.

We will first find the characteristic function $G_\sigma(f)$ of a gaussian distribution of mean 0 and variance σ^2. Use equation (8.5-1) with definition (8.4-10) of the characteristic function:

$$G_\sigma(f) = \int_{-\infty}^{\infty} \frac{1}{(2\pi)^{1/2}\sigma} \exp(-j2\pi vf) \exp(-v^2/2\sigma^2)\, dv \quad (8.6\text{-}1)$$

This is the Fourier transform of a gaussian pulse. We will show that it is gaussian in frequency as well. We observe that in general narrow pulses have broad spectra and broad pulses have narrow spectra. But, we will find that a suitable gaussian pulse is equal in shape to its own spectrum.

We start by noting that $G_\sigma(0)$ is 1 because the integral of the gaussian density is 1. We will differentiate $G_\sigma(f)$ with respect to f, and then integrate by parts. This leads to a differential equation for $G_\sigma(f)$. The first step gives

$$\frac{dG_\sigma(f)}{df} = \frac{1}{(2\pi)^{1/2}\sigma} \int_{-\infty}^{\infty} (-j2\pi v)[\exp(-j2\pi vf)] \exp\left(\frac{-v^2}{2\sigma^2}\right) dv$$

$$(8.6\text{-}2)$$

This can easily be integrated by parts:

$$\frac{dG_\sigma(f)}{df} = j(2\pi)^{1/2}\sigma \int_{-\infty}^{\infty} \exp(-j2\pi vf)\, d\exp\left(\frac{-v^2}{2\sigma^2}\right) \quad (8.6\text{-}3)$$

$$\frac{dG_\sigma(f)}{df} = j(2\pi)^{1/2}\sigma\left\{ [\exp(-j2\pi vf)]\left[\exp\left(\frac{-v^2}{2\sigma^2}\right)\right] \Big|_{-\infty}^{\infty} \right.$$

$$\left. + j2\pi f \int_{-\infty}^{\infty} \exp(-j2\pi vf) \exp\left(\frac{-v^2}{2\sigma^2}\right) dv \right\} \quad (8.6\text{-}4)$$

The first term in (8.6-4) is 0 because the function being evaluated is 0 at both limits. This leaves us with

$$\frac{dG_\sigma(f)}{df} = -(2\pi)^{3/2}\sigma f \int_{-\infty}^{\infty} [\exp(-j2\pi vf)] \exp(-v^2/2\sigma^2)\, dv$$

$$(8.6\text{-}5)$$

Referring to (8.6-1), we obtain a differential equation from (8.6-5):

$$\frac{dG_\sigma(f)}{df} = -(2\pi)^2\sigma^2 f G_\sigma(f)$$

This can be easily solved:

$$\frac{dG_\sigma(f)}{G_\sigma(f)} = -(2\pi)^2\sigma^2 f\,df$$

$$d\ln G_\sigma(f) = -2\pi^2\sigma^2\,d(f^2)$$

$$\ln G_\sigma(f) = -2\pi^2\sigma^2 f^2 + C$$

$$G_\sigma(f) = K\exp(-2\pi^2\sigma^2 f^2) \tag{8.6-6}$$

We know that $G_\sigma(0) = 1$, and so $K = 1$. We have derived the characteristic function of the gaussian distribution, or the voltage spectrum of a gaussian pulse which peaks at time 0 and has energy σ^2. The result is that G_σ is a gaussian pulse:

$$G_\sigma(f) = \exp(-2\pi^2\sigma^2 f^2) \tag{8.6-7}$$

(This can be partially checked by using the energy theorem of Section 3.6.) We note that a broad pulse (large σ^2) has a narrow spectrum, and conversely, as must be.

By rigorous analysis it can be shown that if the *characteristic function* of a sequence of random variables Z_m converges to the characteristic function of a random variable Z as m approaches ∞, then the *distribution* of Z_m converges to the distribution of Z as well. We shall use this to show how we must get a gaussian distribution from filtering a random pulse train. This is the central limit theorem.

Consider a pulse whose area X is randomly and with probability $1/2$ equal to $+1$ or -1. The random variable X has a mean of 0 and a variance of 1. Its characteristic function $G_X(f)$ will be

$$G_X(f) = E[\exp(-j2\pi Xf)] = \tfrac{1}{2}\exp(-j2\pi f) + \tfrac{1}{2}\exp(j2\pi f) = \cos 2\pi f \tag{8.6-8}$$

Now suppose that a sequence of such randomly positive and negative pulses with each pulse area independent of all the others forms the input of a particular low-pass filter. The filter has as its output the time integral of the input over the past T seconds, during which time m pulses reached the input. This *ideal integrator* is a pecular and special type of low-pass filter, but the final result holds for any low-pass filtering. The output in this ideal case is the excess of positive pulses over negative pulses in m pulses.

Let Y_m be the integrated output of the m pulses over the time T. Because the input pulses are randomly and independently positive or negative, Y_m is the sum of m independent random variables with the same distribution as X. So, the random variable Y_m has the characteristic function

$$G_{Y_m}(f) = (\cos 2\pi f)^m \qquad (8.6\text{-}9)$$

Here we use the multiplication result of equation (8.4-15) repeatedly. Because Y_m consists of m random pulses of variance unity, the variance of Y_m will be $m \cdot (1) = m$. Let us normalize Y_m to have variance 1 instead of variance m. The resulting random variable is

$$Z_m = Y_m/m^{1/2}$$

whose characteristic function is

$$G_{Z_m}(f) = E\{\exp[-j2\pi(Y_m/m^{1/2})f]\} = G_{Y_m}(f/m^{1/2}) = [\cos(2\pi f/m^{1/2})]^m$$
$$(8.6\text{-}10)$$

As $m \to \infty$, we can use the series expansion for the cosine up to terms of order f^2:

$$\cos(2\pi f/m^{1/2}) = 1 - \tfrac{1}{2}(2\pi f/m^{1/2})^2 \qquad (8.6\text{-}11)$$

$$G_{Z_m}(f) = [1 - 2\pi^2(f^2/m)]^m \qquad (8.6\text{-}12)$$

As m approaches infinity the expression of (8.6-12) approaches

$$G_1(f) = \exp(-2\pi^2 f^2) \qquad (8.6\text{-}13)$$

Hence from (8.6-7) the distribution of the normalized averages Z_m converges to a gaussian of mean 0 and variance 1 as the number of pulses m becomes very large.

We have derived the central limit theorem only for a random variable which takes on values ± 1 with probability $1/2$ for each. But, we can use characteristic functions to derive the result for *any* random variable with a variance. The derivation is very similar. We start with a random variable X with $E(X) = 0$. We use the series expansion for $\exp(-j2\pi fX)$ out to terms in X^2. The rest of the derivation is very similar to what we have just done. We shall not write this out in general here.

In this chapter we have studied the distribution of a noise or signal voltage at a single instant of time. But, often we are interested in random voltages at two different instants of time, or we are interested in an entire random waveform. To describe these, we need to understand multidimensional probability distributions. We will also find that gaussian distributions arise. But, instead of having only means and variances to deal with, there is a new parameter, the *correlation*.

Problems

1. Use the energy theorem of Section 3.6 to partially check (8.6-7). We are to show that the integral of the square of the original gaussian pulse equals the

integral of the square of its characteristic function. Try to do this without evaluating either integral.

2. What happens to the characteristic functions $G_{Y_m}(f)$ in (8.6-9) if the original pulse areas were 0 and 2 instead of -1 and $+1$?

3. Find the Fourier transform of $\exp(-\pi v^2)$. [This is the best way of remembering (8.6-7).]

4. Derive the central limit theorem for pulses whose areas have expected value 0 and finite variance σ^2. You may find it useful to use

$$\exp(-j2\pi Xf) \doteq 1 - j2\pi Xf - 2\pi^2 X^2 f^2$$

when Xf is small.

5. In telephone cables containing many twisted pairs, repeaters are placed every mile or so. This means that there may be 20 or more repeaters in a given cable run between two regional toll offices of a major metropolitan area. At each repeater, the relative positions of the different pairs within the cable are randomly interchanged for the cable run to the next repeater. This is done to help combat crosstalk, or inductive pick-up between closely spaced pairs. Why does this combat such crosstalk?

9

Random Processes and Gaussian Signals and Noise

In the previous chapter, we studied noise as a random variable, the noise voltage at a particular instant of time. But, in communication, we are presented with entire *waveforms*. These are random, but they are not described by a single distribution of values. We need a way of describing their randomness that will provide answers to communication questions that depend on the entire waveform. For example, we may wish to know the actual energy in a noise waveform over some period of time. This is a random variable, but it depends on more than just a single value of the waveform.

In this chapter, we will see how to describe such random waveforms. We first study two-dimensional distributions, including in this the important notion of the correlation of two random variables. We then study two- and n-dimensional gaussian distributions. This allows us to consider gaussian random processes, such as band-limited gaussian noise. We will learn what it means for a random process to be stationary or time invariant. As one application of these techniques, we will study the detectability of a spectral line in gaussian noise. We shall also learn how to represent narrowband gaussian noise, and how we can use this representation to double the data rate for a given communication bandwidth.

9.1. Two-Dimensional Random Variables and Their Correlation

We sometimes wish to consider and compare two random variables, such as two rf voltages, which are the samples of a noise waveform at two different times. Or, we may wish to represent one random variable, say an rf voltage, by a pair of random variables, in this case the in-phase and quadrature components with respect to some rf sinusoid or phase reference. Questions arise concerning the correlation or covariance of the two variables, and concerning whether or not the variables are truly independent.

Suppose we have two random variables X and Y and their variances:

$$\sigma_X^2 = E(X^2) - [E(X)]^2 = [E(X - \bar{X})]^2$$
$$\sigma_Y^2 = E(Y^2) - [E(Y)]^2 = [E(Y - \bar{Y})]^2$$

Here, X and Y may represent samples of the same waveform, noise or signal, taken time t apart. Their *covariance* is defined as

$$\text{Cov}(X, Y) = E[(X - \bar{X})(Y - \bar{Y})]$$

This is unaffected by a shift in the means. Using a little algebra, we can derive an alternate expression:

$$\text{Cov}(X, Y) = E(XY) - E(X)E(Y) \qquad (9.1\text{-}1)$$

The second form follows from the fact that $\bar{X}$ and $\bar{Y}$ are constants. We note that the expression for covariance is similar in form to the expression for the variance. It is also related to the definition of autocorrelation and cross-correlation given in Chapter 6. We shall explain this relation in Section 9.4.

We now show that the following simple inequality holds:

$$|\text{Cov}(X, Y)| \le \sigma_X \sigma_Y$$

In order to do this, we must derive Schwartz's inequality:

$$|E(XY)| \le [E(X^2)E(Y^2)]^{1/2} \qquad (9.1\text{-}2)$$

We have already used this in Section 1.3 [equation (1.3-4)] in studying the effective area of antennas.

For any number a, the random variable $(X - aY)^2$ has a nonnegative expectation. Also, $(X - aY)^2$ can be zero only if $X = aY$, when X and Y are proportional random variables. We note that they will not be proportional if they represent samples of ordinary noise waveforms taken at different times. This is because the value of the waveform at one time does not exactly determine its value at any other time.

So we must have

$$E(X - aY)^2 \ge 0 \qquad (9.1\text{-}3)$$

Here equality holds only for the unusual case in which X and Y are proportional. Equation (9.1-3) can be expanded and the expectation of the expanded terms taken. The result is that for any a

$$E(X^2) - 2aE(XY) + a^2 E(Y^2) \ge 0 \qquad (9.1\text{-}4)$$

Let the random variable $Y \neq 0$. This means that Y takes on some nonzero values with positive probability. So, $E(Y^2) \neq 0$. The quadratic equation represented by (9.1-4) will not have two real roots, for then it would be negative for some values of a. If we recall the standard form for the solution of a quadratic equation, we see that we must have

$$[E(XY)]^2 - E(X^2)E(Y^2) \leq 0 \qquad (9.1\text{-}5)$$

This is an alternate form of (9.1-2). When Y is the zero random variable, of course, (9.1-2) holds without any derivation necessary.

We note that equality holds if and only if

$$E(X - aY)^2 = 0 \qquad (9.1\text{-}6)$$

for some a. This means that $X - aY = 0$, and X is proportional to Y. We conclude that Schwartz's inequality is strict unless X and Y are proportional.

In order to obtain the result on covariance that we desire, we simply substitute for X and Y in (9.1-2) the random variables $X - \bar{X}$ and $Y - \bar{Y}$. From the expressions given above for σ_X^2, σ_Y^2, and $\text{Cov}(X, Y)$ we see that this gives

$$|\text{Cov}(X, Y)| \leq \sigma_X \sigma_Y \qquad (9.1\text{-}7)$$

Here equality holds if and only if X and Y are linearly related random variables.

We will define the *correlation* $\rho(X, Y)$ of X and Y as a normalized covariance, assuming σ_X and σ_Y are nonzero:

$$\rho(X, Y) = \frac{\text{Cov}(X, Y)}{\sigma_X \sigma_Y} \qquad (9.1\text{-}8)$$

We see that the correlation is unaffected by a scale change in X and Y. We also note that the numerator, and hence the correlation itself, is unaffected by a shift in the means. By (9.1-7) we have the following result:

$$-1 \leq \rho(X, Y) \leq +1 \qquad (9.1\text{-}9)$$

Equality holds in (9.1-9) with $+1$ if and only if $X = aY + b$ with $a > 0$; the -1 is obtained if $a < 0$. If X and Y represent samples of a nonwhite noise waveform taken very close together, ρ will be close to 1.

If $\rho = 0$, we call the random variables *uncorrelated*. From (9.1-1) we can see that independent random variables are uncorrelated. But, uncorrelated random variables are not necessarily independent. For example, let X be a gaussian random variable of mean 0, such as a sample of a noise waveform at some particular instant of time. Pass X through a rectifier to obtain another random variable $|X|$. In this case, X and $|X|$

will be uncorrelated. For, σ_X and $\sigma_{|X|}$ are nonzero, and ρ is defined. Given any value of x taken on by $|X|$, the values $+x$ and $-x$ are equally likely to have occurred. Hence

$$E(X \cdot |X|) = \tfrac{1}{2}E(X^2) + \tfrac{1}{2}E(-X^2)$$
$$E(X \cdot |X|) = 0 = E(X)E(|X|)$$

We see that because $E(X) = 0$, the covariance of X and $|X|$ will be zero. This means that X and $|X|$ are uncorrelated. But, the value taken on by X determines the value taken on by $|X|$. This is almost the opposite of independence.

A measure of independence more powerful than correlation will be introduced in Chapter 12 when we study the mutual information of two random variables.

The next section studies the characteristic function of a pair of random variables which may be correlated. We will see that although uncorrelated random variables need not be independent, they become so if we average enough of them.

Problems

1. X and Y are random variables each taking values ± 1 with probability $1/2$. With what probabilities should the values $(+1, +1)$, $(+1, -1)$, $(-1, +1)$, and $(-1, -1)$ be taken in order that the correlation between X and Y be $+\tfrac{1}{2}$? The same for correlation $-\tfrac{1}{2}$.

2. Show that if two random variables X and Y taking only two values each are uncorrelated, then they are independent. HINT: we can assume that the values X and Y take are 0 and 1. Why? Use the characterization of independence that says that the joint probabilities are the products of the separate probabilities.

3. Find a way of generating many examples of two random variables which are uncorrelated but not independent. HINT: starting with any random variables X, Y with Y not constant, observe that Y and $X - bY$ are uncorrelated for some unique b, and find that b. What happens when $Y = X^3$?

4. Let $X = X(t_0)$ and $Y = X(t_1)$, where $X(t)$ is a noise waveform. Why would you expect $\rho(X, Y)$ to be close to $+1$ if t_1 is close to t_0? Assume that the power spectrum of $X(t)$ does not have much energy at high frequencies.

9.2. Characteristic Functions of Pairs of Random Variables

Just as in the one-dimensional case, we can introduce the characteristic function of a pair (X, Y) of random variables. It is a function of *two* frequencies f and g:

$$G_{(X,Y)}(f, g) = E(\exp[-j2\pi(Xf + Yg)]) \qquad (9.2\text{-}1)$$

If the pair of random variables has a density, this is just the two-dimensional Fourier transform of the density, which we studied in Section 3.10. If X and Y are independent, this is the product of the two one-dimensional characteristic functions of X and Y. Conversely, if $G_{(X,Y)}$ factors into a product of a function of f and a function of g, then X and Y are independent. This is because the two-dimensional characteristic function uniquely determines the two-dimensional distribution, just as the one-dimensional characteristic function determines the one-dimensional distribution. The same is true for the n-dimensional characteristic function of n random variables.

We can filter or sum *pairs* of random variables (X, Y) component-wise, just as we did for one-dimensional random variables in Section 8.6. The two-dimensional characteristic function of the sum of two independent pairs of random variables is the product of the two two-dimensional characteristic functions. Consider m independent pairs of random variables $(X_1, Y_1), \ldots, (X_m, Y_m)$, each with the same distribution as (X, Y). If we sum them as two-dimensional vectors, the resulting characteristic function $G_m(f, g)$ will be

$$G_m(f, g) = [G_1(f, g)]^m = \{E(\exp[-j2\pi(fX + gY)])\}^m \quad (9.2\text{-}2)$$

Here $G_1(f, g)$ is $G_{(X,Y)}(f, g)$, the common characteristic function of any of the pairs (X_i, Y_i). X_i and Y_i need not be independent of each other. But, the different *pairs* (X_i, Y_i) and (X_j, Y_j) are independent of each other, even though Y_i may depend on X_i.

Suppose the random variables X and Y have mean 0. Let $H_m(f, g)$ be the characteristic function of the vector sum of the m pairs (X_i, Y_i), normalized by dividing by $m^{1/2}$. This division is to keep the variances or powers of each component constant, and we can see that the covariance and correlation are preserved as well. The expression is

$$H_m(f, g) = G_m\left(\frac{f}{m^{1/2}}, \frac{g}{m^{1/2}}\right) = \left\{E\left(\exp\left[-j2\pi\left(\frac{fX}{m^{1/2}} + \frac{gY}{m^{1/2}}\right)\right]\right)\right\}^m$$

$$(9.2\text{-}3)$$

As in Section 8.6, we can let $m \to \infty$, and disregard terms like $m^{-3/2}$ or smaller. We use the approximate expansion

$$\exp\left[-j2\pi\left(\frac{fX}{m^{1/2}} + \frac{gY}{m^{1/2}}\right)\right] = 1 - j2\pi\left(\frac{fX}{m^{1/2}} + \frac{gY}{m^{1/2}}\right)$$

$$- 2\pi^2\left(\frac{fX}{m^{1/2}} + \frac{gY}{m^{1/2}}\right)^2 \quad (9.2\text{-}4)$$

Because $E(X)$ and $E(Y)$ are equal to zero, (9.2-3) can be written in the following form:

$$E\left(\exp\left[-j2\pi\left(\frac{fX}{m^{1/2}} + \frac{gY}{m^{1/2}}\right)\right]\right) = 1 - \frac{2\pi^2}{m}(f^2\sigma_X^2 + 2fg\rho\sigma_X\sigma_Y + g^2\sigma_Y^2)$$

$$(9.2\text{-}5)$$

Here $\rho = \rho(X, Y)$ is the correlation of X and Y, which we defined in Section 9.1.

Now let X and Y be uncorrelated, i.e., $\rho(X, Y) = 0$. We have already observed that the summed random variables must remain uncorrelated. But equation (9.2-5) now implies

$$H_m(f, g) = \left(1 - \frac{2\pi^2}{m}(f^2\sigma_X^2 + g^2\sigma_Y^2)\right)^m \qquad (9.2\text{-}6)$$

As m approaches infinity, this approaches

$$H_\infty(f, g) = \exp\left[-2\pi^2(f^2\sigma_X^2 + g^2\sigma_Y^2)\right] \qquad (9.2\text{-}7)$$

We can recognize the form of (9.2-7) as the product of the one-dimensional characteristic functions of two gaussian distributions. These are the limiting distributions of the normalized sums of X_i and Y_i. Hence, in the limit the normalized sums of X_i and Y_i are independent.

We see that although uncorrelated random variables need not be independent, if we add up enough of them and normalize so that the variances of the sums are unchanged, then the sums will be nearly independent. This is true for any finite number of random variables—if they are pairwise uncorrelated, the normalized averages approach a family of independent random variables. These random variables will each be gaussian, because they are normalized averages of independent random variables as in Section 9.1.

We shall exploit this fact in the next section to show how the general two-dimensional gaussian distribution arises in communication. We first show that uncorrelated joint gaussians are independent, and then show that a two-dimensional gaussian distribution is entirely determined by its correlation (if the means and variances are also specified). We must first define the two-dimensional gaussian distribution.

Problems

1. Find the characteristic function of the pair of random variables with the following joint distribution:

$$p(1, 1) = p(-1, -1) = \tfrac{3}{8}$$
$$p(-1, 1) = p(1, -1) = \tfrac{1}{8}$$

2. Find the correlation of the above random variables both directly and by

using the characteristic function. Check that the answers are the same. HINT: first show that always

$$E(XY) = \frac{-1}{(2\pi)^2} \frac{\partial^2 G}{\partial f \, \partial g}\bigg)_{f,g=0}$$

where G is the characteristic function.

9.3. The Two-Dimensional Gaussian Distribution

The derivation in the last section applies to any pair of random variables. Here we consider the properties of a special kind of pair of random variables, each of which is gaussian. We saw why noise was gaussian in Section 9.1. Here we will learn how noise measurements made at two instants of time are distributed.

If X and Y are independent gaussians, then we say that the two-dimensional random variable (X, Y) has a *two-dimensional independent gaussian distribution*. More generally, if X and Y are independent gaussians, and a, b, c, d are any constants, the two-dimensional random variable generated by the two-port linear device D of Fig. 9.1,

$$(U, V) = (aX + bY, cX + dY) \qquad (9.3\text{-}1)$$

is said to have a general *two-dimensional gaussian distribution* or joint

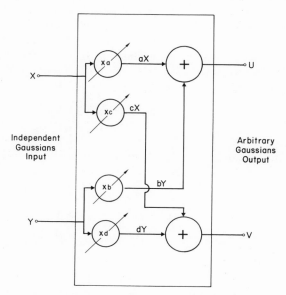

Figure 9.1. Generation of two-dimensional gaussians by memoryless linear device D.

gaussian distribution. Such distributions arise in studying error prob-abilities and in many other ways. Here we note that U and V can each be gaussian separately, and yet not be joint gaussians.

Knowing that (U, V) is of the form (9.3-1) (and knowing the means and variances of X and Y) completely determines the joint distribution of (U, V). We can in principle then answer any probability question about U and V together. For example, if U and V are the inputs to a multiplier, we can, with difficult calculation, find the probability that the multiplier output UV exceeds a given threshold.

We will derive the following important fact: gaussian distributions are completely determined by their means, their variances, and by their *correlation*. This at first seems surprising. For, we shall derive in (9.3-4) a formula for the correlation of U and V which implies that a given correlation value $\rho(U, V)$, and given means and variances, can be ob-tained with infinitely many different linear devices [coefficients a, b, c, d in (9.3-1)].

For example, suppose that the correlation desired between U and V is to be zero. The eight linear devices

$$(U, V) = (\pm X, \pm Y) \quad \text{and} \quad (\pm Y, \pm X)$$

obviously produce uncorrelated (because independent) (U, V). But, there are other devices which do so. These others will have crosstalk between X and Y inputs, that is, U will depend on both X and Y, and V will depend on both X and Y. In spite of this crosstalk, we shall show that all uncorrelated gaussian random variables are independent. We will use this to show that a two-dimensional gaussian distribution with given means and variances is completely determined by its correlation.

What are the means, variances, and correlation of the random variables (U, V)? The means add linearly, as we expect:

$$\begin{aligned} E(U) &= aE(X) + bE(Y) \\ E(V) &= cE(X) + dE(Y) \end{aligned} \tag{9.3-2}$$

To compute the variances, we can assume that X and Y, and so U and V, have zero means. We calculate

$$\begin{aligned} \sigma_U^2 &= E(U^2) = E(a^2X^2 + 2abXY + b^2Y^2) \\ \sigma_V^2 &= E(V^2) = E(c^2X^2 + 2cdXY + d^2Y^2) \end{aligned}$$

Since X and Y are independent and zero mean, $E(XY) = \bar{X}\bar{Y} = 0$, and the required formulas are

$$\begin{aligned} \sigma_U^2 &= a^2\sigma_X^2 + b^2\sigma_Y^2 \\ \sigma_V^2 &= c^2\sigma_X^2 + d^2\sigma_Y^2 \end{aligned} \tag{9.3-3}$$

These equations describe the output powers of a two-port linear device (or network) without memory, when the inputs are independent or uncorrelated noises of zero mean.

We can find the covariance of U and V:

$$\text{Cov}(U, V) = E[(aX + bY)(cX + dY)]$$
$$\text{Cov}(U, V) = ac\sigma_X^2 + bd\sigma_Y^2 \qquad (9.3\text{-}4)$$

We easily see from Equations (9.3-3) and (9.3-4) that the same variances and covariances can be obtained with infinitely many different a, b, c, d.

The correlation between U and V [see (9.1-8)] is

$$\rho(U, V) = \frac{ac\sigma_X^2 + bd\sigma_Y^2}{[(a^2\sigma_X^2 + b^2\sigma_Y^2)(c^2\sigma_X^2 + d^2\sigma_Y^2)]^{1/2}} \qquad (9.3\text{-}5)$$

The condition that U and V be uncorrelated is

$$ac\sigma_X^2 + bd\sigma_Y^2 = 0 \qquad (9.3\text{-}6)$$

We note that (9.3-3) through (9.3-6) depended only on the fact that X and Y are uncorrelated, not on the fact that they are gaussian or independent.

We will now show that uncorrelated joint gaussians are independent. Let (U, V) be given by (9.3-1), with a, b, c, d chosen so that $\rho(U, V)$ as given by (9.3-5) is zero. Figure 9.2 shows two block diagrams which have independent gaussians (X, Y) as their input and the same uncorrelated joint gaussians (U', V') as their output. In each we take a normalized average, as in Section 9.2. In one case we take normalized averages of X and Y; in the other case we take normalized averages of U and V. We can average over many inputs X and Y to give two gaussian variables X' and Y', with the same distributions as X and Y. We can then input these

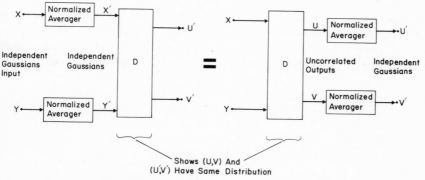

Figure 9.2. Independence of uncorrelated gaussians (D is the device of Fig. 9.1 with $ac\sigma_X^2 + bd\sigma_Y^2 = 0$).

to the linear device D. Or, we can average over U and V to give two gaussian variables U' and V' for the final output. The two block diagrams perform the identical operation, because D is a linear device. We note here that (U, V) has the same distribution as (U', V'). This is because (U, V) and (U', V') are the outputs of the same linear device D with identically distributed inputs. The input distributions are identical because the pairs (X, Y) and (X', Y') are independent gaussians with the same means and variances.

If we average *before* the device D we input two independent gaussian variables (X', Y') and get out two uncorrelated gaussian variables (U', V'). But, if we average *after* the linear device, Section 9.2 tells us that (U', V') must be *independent*, not merely uncorrelated. This holds because they are normalized averages of uncorrelated random variables. We conclude that U' and V' are independent. And, they have the same joint distribution as U and V, the original pair of uncorrelated gaussians. So, U and V are independent as well. Uncorrelated joint gaussian random variables are independent.

The correlation of gaussian random variables of given means and variances completely determines the distribution, at least if the correlation is zero. What about other values of the correlation? The correlation of a gaussian pair of random variables (U, V) determines the distribution in general. Let us show this. We run the pair of correlated gaussians through a peculiar device we can call a *decorrelator*. This is a memoryless two-port linear device that produces uncorrelated outputs (X, Y) as in Fig. 9.3. Linear devices, by the very definition of the joint gaussian property, must preserve the joint gaussian property. So, the outputs are not merely uncorrelated but independent.

We can run these independent gaussians (X, Y) through the *inverse* linear device, to obtain the original gaussian random variables (U, V). The linear device, and its inverse, depend only on the original correlation and variances. So, every gaussian pair with a given correlation and given variances can be obtained as the output of the same linear device, the inverse decorrelator, with independent gaussian inputs of the same means and variances. But, this linear device completely determines the output

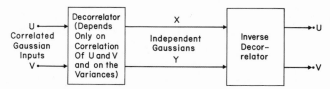

Figure 9.3. Uniqueness of the correlated two-dimensional gaussian distribution.

distribution. We conclude that all we need to know about joint gaussians to answer any probability questions are their means, variances, and correlation. The same result is true for any number of jointly gaussian random variables.

We now have some familiarity with two- and even n-dimensional distributions, gaussian or not. In the next section, we will see how to put these multidimensional distributions together to obtain a statistical description of a random waveform. We need to do this before we can rigorously describe noises or signals as random phenomena.

Problems

1. Show that if U and V are gaussian of mean 0, variances σ_U^2 and σ_V^2, and correlation ρ, their joint density function is

$$P(u, v) \, du \, dv = \frac{1}{2\pi(1 - \rho^2)^{1/2}\sigma_U\sigma_V}$$
$$\times \left\{\exp\left[-\frac{1}{2(1 - \rho^2)}\left(\frac{u^2}{\sigma_U^2} - 2\rho\frac{uv}{\sigma_U\sigma_V} + \frac{v^2}{\sigma_V^2}\right)\right]\right\} du \, dv$$

   (The *Jacobian* tells us what to multiply the element of area $dx\,dy$ by to get the new element of area $du\,dv$ where u and v are functions of x and y. Use the fact that the Jacobian of a linear transformation is the determinant.) How should the formula be interpreted if $\rho = \pm 1$? Conclude that uncorrelated gaussians are independent, and that the joint distribution of a gaussian pair of random variables (U, V) is completely determined by $\bar{U}$, $\bar{V}$, σ_U^2, σ_V^2, and ρ. This formula is not very useful for calculations. HINT: assume $\sigma_X^2 = \sigma_Y^2 = 1$ in (9.3-1).

2. Let X and Y be independent gaussians. Design infinitely many different linear devices
$$U = aX + bY$$
$$V = cX + dY$$
   which produce an output (U, V) with U and V independent and of variance 1. You may assume that X and Y themselves each have mean 0 and variance 1.

3. Show that if U and V are correlated random variables of positive variances and with correlation not ± 1, then U and V can be passed through a memoryless linear device (which depends only on the variances and correlation) to produce random variables X and Y of positive variances which are uncorrelated. Show that this linear device is invertible by finding its inverse.

4. Find the characteristic function of a two-dimensional gaussian of means 0, variances σ_X^2 and σ_Y^2, and correlation ρ. Use this to show that normalized averages of any two-dimensional distribution (with variances) approach a two-dimensional gaussian of the same variances and correlation. HINT: rederive (9.2-7) without assuming $\rho = 0$, and use the fact that normalized averages of a gaussian distribution form a gaussian of the same variances and correlation.

9.4. Stationary Random Processes and Their Autocorrelation Functions

In the preceding section, we learned about two- and n-dimensional random *variables*. Here we use what we have learned to study random waveforms or *processes*. These can be thought of as noise. But, even signals can be random.

Random processes are waveforms $X(t)$ whose entire time history is probabilistic. We can think of $x(t)$ as the "value" taken on by $X(t)$, or as a single function chosen from the ensemble $X(t)$ of possible functions. We can also define expected values over the ensemble of functions $X(t)$.

In Section 2.5, where we studied noise as a signal, we represented a random waveform $V = V(t)$ corresponding to band-limited white noise as a sum of sinusoids of random amplitude and phase, with equal expected power at every frequency. By replacing the sum in (2.5-1) by a suitable kind of integral, the approach of Section 2.5 can be made fully rigorous. We shall not pursue such an approach further in this book. A different but related way of defining random waveforms which is suggested by the concept of signal space in Section 11.1 is to specify the coefficients of the random function when expanded in a suitable set of orthonormal functions. These coefficients are analogous to Fourier coefficients and are random variables themselves. A set of coefficient values then corresponds to a random infinite series, or to an infinite series with random coefficients, and thus defines a random function or waveform. All approaches to defining random functions are equivalent in the mathematical sense, but some are more convenient for some purposes and some for other purposes.

Instead of choosing an entire $X(t)$ waveform at once, we may be given the n-dimensional joint distributions of the values $[X(t_1), X(t_2), \ldots, X(t_n)]$ at n fixed instants of time. The random process can also be defined by these joint distributions, for all n and all times $t_1, t_2, \ldots, t_n$. A *gaussian* random process is one for which all these *joint* distributions are gaussian. For band-limited processes, we can restrict the t_i to be sampling instants. We shall make use of this in Section 11.1, where we study signal space.

The distribution of a gaussian random process $X(t)$ is completely determined by its *autocorrelation function*

$$C(s, t) = E(X(s)X(t))$$

(Here we have assumed $X(t)$ has mean 0 for all t.) We will relate this below to the definition of autocorrelation given in Section 6.1.

Knowing only the autocorrelation $C(s, t)$ and the fact that we have a gaussian process enables us in principle to answer any probability questions about the waveform $X(t)$. This is because an n-dimensional gaussian distribution is completely determined by the correlation values. There are $n(n + 1)/2$ different ones of these, including the variances.

For example, the autocorrelation $C(s, t)$ of a gaussian process determines the probability that the energy in $X(t)$ over 1 second is at least 1 joule. This can be written as

$$P\left(\int_0^1 X^2(t)\, dt\right) \geq 1$$

Such questions occur in studying fluctuations in measured noise power and in finding error probabilities in communication systems with non-linear detection. If the noise is not gaussian, these problems become much more difficult, because we need to know more about the noise than the autocorrelation and power alone.

In Section 6.5 we defined a stationary signal as one from which one cannot tell absolute time. A stationary random process could be defined as one which produces waveforms which are stationary in this broad sense. It is customary to require that stationary random processes produce waveforms from which not even *phase* or relative time information can be extracted. This is the same as requiring that the joint probability distributions at times $t_1, t_2, \ldots, t_n$ depend only on the relative differences $t_2 - t_1, t_3 - t_1, \ldots, t_n - t_1$.

There is no universally accepted name for random processes which produce waveforms that are stationary in the broader sense of Section 6.5. But, such processes arise in communication. For example, a sequence of pulses at a fixed interval of repetition with peak amplitudes randomly ± 1 is stationary in this broader sense. It is not stationary in the narrower sense, because the amplitudes *between* the pulse peaks or sampling instants will be distributed differently from the amplitudes *at* the sampling instants. If the pulses exhibit no intersymbol interference, the possible amplitudes at the sampling instants are only ± 1. At other instants, many more values are possible.

If $X(t)$ is a stationary noise or signal (in the narrower sense), its autocorrelation $C(s, t)$ depends only on the differences $\tau = s - t$, because the joint distribution of $X(s)$ and $X(t)$ depends only on $s - t$. And, for a gaussian process of zero mean, the opposite is true: if

$$C(s, t) = C(s - t) = C(\tau)$$

then $X(t)$ is stationary. This is because the distribution of n jointly gaussian random variables of zero mean depends only on the covariances $E(X(t_i)X(t_j))$. And, the covariances are $C(t_i - t_j)$.

For stationary noise, the power spectral density $P(f)$ (see Section 6.6) is defined as the Fourier transform of the autocorrelation:

$$P(f) = \int_{-\infty}^{\infty} C(\tau) e^{-j2\pi f\tau}\, d\tau \tag{9.4-1}$$

This gives the density of the power in the random signal $X(t)$ at frequency f. In Section 6.6, we defined the autocorrelation as

$$C(\tau) = \lim_{T \to \infty} \frac{1}{T} C_T(\tau) \qquad (9.4\text{-}2)$$

Here $C_T(\tau)$ is the autocorrelation of the actual noise waveform $X(t)$ truncated to be 0 if $|t| > T/2$. $C(\tau)$ seems to be a random variable, because it may depend on the actual $x(t)$ observed. Is it really random?

There is a property of stationary random processes called the *ergodic property* or *ergodicity*. The definition is that the expectation of any function of the values of the random waveform at a number of instants of time can be computed as a time average, from any *particular* waveform that happens to occur. This means that if

$$F\big(X(t_1), X(t_2), \ldots, X(t_n)\big)$$

is such a function, its expected value over all functions in the ensemble can be computed from a single observed $x(t)$. The formula is

$$E\big(F(X(t_1), X(t_2), \ldots, X(t_n))\big)$$
$$= \lim_{T \to \infty} \frac{1}{T} \int_{-T/2}^{T/2} F\big(x(t_1 + \tau), \ldots, x(t_n + \tau)\big)\, d\tau \qquad (9.4\text{-}3)$$

We say that (9.4-3) expresses the fact that space averages (expectations over the joint or ensemble distribution of the $X(t_i)$) are equal to time averages for a particular function $x(t)$ in the ensemble.

Nonergodic processes have a kind of memory. If they begin in one state, they forever exhibit behavior determined by that state. Ergodic processes, on the other hand, go through all possible states in the same predictable proportion no matter how they begin.

There are stationary processes which are not ergodic. Not even a stationary gaussian process need be ergodic. In spite of this, every stationary gaussian process whose power spectrum has no delta function portion, and which is likely to be met in practice, will be ergodic. But, there is no simple condition that is exactly equivalent to ergodicity for stationary gaussian processes.

For a stationary ergodic process, the mean or time average

$$\frac{1}{T} \int_{-T/2}^{T/2} x(t)\, dt$$

approaches the ensemble mean value $E(X(t))$. Since the process is stationary, this mean is independent of time t. For this reason, we sometimes say that a random process with $E(X(t)) = 0$ has no dc. Or, we may subtract $E(X(t))$ from $x(t)$ and say we have subtracted off the dc, even when $X(t)$ is not ergodic.

For an ergodic process, we see from (9.4-3) that the limit or time average in (9.4-2), which seems to depend on the $x(t)$ that actually occurs, converges to the space average

$$C(\tau) = E(X(t)X(t + \tau)) \tag{9.4-4}$$

The limit is the same for every observed waveform $x(t)$. So, $C(\tau)$ is not a random quantity for ergodic processes. The two definitions of autocorrelation, the one in Section 6.6 and the one here, coincide for stationary ergodic noises and signals.

Gaussian noise is easy to obtain by taking averages of a sufficient number of independent random processes. This follows from Sections 9.2 and 9.3. For example, impulse noise and crosstalk in telephone circuits produce gaussian noise if there are enough noise sources adding together and no one noise source is dominant. And, gaussian noise passed through *any* linear device, even one with memory, still remains gaussian noise. In general, the filtered noise will have a different autocorrelation, and a different spectrum (if stationary).

It is also worth noting that stationary noise of any power spectral density can be obtained by passing stationary white noise through a time-invariant filter whose transfer function has magnitude squared equal to the desired spectral density. This follows from the results of Section 3.7 on linear networks and of Section 6.6 on the autocorrelation of a stationary signal. If the resulting output noises are all gaussian no matter what the linear device or filter may be, we call the input noise white *gaussian* noise. We have met white gaussian noise already as thermal noise and as the non-dc component of shot hoise. Because white noise has infinite average power, it does not have sample waveforms that are finite in any sense. We must filter the white noise to obtain a random noise with a finite waveform. We shall not pursue this here.

In the next section, we will study the accuracy with which we can measure the spectrum of gaussian noise which is not white. We shall look for a spectral line in a flat background.

Problems

1. A random noise which is broad-sense stationary but not narrow-sense stationary has relative time or phase information in its observed waveforms. We would still like the time average $C(\tau)$ as given by (9.4-2) to be equal to something which involves the space average $C(\tau)$ as given by (9.4-4). What would you guess is the appropriate formula?

2. There is a stationary gaussian process $X(t)$ with spectrum $P(f) = \delta(f)$. Describe it in another way. Is the sample autocorrelation function of (9.4-2) equal to the $C(\tau)$ obtained by inverse Fourier transforming (9.4-1)? What has gone wrong?

3. Noise in a rural telephone system can be caused by power lines lying near the telephone circuits. This noise varies from one circuit to the next, depending on how the circuits were laid out. Considering the layout as random as well as the electrical interference, we have a random noise which is stationary. Is it ergodic? What about the electrical noise between two specific telephones on different coasts of the U.S. connected by the national telephone network for frequent long-distance calls?

9.5. Detection of Spectral Differences

In Section 2.5 we studied noise as a signal, without the advantage of our study of probability. The noise had a flat power spectrum, and so $P(f)$ was independent of frequency. Many noises which we regard as signal are not flat. Spectral lines or salients or regions of enhanced spectrum can be caused by nonthermal radiation due to molecules in interstellar space. These are called molecular lines. In two-dimensional range-doppler radar mapping (see Section 12.8), spectral peaks can be caused by enhanced reflectivity of the target at locations corresponding to various doppler shifts. The Cyclops report (Reference 3 of Chapter 2), which we mentioned in Section 2.4 where we studied the detection of faint sources, is concerned with the efficient detection of spectral lines at un-known locations in frequency. Spectrum analyzers searching for radio frequency interference or RFI have the same problem. And, in frequency-shift keying for data transmission, we also use a form of spectrum analyzer as the detector. Detecting peaks in spectra is impor-tant. Section 10.2 of Reference 1 studies in detail the application to frequency-shift keying for digital communication.

Here we shall consider a signal whose spectrum is flat or white and of known amplitude except for a very narrow spectral line at unknown frequency within a given frequency range (see Fig. 9.4). The flat part is

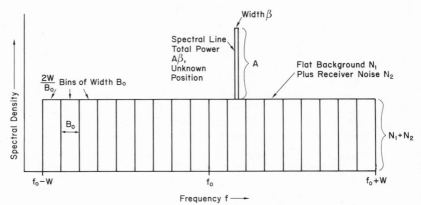

Figure 9.4. Narrow spectral line in background noise.

called the *background*. We seek to locate the line within a frequency bin of width B_0. B_0 can be much greater than the line width, because of the difficulty of building very high resolution spectrum analyzers. We shall assume that this is the case.

The spectrum analyzer is said to *locate* the line if it finds a bin such that there is a low probability of the bin not being the correct one. For simplicity of the analysis, we will use as our estimate of the probability of not being correct the probability of a single-comparison error. This will occur if the power measured in the correct bin is less than the power measured in a fixed incorrect bin. We will find the probability of a single-comparison error and use that as if it were the true error probability. The actual error probability of course will be greater than this and will depend on the number of bins. But, unless the number of bins is extremely large, the conclusions will not depend very much on the number of bins.

In finding error probability, we shall also make the simplifying assumption that the amplitude of the spectral line is known. This may not hold in practice in many applications, but it permits us to analyze the situation more easily.

We will assume that the search is conducted between frequencies $f_0 - W$ and $f_0 + W$, where $f_0 > W$. Here W is a multiple of B_0, and there are $2W/B_0$ frequency bins. The line itself has a flat power spectrum of amplitude A and width β. Here β is assumed much less than B_0, so that we can assume that the spectral line does not affect the power detected in more than one bin. And, the observing bandwidth B is assumed much smaller that the bandwidth β of the spectral line. In a reasonable system, B_0 might be 1 kHz, the line width β a few Hz, and B a few hundredths of a Hz.

Let the flat background have one-sided spectral density N_1. Let the receiver noise density be N_2, so that the total observed spectrum has N_2 added to it at all frequencies. This happens because independent noise powers add, and so spectral densities add as well. The *average* or expected power in the correct bin is the power $A\beta$ in the bandwidth β of the line, plus the noise power $(N_1 + N_2)B_0$ due to the background and to the receiver, which occurs over the entire bandwidth B_0 of the bin. The result is

$$P_0 = A\beta + (N_1 + N_2)B_0 \qquad (9.5\text{-}1)$$

We shall assume that the line power $A\beta$ is much smaller than the second or noise term $(N_1 + N_2)B_0$ above. The result of this is that the mean-square power fluctuation in the correct bin is due mostly to the variance or fluctuation of the second term in (9.5-1). This is not essential, but it simplifies the calculations. Because we have assumed that the signal power $A\beta$ is small compared with the noise power, this is also the fluctuation for an incorrect bin. Because

$$B \ll B_0$$

then by (2.5-13), this fluctuation P_b^2 is

$$P_b^2 = (N_1 + N_2)^2 2BB_0 \qquad (9.5\text{-}2)$$

Here $B = 1/2T$ is half the reciprocal of the observing or integration time T. We note that the method of derivation in Section 2.5 did not use probability. We could now rederive the results there more rigorously. However, the previous answers are correct, and we shall accept them.

For any one incorrect bin, we have from (2.5-14) and (9.5-2)

$$P_0' = (N_1 + N_2)B_0 = P_n \qquad (9.5\text{-}3)$$

$$(P_b')^2 = (N_1 + N_2)^2 2BB_0 \qquad (9.5\text{-}4)$$

Here P_0' is the noise power P_n, or the average power of the incorrect bin, and $(P_b')^2$ its mean-square fluctuation.

How do we find the probability of mistaking the correct bin for a given incorrect one in a one-on-one comparison? Here the averaging time $T = 1/2B$ is large compared to the reciprocal bandwidth or "decorrelation time" of the noise and of the signal of spectral width β, because $B \ll \beta \ll B_0$. So, the central limit theorem of Section 8.6 implies that the power outputs of the correct and incorrect bins are joint gaussian random variables. They are independent, because the noises in disjoint spectral regions are independent. We shall formally derive the fact that the noises must be uncorrelated in Section 11.1. Because they are gaussian, they will be independent. We note that we have been using this property often but implicitly, ever since we defined the power spectral density: output signals from different bins of a spectrum analyzer are uncorrelated random variables.

The mean P_0 of the correct output is given by (9.5-1). The mean P_0' of the incorrect output is given by (9.5-3). The mean of the difference of outputs, correct minus incorrect, is given by

$$\mu = A\beta \qquad (9.5\text{-}5)$$

Because the random variable with this mean is a sum of two independent gaussians, it will be gaussian. The variance σ^2 of the difference is the sum of the variances, because the random variables are independent. Using (9.5-2) and (9.5-4), this becomes

$$\sigma^2 = 4BB_0(N_1 + N_2)^2 \qquad (9.5\text{-}6)$$

We will make a single-comparison error if the gaussian random variable of mean (9.5-5) and variance (9.5-6) is less than 0. By subtracting the mean and scaling by the square root of the variance, we see that this event has the same probability as the probability p that a gaussian random variable of mean 0 and variance 1 be less than $-L$, where L is easily found as

$$L = A\beta/2(N_1 + N_2)(BB_0)^{1/2} \tag{9.5-7}$$

From Section 8.5

$$p = \frac{1}{(2\pi)^{1/2}} \int_{-\infty}^{-L} \exp\left(\frac{-v^2}{2}\right) dv$$

$$p = \frac{1}{(2\pi)^{1/2}} \int_{L}^{\infty} \exp\left(\frac{-v^2}{2}\right) dv = 1 - \Phi(L) \tag{9.5-8}$$

Here Φ is the standard gaussian distribution of mean 0 and variance 1.

We see that p and L determine each other. Small p corresponds to large L, to small B, or to long observing time T. This of course must be. When $L = 3$ (a three sigma error event), $p = 0.0013$ from Table 8.1.

The performance criterion can be defined by giving a required L. Equation (9.5-7) can be used to determine the required observing bandwidth B. The result is

$$B = A^2\beta^2/4(N_1 + N_2)^2 B_0 L^2 \tag{9.5-9}$$

Here $A\beta$ is the *signal* power P_s in the spectral line and $(N_1 + N_2)B_0$ is the *noise* power P_n in a bin of bandwidth B_0. We can write (9.5-9) as

$$B = \left(\frac{P_s}{P_n}\right)^2 \cdot \frac{B_0}{4L^2} \tag{9.5-10}$$

The derivation is valid for weak signals where $P_s \ll P_n$. This is usually the case in spectrum analyzers—receiver or background noise predominates. In using (9.5-10), we should remember that P_n depends on B_0 via (9.5-3).

In the next section, we will study gaussian noise of a narrow spectrum, but there will be no background noise level. The noise is called *narrowband gaussian noise*. It has a narrow spectrum concentrated symmetrically around a carrier.

Problems

1. Suppose that receiver frequency resolution is doubled. That is, the bins are made half as wide. What does this do to the required observation time? (Here the error probability p is held constant, and A, β, and $N_1 + N_2$ are the same in either case.) What does this say about the resolution we should strive for? What might this do to the cost of the instrument? How might this conclusion be affected if we consider the actual error probability and not merely the single-comparison error probability?

2. Suppose the spectral line originates from a point source in the sky, and

$N_1 = 0$. The effective area of the receiving antenna is doubled without changing the receiver noise temperature. What does this do to the line amplitude A? What does this do to the required observation time? (Compare Problem 2 of Section 2.5.) This is a common phenomenon in incoherent detection or power detection or detection without phase reference.

3. A monostatic planetary radar (same antenna transmitting and receiving) is used to map Venus by measuring returns from particular range-doppler cells. Eight hours are required to make a good map, and surface noise or speckle can be ignored. The effective antenna area is tripled. How much time is needed to make a good map now? (This means that we want to keep the probability of detecting a small bright feature or spectral salient in the spectrum the same. Speckle, which we studied in Section 7.7, is becoming important for some earth-based planetary radars because the integration times needed to average out receiver noise are no longer long enough to average out the speckle. The conclusions of this problem are changed with speckle present.)

9.6. Representation of Narrowband Gaussian Noise

In connection with amplitude or phase modulation of a carrier of frequency ω_0, it is convenient to think of noise components $n_1(t)2^{1/2} \cos \omega_0 t$ and $n_2(t)2^{1/2} \sin \omega_0 t$. We will assume for simplicity that the noise power is distributed symmetrically around the frequency $\omega_0 = 2\pi f_0$. This would be true for white noise passed through a band-pass filter whose frequency response was symmetrical about ω_0.

We will show that if the noise is stationary gaussian, the two noise components n_1 and n_2 are stationary gaussian and independent. This means more than that $n_1(t)$ and $n_2(t)$ are independent random variables for every fixed t. Independence of n_1 and n_2 as random functions (or noises or processes) means that knowing the total time history of n_1 forever tells us nothing about the noise waveform n_2, no particular value, nor anything else about it.

The results of the preceding section will be used to derive the independence result. We will rely on the fact that a joint gaussian distribution of given means and variances is determined solely by the correlation. The two baseband noises n_1 and n_2 will have the same spectrum as one sideband of the spectrum of $n(t)$, translated back to dc. Let us derive all these facts.

First consider the circuit of Fig. 9.5. It is a way of decomposing a band-pass waveform $x(t)$. This is one whose energy is confined to the spectral region

$$f_0 - B \leq f \leq f_0 + B \qquad (9.6\text{-}1)$$

Here f_0 is a carrier frequency and the band limit B satisfies

$$B < f_0 \qquad (9.6\text{-}2)$$

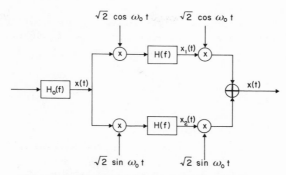

Figure 9.5. Decomposition of band-pass waveform.

The filter $H_0(f)$ is an ideal band-pass filter. This means

$$H_0(f) = \begin{cases} 1, & |f - f_0| \leq B \\ 0, & |f - f_0| > B \end{cases} \tag{9.6-3}$$

The filter $H(f)$ is an ideal low-pass filter:

$$H(f) = \begin{cases} 1, & |f| \leq B \\ 0, & |f| > B \end{cases} \tag{9.6-4}$$

Figure 9.5 decomposes a band-pass signal $x(t)$ into two low-pass signals $x_1(t)$ and $x_2(t)$:

$$x_1(t) = [x(t)2^{1/2} \cos \omega_0 t]_{\text{low passed}}$$
$$x_2(t) = [x(t)2^{1/2} \sin \omega_0 t]_{\text{low passed}} \tag{9.6-5}$$

From these, $x(t)$ is reconstructed as

$$x(t) = x_1(t)2^{1/2} \cos \omega_0 t + x_2(t)2^{1/2} \sin \omega_0 t \tag{9.6-6}$$

We shall explain why this recovers $x(t)$.

Consider the part of Fig. 9.5 after the band-pass filter $H_0(f)$. It appears to be a time-*variant* filter, because $\cos \omega_0 t$ and $\sin \omega_0 t$ are time variant. But, we will show that this part is identical to the time-*invariant* filter $H_0(f)$. This will demonstrate (9.6-6), because if $x(t)$ is already confined to spectral region (9.6-1), passing it through another ideal band-pass filter leaves it unaltered.

We can verify that this filter gives $H_0(f)$ by verifying that it produces the same output as $H_0(f)$ for sinusoidal inputs

$$\cos (2\pi ft + \phi) \tag{9.6-7}$$

Every signal can be represented via the Fourier transform as a linear sum of cosines in arbitrary phases, with coefficients depending on frequency $\omega = 2\pi f$. And, Fig. 9.5 is a linear circuit and so is the circuit with transfer function $H_0(f)$. So, if the two circuits agree on sinusoidal inputs (9.6-7) for all ω and ϕ, they agree for all input waveforms. Verification for inputs of the form (9.6-7) is a routine verification. Both filters give 0 output if

$$|f - f_0| > B \qquad (9.6\text{-}8)$$

and both give (9.6-7) unaltered if

$$|f - f_0| < B \qquad (9.6\text{-}9)$$

We also note that if $x_1(t)$ and $x_2(t)$ are low-pass waveforms, and $x(t)$ is *defined* by (9.6-6), $x(t)$ is band pass. And, $x_1(t)$ and $x_2(t)$ can be recovered from $x(t)$ via (9.6-5). In other words, a band-pass waveform $x(t)$ has only one representation of the form (9.6-6) with $x_1(t)$ and $x_2(t)$ low pass.

Let us apply this to stationary gaussian noise $n(t)$ of mean 0 which is band limited by (9.6-1) and whose spectrum is symmetric about the carrier $\omega_0 = 2\pi f_0$. We have the band-pass decomposition

$$n(t) = n_1(t)2^{1/2}\cos\omega_0 t + n_2(t)2^{1/2}\sin\omega_0 t \qquad (9.6\text{-}10)$$

Here $n_1(t)$ and $n_2(t)$ are low-pass gaussian processes. They are gaussian because they were obtained by the linear (although time-varying) operation (9.6-5) on the gaussian process $n(t)$.

It is somewhat surprising that $n_1(t)$ and $n_2(t)$ are (narrow-sense) stationary noises. This is in spite of the time-varying operation (9.6-5). In fact, the noises are stationary when considered as a *two*-dimensional process. This means that the joint *four*-dimensional distribution of $\big(n_1(t),$ $n_2(t)\big)$ and $\big(n_1(s),\ n_2(s)\big)$ depends only on the time difference $s - t$. Furthermore, n_1 and n_2 are *independent* noises. Finally, n_1 and n_2 each have half the power n has, with spectrum the same as the spectrum of one sideband of n, but translated back to dc.

We shall demonstrate these facts. We create noises $m_1(t)$ and $m_2(t)$ that are independent stationary gaussian processes of the spectral density and power claimed for n_1 and n_2. We will use m_1 and m_2 to generate a random noise process by (9.6-6), which we call $m(t)$:

$$m(t) = m_1(t)2^{1/2}\cos\omega_0 t + m_2(t)2^{1/2}\sin\omega_0 t \qquad (9.6\text{-}11)$$

We see that $m(t)$ is a gaussian process. We will show that it is the same process (has the same statistical properties) as the original $n(t)$. To show this, we will show that it has the same autocorrelation function as $n(t)$, and then use the results of Section 9.4. Since we have observed that n_1 and n_2 are *uniquely* determined via (9.6-10), the n_1 and n_2 that are derived from $n(t)$ have the same statistical properties, including independence, spectral shape, and power, as m_1 and m_2. This is what we wanted to demonstrate.

Let $C(\tau)$ be the common autocorrelation of $m_1(t)$ and $m_2(t)$:

$$C(\tau) = E(m_1(t)m_1(t + \tau)) = E(m_2(t)m_2(t + \tau)) \qquad (9.6\text{-}12)$$

What is the autocorrelation function $C_0(s, t)$ of $m(t)$? Here we write $C_0(s, t)$, because we do not yet know that $m(t)$ is stationary. The autocorrelation is

$$C_0(s, t) = E(m(s)m(t))$$
$$= E((m_1(s)2^{1/2} \cos \omega_0 s + m_2(s)2^{1/2} \sin \omega_0 s)$$
$$\times (m_1(t)2^{1/2} \cos \omega_0 t + m_2(t)2^{1/2} \sin \omega_0 t)) \quad (9.6\text{-}13)$$

By independence of m_1 and m_2, the cross terms in (9.6-13) are 0. We are left with

$$C_0(s, t) = 2 \cos \omega_0 s \cos \omega_0 t E(m_1(s)m_1(t))$$
$$+ 2 \sin \omega_0 s \sin \omega_0 t E(m_2(s)m_2(t))$$
$$= C(s - t)(2 \cos \omega_0 s \cos \omega_0 t + 2 \sin \omega_0 s \sin \omega_0 t)$$
$$C_0(s, t) = 2 \cos \omega_0 (s - t)C(s - t)$$
$$(9.6\text{-}14)$$

Here $C(\tau)$ is the autocorrelation of m_1 or m_2.

So $C_0(s, t)$ depends only on $\tau = s - t$. Because $m(t)$ is gaussian and has an autocorrelation that depends on τ only, it is stationary. Its autocorrelation $C_0(\tau)$ is

$$C_0(\tau) = 2 \cos \omega_0 \tau C(\tau) \qquad (9.6\text{-}15)$$

When $\tau = 0$, (9.6-15) says that the power $C_0(0)$ in m is twice the power in either m_1 or m_2. Each component m_1 and m_2 carries half the power in m, as must be.

What does (9.6-15) imply for the spectrum $P_0(\omega)$ of $m(t)$ in terms of the spectrum $P(\omega)$ of $m_1(t)$ or $m_2(t)$? We can use the Fourier cosine transform here, because $C_0(\tau)$ is symmetrical in τ:

$$P_0(\omega) = \int C_0(\tau) \cos \omega\tau \, d\tau$$

$$= \int C(\tau)2 \cos \omega_0\tau \cos \omega\tau \, d\tau$$

$$= \int C(\tau)[\cos (\omega - \omega_0)\tau + \cos (\omega + \omega_0)\tau] \, d\tau$$

$$P_0(\omega) = P(\omega - \omega_0) + P(\omega + \omega_0) \qquad (9.6\text{-}16)$$

We see that the spectrum of $m(t)$ is the spectrum of $m_1(t)$ or $m_2(t)$, but moved out to be centered as sidebands around $\pm\omega_0$. Because $P(\omega)$ is symmetric about $\omega = 0$, the two sidebands must have spectra symmetric about $\pm\omega_0$. By definition of the noises m_1 and m_2, $P_0(\omega)$ was the spectrum of $n(t)$ as well. So, $m(t)$ and $n(t)$ have the same spectrum, and so have the same autocorrelation and the same statistical properties. And, the pairs (m_1, m_2) and (n_1, n_2) have the same statistical properties as well because they were derived from $m(t)$ and $n(t)$ by the same circuits. So, n_1 and n_2 have the properties we claimed, and we have completed the derivation of narrowband gaussian noise.

The signal decomposition (9.6-6) permits us to send independent "in phase" and "quadrature" signals $x_1(t)$ and $x_2(t)$, each of bandwidth B, over a white gaussian channel of rf bandwidth $2B$, using total power P. The in-phase and quadrature channel each will have exactly the same performance as if $x_1(t)$ alone or $x_2(t)$ alone were sent with power $P/2$. (We note that the noises on the two channels are independent, and so there is no reason not to receive them as independent channels.) Here we have a way of doubling available bandwidth without decreasing signal-to-noise performance. In-phase and quadrature channels are often used in communication satellite systems and in other applications.

Figure 9.6 shows a typical waveform $n(t)$ of narrowband gaussian noise. The bandwidth of the gaussian processes $n_1(t)$ or $n_2(t)$, which is half the bandwidth $2B$ of $n(t)$ around the carrier, is a few percent of the center or carrier frequency f_0. We can plainly see the low-pass envelope

$$[n_1{}^2(t) + n_2{}^2(t)]^{1/2}$$

To make the components $n_1(t)$ and $n_2(t)$ separately available requires a demodulator built as in Fig. 9.5.

We can make a rough determination of B/f_0 by dividing the period of the carrier by the average fluctuation period (wavelength) of the envelope. The base bandwidth in Fig. 9.6 relative to the carrier as estimated in this way is about 3%. This agrees with parameters of the experimental setup used to generate the waveform.

In the next chapter, we use the noise theory developed in this and preceding chapters to study error probabilities in signaling. We will also

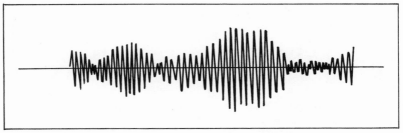

Figure 9.6. Narrowband gaussian noise.

consider shaping pulses to match the propagation characteristics of channels, particularly those that do not pass dc.

Problems

1. Use the results of this section to derive expression (4.6-2) for the signal-to-noise ratio in amplitude modulation:

$$\frac{S}{N} = \frac{\text{transmitted power}}{N_0 B}$$

(HINT: break the total rf noise up into an in-phase and a quadrature component.) Do this also for expression (4.6-9) for the signal-to-noise ratio in low-index phase modulation:

$$\frac{S}{N} = \frac{(\overline{x^2 \phi_d^2})(\text{transmitted power})}{N_0 B}$$

HINT: let the carrier have peak amplitude v_0. We demodulate by multiplying equation (4.4-2) by $2^{1/2} \sin \omega_0 t$.

2. Show that the part of Fig. 9.5 after the band-pass filter $H_0(f)$ on the left gives $H_0(f)$ by verifying the trigonometric identities involved.

3. What happens if we demodulate $n(t)$ into $n_1(t)$ and $n_2(t)$ with a phase error ϕ? This means that $\sin(\omega_0 t + \phi)$ and $\cos(\omega_0 t + \phi)$ are used instead of $\sin \omega_0 t$ and $\cos \omega_0 t$ before the low-pass filter. Does $n(t)$ have any information concerning the *carrier* phase?

4. Find stationary gaussian processes $n_1(t)$ and $n_2(t)$ which give independent random variables for the same t but where the two processes are not independent. HINT: let $n_1(t)$ be band-limited white gaussian noise, and let $n_2(t)$ be a suitable time shift of $n_1(t)$.

10

Some Aspects of Data Transmission

We have seen in earlier chapters that the sampling theorem enables us to represent band-limited signals by a sequence of samples and that a sample can be transmitted with any desired degree of fidelity as a discrete number by means of a pulse or a group of pulses having discrete amplitudes. We have seen that the transmission of sequences of numbers, or data, has an increasing application in telephony and television as well as in the transmission of alphanumeric data between computers, between terminals, or between terminals and computers. It is also universally used for telemetry from space.

We have already discussed signals and noise, the optimization of signal-to-noise ratio, and the probability of errors caused by gaussian noise in the baseband transmission of sinc and rectangular pulses. There are, however, other important considerations in data transmission, including shaping the spectra of data signals, especially near dc, and errors in their rf transmission. We will discuss these matters in this chapter, as they apply to *short pieces* of data. The next chapter studies how well we can do in reducing errors when we treat *long blocks* of data as a single unit for purposes of transmission and reception.

10.1. Errors in the Transmission of Baseband Pulses

In Section 8.5 we discussed errors due to noise in the transmission of two-level or multilevel pulses. Here we will review the results briefly.

For two-level transmission of pulses of amplitude $\pm v_0$, the probability of error P_e is

$$P_e = \frac{1}{(2\pi)^{1/2}} \int_{x=v_0/\sigma}^{\infty} \exp\left(\frac{-x^2}{2}\right) dx$$

$$P_e = 1 - \Phi\left(\frac{v_0}{\sigma}\right) \tag{10.1-1}$$

Here Φ is the gaussian distribution function. When the probability of error is small

$$P_e \doteq \frac{1}{(2\pi)^{1/2}(v_0/\sigma)} \exp\left(\frac{-v_0^2}{2\sigma^2}\right) \qquad (10.1\text{-}2)$$

Here σ^2 is the mean-square noise voltage at the sampling time at which the signal voltage is plus or minus v_0. In Section 7.4 we saw that for a proper choice of filters, this signal-to-noise ratio S/N will be

$$\frac{S}{N} = \frac{v_0^2}{\sigma^2} = \frac{2E_0}{N_0} = \frac{2TP_0}{N_0}$$

where E_0 is the energy of the transmitted pulse, P_0 is the power of the transmitted pulse stream, N_0 is the noise power density (for positive frequencies), and T is the time allotted per pulse. In Sections 7.2 and 7.5 we saw that this result can be attained without intersymbol interference with sinc pulses or rectangular pulses. And, in Section 7.4, we saw that this ratio holds for any pulse shape.

In Fig. 8.4 the error rate was plotted vs. v_0^2/σ^2.

We should note that this graph depends on the noise being additive gaussian noise. The error will then be given correctly in terms of v_0^2/σ^2 regardless of noise spectrum or pulse spectrum. But, the relation

$$v_0^2/\sigma^2 = 2E_0/N_0$$

holds only if the noise is white and then only if we optimize the receiving filter as in Section 7.4. If we do not optimize the filter, v_0^2/σ^2 will be smaller than $2E_0/N_0$.

This discussion has neglected the spectral occupancy of the transmitted signal. The next section begins to treat this aspect.

Problems

1. Suppose when we transmit plus or minus sinc pulses of bandwidth B, add white noise, and receive them with a receiver of bandwidth B the error rate is one error in 10^9 pulses. If we double the bandwidth of the receiver while keeping the signaling rate the same, what will the error rate be? (Assume gaussian noise.)

2. A data system is required to transmit 10 binary pulses a second. The receiver noise temperature is 72.5°K. The path loss is 200 dB. What transmitter power is required for an error rate of 1 in 10^8? How does this power depend on the data rate, the receiver noise temperature, and the receiving antenna effective area?

10.2. Shaping the Spectra of Data Signals

The Fourier transform of a sinc pulse is constant within its bandwidth, and the Fourier transform of a rectangular pulse is nearly constant at low frequencies. If $X(f)$ is the Fourier transform of the pulse shape, the energy spectrum is $X(f)X^*(f)$. This is nearly constant near 0 frequency (dc) for rectangular pulses.

A simple-minded representation of a binary data signal is a sequence of such pulses that are randomly positive or negative. We see from Section 8.3 that the power spectrum of such a sequence will be the energy spectrum of the individual pulses divided by T, the pulse spacing. Thus the power spectrum of these random data signals will also be constant near zero frequency.

Now, suppose that we instead use the original pulses linearly filtered by $H(f)$. The results of Section 8.3 will hold for a random input in the sense that the power spectrum will be $H(f)H^*(f)$ times the power spectrum $P(f)$ for the original pulses. In this way we can attempt to shape the spectrum, or to get any desired power spectrum near dc.

There are other ways of getting a spectrum shaped to our needs. In later sections we will find ways of encoding (rather than filtering) a random sequence of binary digits so as to shape the spectrum $P(f)$. Memory is used so that the encoding of the present pulse depends on the previous pulses. Here we will ask, why should we be so concerned about the transmitted power spectrum?

We might want to transmit a data signal over a telephone circuit that does not pass low frequencies. Such a circuit might be a wire circuit including transformers, which do not pass dc. Or, we might have a single-sideband radio circuit in which we filter low frequencies out of the signal in order to provide a narrow undisturbed channel for transmission of the carrier needed in the demodulation of the signal at the receiving end. For whatever reason, we may need to generate a power spectrum with no dc.

But, what happens if we eliminate low frequencies from a data signal whose spectrum has some substantial energy at zero frequency? Qualitatively, we expect the voltage midway between the peak positive pulse voltage and the peak negative pulse voltage to wander up and down, depending on the pulse pattern. Figure 10.1 shows this for sinc pulses. The wandering will result in errors in discriminating positive from negative pulses. These errors are beyond those that result from the loss of the signal power that was filtered out near zero frequency. We may amplify the signal after filtering so that it is back to its previous power level. But, we will still have intersymbol interference introduced by the filtering. This causes errors, even in the absence of noise. And, the performance becomes more sensitive to the timing of the sampling operation when there is intersymbol interference. See Reference 2 for a more detailed discussion.

One remedy for the intersymbol interference caused by filtering out the dc is to generate a *stream* of pulses whose power spectrum goes to

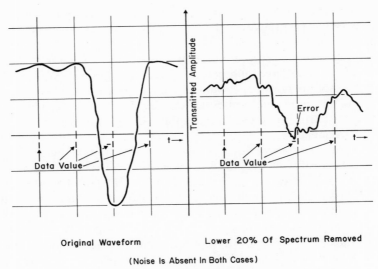

Original Waveform   Lower 20% Of Spectrum Removed

(Noise Is Absent In Both Cases)

Figure 10.1. Intersymbol interference caused by removing dc.

zero at zero frequency in the first place. Then, failure to transmit frequencies near zero will not seriously distort the received train of pulses. We will see in the next section how to do this without expanding the bandwidth required for transmission. There will be a power penalty, but one which we can ordinarily afford in wire communication, where we may not be able to use methods of transmission that require expanded bandwidth.

Problems

1. A data signal with dc suppressed is sent through an unknown number of single-sideband telephone links, each of which may shift all frequencies in the signal by an unknown small frequency. Can you suggest a way of recovering a true carrier frequency by means of which the data signal can be correctly interpreted? You may have to transmit something besides the data signal.

2. Is there a way in which information about both the carrier and the pulse rate can be sent along with a data signal? Assume the data spectrum goes to 0 at the upper limit $B = N/2$ as well as at 0 (we will see how to do this in Section 10.4). You may have to transmit an additional signal beyond that in Problem 1.

3. Draw a sinc pulse in which the lower 10% of the frequencies have been removed. Does the filtered pulse go to zero at the sampling instants as the original pulse did?

4. Suggest a method based on the Nyquist criterion of Section 5.4 for obtaining pulses with *no* energy up to frequency B, with top frequency $2B$, and with no

intersymbol interference. What is the time domain form of the pulse? (HINT: refer to Problem 4 of Section 7.1.) Sketch your answer.

10.3. Bipolar Encoding

Bipolar encoding is a very simple means for encoding a sequence of random binary digits into a sequence of pulses whose power spectrum is zero at zero frequency. Because we encode with memory, the pulses are not independent, and the transmitted power spectrum is not the number of pulses per second times the pulse energy spectrum.

In bipolar encoding we represent 0 as an "impulse" of area or strength 0, (that is, as 0), and 1 alternately as an impulse or delta function of strength +1 and −1. Here the last encoding of an input 1 must be remembered. Suppose that 0's and 1's have equal and independent probabilities of 1/2. What is the *conditional* probability of various characters following a particular character? Bipolar encoding can easily be seen to have the following conditional probabilities of transmission:

Following a zero	Following a +1	Following a −1
$p(0) = 1/2$	$p(0) = 1/2$	$p(0) = 1/2$
$p(1) = 1/4$	$p(1) = 0$	$p(1) = 1/2$
$p(-1) = 1/4$	$p(-1) = 1/2$	$p(-1) = 0$

A simple argument shows that the probabilities for the second or further character beyond a given character are independent of that character and are therefore

$$p(0) = 1/2$$

$$p(1) = 1/4$$

$$p(-1) = 1/4$$

We shall find the autocorrelation of this impulse train when we have N impulses per second. For $\tau = 0$ the autocorrelation impulse strength will be

$$N[(1/2)(0)(0) + (1/4)(1)(1) + (1/4)(-1)(-1)] = N/2$$

Here N is the number of bits to be transmitted in 1 sec, or the number of impulses per second if we count impulses of strength 0. For $\tau = \pm T$ the strength of the autocorrelation impulse will be

$$N[(1/4)(1/2)(1)(-1) + (1/4)(1/2)(-1)(1)] = -N/4$$

For $\tau = \pm nT$, n an integer ≥ 2, independence guarantees that the autocorrelation impulse strength is 0. Hence

$$C(\tau) = (N/2)[-(1/2)\delta(\tau + T) + \delta(\tau) - (1/2)\delta(\tau - T)] \quad (10.3-1)$$

The power spectrum $P(f)$, which is the Fourier transform of this, is

$$P(f) = (N/2)[-(1/2)e^{j2\pi fT} + 1 - (1/2)e^{-j2\pi fT}]$$

$$P(f) = (N/2)[1 - \cos(2\pi fT)] \tag{10.3-2}$$

This is 0 at dc, and has 0 slope there. So, we can say that there is no dc.

If we actually transmitted, not impulses, but pulses specified by $H(f)$, the power spectrum $P_1(f)$ would be, as in Section 8.3,

$$P_1(f) = N\left[\frac{1 - \cos(2\pi fT)}{2}\right]H(f)H^*(f) \tag{10.3-3}$$

We have generated this spectrum without having to use a filter which passes only the higher frequencies or without having to generate pulses with no energy at dc. But, what is the performance of bipolar encoding?

Suppose first we simply send out randomly $+$ or $-$ pulses, without bipolar encoding. Then, according to (7.4-10), we would have at the point at which the optimally filtered pulses are detected

$$v_p^2/\overline{v_n^2} = 2E_0/N_0 = 2P_0T/N_0 \tag{10.3-4}$$

This does not depend on the pulse shape. Here v_p is the pulse height after filtering at the receiver, E_0 is transmitted energy per pulse, P_0 is power of the transmitted pulse stream, T is pulse spacing, $\overline{v_n^2}$ is the noise power, and N_0 is the power spectral density of the additive noise. The level spacing of Section 8.5 is $2v_0 = 2v_p$. Hence

$$v_0^2/\overline{v_n^2} = 2P_0T/N_0 \tag{10.3-5}$$

But, in the bipolar pulse stream, the level spacing is $v_0 = v_p$, not $2v_p$, and hence we need twice the pulse voltage for approximately the same error rate, or four times the energy per pulse. But, since we are equally likely to transmit a 0 as we are to transmit a plus or minus pulse, we will transmit only half this energy per unit time (at least for sinc pulses). So, we see that we need $\frac{1}{2} \times 4 = 2$ times the power for the bipolar case as we do for the binary case. Actually, we need a little more. The probability of making an error is higher in the bipolar than in the two-level case because in the two-level case the probability of error is P_e, as given by (10.1-1). In bipolar encoding, the probability of making an error when the pulse is positive or negative is P_e; the probability of making an error when the pulse is zero is $2P_e$. So, the average probability of making an error in bipolar is $(3/2)P_e$.

Thus we pay a little more than a 3 dB power price for bipolar encoding. In return, because we use pulses with no intersymbol interference, we are able to send the signal without intersymbol interference through a system that does not transmit frequencies near zero. And,

Figure 10.2. Bipolar encoder as finite-state machine. State I = "Last 1 encoded as −1;" State II = "Last 1 encoded as +1." In the pairs of numbers a, b over each arrow, a is the input, b is the output ($a = 0$ or 1; $b = 0, +1,$ or -1.

(10.3-3) shows that we do not need any *more* bandwidth. Systems which use a subcarrier to translate the data spectrum away from dc require a higher top frequency, but they do not require more signal power. We note that, in wire communication, power is often not the problem; bandwidth is.

It is instructive to think of a bipolar encoder as a *finite-state machine*. The encoder of Fig. 10.2 has two states which we will call I and II. Its operation can be completely described by specifying the output and the new state when there is a given input in a given state.

When the machine is in state I, inputting a 0 results in an output of 0 and no change of state. Inputting a 1 results in an output of +1 and a change to state II. When the machine is in state II, inputting a 0 results in an output of 0 and no change of state. Inputting a 1 results in an output of −1 and a change to state I.

The concept of states is important in complicated encoding, such as described in Section 10.5. It will also occur when we study scramblers in Section 10.6 and in convolutional encoding for error protection in Chapter 12.

The next section considers another way of removing the low frequencies from a transmitted data stream.

Problems

1. A random binary stream of pulse rate N is divided into two streams of even and odd numbered digits, each of rate $N/2$. These two streams, interleaved, form the original stream. Each stream is separately bipolar encoded with impulses. Are the two streams independent? The resulting pulses are interleaved to form an output of N pulses per second. How does the resulting power spectrum differ from that for ordinary bipolar encoding? (HINT: the power spectrum of the sum of independent processes is the sum of the spectra.) What happens if we use sinc pulses with no intersymbol interference at the rate of N pulses per second?

2. What fraction of the power in a bipolar-encoded pulse stream using sinc pulses is at frequencies below 200 Hz, if there are $N = 8000$ bits or pulses to be transmitted per second?

3. We can try to generalize bipolar encoding to three levels by letting

$$0 \to 0, \qquad 1 \to \pm 1 \text{ alternately}, \qquad 2 \to \pm 2 \text{ alternately}$$

What is the finite-state machine? Can you suggest what difficulty we might get into if we try this for data in which 0, 1, and 2 occur independently and with equal probabilities?

4. In discussing the power required for bipolar transmission, why did we have to restrict the discussion to sinc pulses? Show that we need twice the power for *any* filtered pulses with no intersymbol interference (all we really need is that the filtered pulse goes to zero at the time T at which the peak of the next pulse comes along).

10.4. Partial-Response Encoding

Partial-response encoding is another way to make the power spectrum zero at zero frequency.

In the usual form of partial-response encoding the input is a binary digital signal of 0's and 1's occurring T seconds apart. Encode by the following method: if the input is a 0, output two successive impulses, $2T$ seconds apart, of strengths $-\frac{1}{2}$ and $+\frac{1}{2}$. If the input is a 1, output two successive impulses, again $2T$ seconds apart, but of strengths $+\frac{1}{2}$ and $-\frac{1}{2}$. Output impulses occurring at the same time are additive so the final output stream will consist only of impulses of strengths -1, 0, and $+1$, with T seconds between impulses. (This will be so except for the first few transmitted pulses.) We may use the notation

$$0 \rightarrow -\tfrac{1}{2},\, 0,\, +\tfrac{1}{2}$$

$$1 \rightarrow +\tfrac{1}{2},\, 0,\, -\tfrac{1}{2}$$

Here successive symbols correspond to pulses T seconds apart.

We see that the result is the same as if we passed a sequence of $-1, +1$ impulses through a linear network with an impulse response

$$\tfrac{1}{2}[\delta(t) - \delta(t - 2T)]$$

The transfer function of such a network is the Fourier transform $H_1(f)$ of the time response. This is

$$H_1(f) = \int_{-\infty}^{\infty} \tfrac{1}{2}[\delta(t) - \delta(t - 2T)]e^{-i2\pi fT}\, dt$$

$$= \tfrac{1}{2}(1 - e^{-i4\pi fT})$$

$$= je^{-i2\pi fT}(e^{i2\pi fT} - e^{-i2\pi fT})/2j$$

$$H_1(f) = je^{-i2\pi fT} \sin (2\pi fT) \tag{10.4-1}$$

The power spectrum $P(f)$ of this network excited by the random train of impulses of strengths ± 1 will be $H_1(f)H_1^*(f)$ times the power spectrum which is everywhere N, if there are $N = 1/T$ impulses per second. This is the spectrum of partial-response encoding of random ± 1 data.

Since

$$H_1(f)H_1{}^*(f) = \sin^2(2\pi fT) = \tfrac{1}{2}[1 - \cos(4\pi fT)]$$

we see that $P(f)$ is given by

$$P(f) = NH_1(f)H_1{}^*(f) = (N/2)[1 - \cos(4\pi fT)] \qquad (10.4\text{-}2)$$

Since $P(0) = 0$ and the derivative $P'(0) = 0$, there is no dc component. This is the same as for the bipolar encoding of Section 10.3, but the spectrum is different from the bipolar spectrum (see Problem 1 of Section 10.3). The spectrum for pulses whose shape is specified by some $H(f)$ rather than impulses can be obtained by multiplying $P(f)$ in (10.4-2) by $H(f)H^*(f)$.

We see from (10.4-2) that the partial-response spectrum $P(f)$ is also 0 at the *upper* edge of the Nyquist sampling band. Why is this useful? We note that the receiver filter is matched to the pulse spectrum, not to the overall transmitted spectrum, so that it does not go to zero at the upper edge of the band. Thus any errors in the phase or amplitude of the receiver filter characteristic at the upper edge of the Nyquist band ($f = 1/2T$) will have little effect because the signal power goes to zero at this frequency. Or, we may wish to insert information about the carrier or pulse rate at the upper edge of the band.

There is a 3+ dB power penalty in partial-response encoding owing to removal of dc, as in Section 10.3. The reasoning is the same as before.

We can also get the *bipolar* spectrum of (10.3-2) by encoding input data as follows:

$$0 \rightarrow -\tfrac{1}{2}, +\tfrac{1}{2}$$
$$1 \rightarrow +\tfrac{1}{2}, -\tfrac{1}{2}$$

This is a different encoding, but the spectrum will be the same as the bipolar spectrum. The calculation is similar to the one for partial-response encoding. In this case, as in bipolar, the spectrum has the disadvantage of a *maximum* at the upper edge of the Nyquist band.

In the next section, we will see how to encode data which takes on more than two values.

Problems

1. Make the partial-response encoding of

$$1, 1, 1, 1, 1, 1, \text{etc. forever}$$

Can you decode the rest of such a message if you miss the first few received pulses?

2. Encode

$$0, 0, 0, 0, 0, 1, 1, 1, 1, 1, \text{ etc.}$$

Can you decode this message if you miss the first few received pulses?

3. Can you suggest a way of preprocessing the digits to be encoded so that the partial-response decoding can be carried out unequivocally? Compare with bipolar encoding.

4. Show that the bipolar spectrum is obtained for the encoding $0 \rightarrow -\frac{1}{2}, +\frac{1}{2}$, $1 \rightarrow +\frac{1}{2}, -\frac{1}{2}$. Can you suggest a heuristic explanation for the fact that putting the 0 in between produces the partial-response encoding spectrum?

5. We have found the message spectra for bipolar and partial-response encoding when the message symbols are chosen randomly and with equal probabilities. Will these be the spectra when the message symbols are chosen *periodically*? Give an example. (Review Problem 1 of Section 6.7, on the relation of the Fourier series of a periodic waveform to its Fourier transform.)

6. What fraction of the power in a partial-response-encoded pulse stream using sinc pulses is at frequencies below 200 Hz if there are $N = 8000$ bits to be transmitted per second? (Compare Problem 2 of Section 10.3.) How does this compare with bipolar encoding?

10.5. Many-Level Partial-Response Encoding

In partial-response encoding as described in Section 10.4, we start with a signal stream consisting of the binary digits or bits 0 and 1. Let us generalize this to an encoding of the digits which have m values (for binary, $m = 2$). Thus the input to the encoder is a sequence $x_0, x_1, \ldots$, where each x_i is one of the m numbers $0, 1, \ldots, m - 1$. The output of the encoder will be a sequence $y_0, y_1, \ldots$, where each y_i is one of the $(2m - 1)$ numbers $0, \pm 1, \ldots, \pm(m - 1)$.

From the sequence of input digits we derive a sequence of states s. We assume $s_{-1} = 0$. The rule for generating the successive states is the congruence

$$s_n = (s_{n-1} + x_n) \text{ (modulo } m) \tag{10.5-1}$$

Here $n = 0, 1, 2, \ldots$. By modulo m we mean to subtract m from $(s_{n-1} + x_n)$ as many times as we can and still have a nonnegative remainder; that remainder is then s_n.

The output digits y are given by subtracting successive states, but *not* modulo m:

$$y_n = s_n - s_{n-1} \tag{10.5-2}$$

These lie between $-(m - 1)$ and $(m - 1)$ inclusive because the s_i are between 0 and $m - 1$. The relation (10.5-1) is a congruence; relation (10.5-2) is an ordinary equation.

Table 10.1. Partial Response Encoding for $m = 4$

Index (n)	-1	0	1	2	3	4	5	6	7	8	9	10	11
Input (x)	*	3	3	2	1	0	2	1	2	0	3	3	2
States (s)	0	3	2	0	1	1	3	0	2	2	1	0	2
Output (y)	*	3	-1	-2	1	0	2	-3	2	0	-1	-1	2

The example shown in Table 10.1 assumes $m = 4$. We see that the output of the encoder is simply an encoding of the states s in which

$$0 \rightarrow 0, \quad 0$$
$$1 \rightarrow 1, -1$$
$$2 \rightarrow 2, -2$$
$$3 \rightarrow 3, -3$$

Here we shift and add as in the previous section. The output has $2m - 1$ levels. In this case, $m = 4$ and $2m - 1 = 7$. The above correspondence of input to output is analogous to bipolar encoding. To obtain partial-response encoding, we split the input into two streams and encode each separately.

Decoding is very easy. To go from y_n to x_n, we use the formula

$$x_n = y_n \quad \text{if} \quad y_n \geq 0 \tag{10.5-3}$$

$$x_n = m + y_n \quad \text{if} \quad y_n < 0 \tag{10.5-4}$$

This follows for (10.5-1) and (10.5-2). We have

$$y_n = (s_{n-1} + x_n) \; (\text{modulo } m) - s_{n-1} = (s_{n-1} + x_n - s_{n-1}) \; (\text{modulo } m)$$
$$= x_n \; (\text{modulo } m)$$

Hence, either $y_n = x_n$, or $y_n = x_n - m$ so that $x_n = y_n + m$. We can see which expression holds by noting that

$$0 \leq x_n \leq m - 1$$

For many-level partial-response encoding, we wish to encode the states as pairs $s_n, 0, -s_n$ instead of as pairs $s_n, -s_n$, so we merely divide the input stream into n odd and n even and encode and decode the two streams separately. The two streams are interleaved for transmission. Many-level partial-response encoding agrees with ordinary partial-response encoding when $m = 2$, but the two-level input bit stream has to be preprocessed.

If the input characters x_n are chosen randomly and with equal probabilities, the states s_n will be random and have equal probabilities. The partial-response encoding of a state is equivalent to applying to an impulse whose amplitude is equal to the numerical value of the state a network for which

$$H_1(\omega)H_1^*(\omega) = \tfrac{1}{2}(1 - \cos 2\omega T)$$

This is as Section 10.4. From this, the spectrum of many-level partial-response encoding can be derived. The shape is the same as that of the spectrum given in Section 10.4, but the amplitude depends on the number of states m.

These last three sections have described ways to remove frequencies near dc in data transmission by enhancing the randomness inherent in the data. We now study encoders with longer memory. They operate on nonrandom data not so much to take out low frequencies as to put in high frequencies. We may wish to do this to whiten the short-term spectrum or to synchronize the symbols more easily.

Problems

1. Encode the three-level signal

$$0, 1, 2, 2, 1, 0, 1, 2$$

    Check by decoding that the original signal is obtained.

2. Decode the five-level signal

$$0, 1, 2, -2, -1, 4, -1, 0, 1$$

    Check by encoding that the above sequence is obtained.

3. How should the partial-response encoding of Section 10.4 have its input preprocessed to agree with the partial-response encoding of this section when $m = 2$? (Compare Problem 3 of Section 10.4.)

4. The encoder for many-level partial-response encoding is a finite-state machine. Draw the state diagram for $m = 4$.

5. With what probabilities are the different $(2m - 1)$ possible y values taken on, knowing that the m states are equally likely and independent of each other? Are successive y's independent?

6. Derive the power spectrum for many-level partial-response encoding with $m = 5$. [HINT: subtract $(m - 1)/2$ from each state, and observe that this cannot affect the answer.] How does the answer depend on m for large m? [Review equation (4.6-18).]

10.6. Randomizing the Data Source

In calculating spectra in the preceding sections, we have assumed a random binary data stream. Actual data streams are not random, and they may have large strings of zeros or ones, or other unfortunate

(a)

(b)

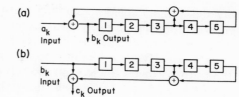

Figure 10.3. (a) Five-stage scrambler and (b) five-stage descrambler.

patterns of digits. A long run of the same digit may cause malfunction in digital circuits intended to extract the pulse rate, and peculiar sequences may do other mischief.

For this reason, *scramblers* are often interposed between a binary source and a data transmission system. Such scramblers make the data source seem random. They provide a prewhitening of the spectrum of the data. This makes the input spectrum, especially the short-term spectrum, flatter or whiter. Here we will explain a typical form of self-synchronizing scrambler. It is based on a shift register, another instance of finite-state machine. The scrambler is a *feedback* shift register and the descrambler is a *feed-forward* shift register. Shift registers will also occur in Chapter 12 in connection with coding of data for error protection. We shall explain all these terms.

Figure 10.3, taken from Reference 3, shows a five-stage scrambler and its associated descrambler. Each cell in the registers represents a unit of delay; $\oplus$ denotes modulo-two addition of the binary data ($0 + 0 = 1 + 1 = 0$; $0 + 1 = 1 + 0 = 1$). If a_k is the input to the scrambler and b_k the scrambled output, then we will have for all k

$$b_k = a_k \oplus b_{k-3} \oplus b_{k-5} \tag{10.6-1}$$

There is feedback because output b_k depends on previous outputs b_{k-3} and b_{k-5}. Similarly, the output of the descrambler will be

$$c_k = b_k \oplus b_{k-3} \oplus b_{k-5} \tag{10.6-2}$$

No c term is on the right-hand side of (10.6-2). This means there is no feedback.

Since (10.6-1) holds,

$$c_k = (a_k \oplus b_{k-3} \oplus b_{k-5}) \oplus b_{k-3} \oplus b_{k-5}$$

$$c_k = a_k \tag{10.6-3}$$

This is so because $b \oplus b = 0$ for any b. So, the descrambler lives up to its name—it reverses the operation of the scrambler.

How do we start the descrambler? Some communication devices need a "T_0 machine" to synchronize operations, or to find T_0 at both ends. But this is not necessary here. The descrambler is self-synchronizing. This is why this scrambling–descrambling method is often preferred. And, the self-synchronizing property also permits recovery from channel bit errors and insertion or deletion of bits. Let us see why these things are so.

If we receive five bits in a row correctly, then all five stages in Fig. 10.3b are correctly loaded, no matter what the past history was. Time zero (T_0) need not be known at the descrambler. When the next bit comes in, the output will be correct, because the descrambler will be in the correct state. In the absence of further channel errors or bit slippage, descrambling proceeds correctly according to (10.6-2), since the register remains in the correct state.

Shift registers are very useful in communication in many ways. Reference 4 shows this, and derives many useful properties. Pages 23–59 and p. 88 are relevant to this section.

This completes our study of representing or encoding digital data to match channel transfer function or spectral characteristics. In Chapter 11, we will study encoding of data to combat channel noise. In the next section, we examine errors in digital data transmission over an rf link.

Problems

1. If a single error is made in transmission in Fig. 10.3, how many characters can be received in error? What does this do to the average bit error probability at low error rates?

2. Describe the output of the scrambler when eight consecutive 0's are input and when eight consecutive 1's are input. Assume the state of the encoder is 10000 in the first case and 01111 in the second. Explain the relation between the two answers. How could this help synchronize symbols? (Refer to Problem 4 of Section 7.5.)

3. Represent the feed-forward descrambler of Fig. 10.3 as a finite-state machine with 32 states. Use a few brief words and no figures.

4. Show that if the input to the scrambler consists of random data (inputs equally likely and independent), then so does the output. Hence one cannot hope to deduce what the scrambler is from merely observing the output when random data are input.

10.7. Phase Shift Keying: Errors Due to Noise

How can we transmit baseband pulses over a radio link?

One possibility is single-sideband transmission. If we have suppressed the spectrum of the baseband signal at zero frequency, as in preceding sections, we can insert the carrier necessary for demodulation and filter it out at the receiver.

Commonly, however, single-sideband transmission is not used in microwave systems. Rather, phase-shift keying (psk) is used. The phase of a carrier is shifted to one of m equally spaced angular positions to send one of m digits or symbols. We studied m-amplitude baseband signaling in Section 8.5.

If $m = 2$ in phase-shift keying, this is equivalent to an amplitude modulation (by $+1$ or -1). The most common forms of phase modulation for radio links are two-phase modulation and four-phase modulation. We shall study m-phase modulation from the standpoint of information theory in Sections 11.7 and 12.4. Here we shall find the error probability in two-phase and four-phase modulation.

Because the baseband pulses which amplitude modulate the carrier are rectangular, the baseband and rf bandwidths in two-phase modulation will be infinite. How shall we demodulate this spectrum? If we use an ordinary am demodulator to obtain baseband pulses, we multiply the received signal by the carrier $2 \cos \omega_0 t$ and low-pass filter. This is as in Section 4.6.

The baseband pulses are detected by integrating over the known pulse period and sampling. Section 7.5 showed that this is optimum.

However, according to Section 7.5, we can regard this system in another way. We can think of sending a peculiar kind of baseband pulse which is equal to $\pm 2 \cos \omega_0 t$ over the pulse period. Section 7.4 showed that the optimum detection for this can be achieved in the time domain by matched filtering. If we use the correlation receiver, this is the same as multiplying by $2 \cos \omega_0 t$ and integrating over the pulse period. We can assume that there was a low-pass filter before the integration, because the double-frequency term integrates to 0 anyway. But, what we then have is the same as the am detector followed by the integrator.

So, as we saw in Problem 2 of Section 7.5, am demodulation followed by ideal baseband detection is optimum for two-phase modulation. It must give the same signal-to-noise ratio as the optimum detection of Section 7.4. From equation (7.4-10), this signal-to-noise ratio will be

$$\frac{S}{N} = \frac{2E_0}{N_0} = \frac{2TP_s}{N_0}$$

Here E_0 is the energy per pulse and there are $N = 1/T$ pulses per second. Each phase 0 or π lasts T seconds. If we write $B = 1/2T$, where B is the baseband bandwidth that would be obtained for sinc pulses, the signal-to-noise ratio becomes

$$\frac{P_s}{N_0 B} = \frac{v_0^2}{\sigma^2} \tag{10.7-1}$$

This is as in equation (8.5-14) for baseband sinc pulses. Here $v_0\sqrt{2}$ is the filtered pulse amplitude and σ^2 is the noise variance. The pulse is $v_0 2^{1/2} \cos \omega_0 t$ centered at $t = 0$ and of duration T. The signal power is v_0^2. The noise variance or power is $N_0 B$ after matched filtering by the $2^{1/2} \cos \omega_0 t$ pulse and low-pass filtering. This is because of Section 9.6, which showed that the rf noise power $\sigma^2 = 2N_0 B$ is equally divided between the sine and cosine channels.

Thus, in terms of the (one-sided) rf noise power density N_0 and the rf signal power P_s, the error probability P_e in ideal two-phase transmission is that given in Section 8.5. For large signal-to-noise ratios we saw in equation (8.5-12) that this probability P_e is approximately

$$P_e \doteq \frac{1}{(2\pi)^{1/2}(v_0/\sigma)} \exp\left(\frac{-v_0^2}{2\sigma^2}\right) \qquad (8.5\text{-}12)$$

It is easy to see that we can send two independent baseband binary pulse streams of the same pulse rate without broadening or changing the rf spectrum. To do this we multiply the first by the rf wave $2^{1/2} \cos \omega_0 t$ and the second by the rf wave $2^{1/2} \sin \omega_0 t$. We saw this in Section 9.6. We recover the first baseband pulse stream by multiplying the composite rf signal by $2^{1/2} \cos \omega_0 t$ and low-pass filtering, and the second by multiplying the composite signal by $2^{1/2} \sin \omega_0 t$ and low-pass filtering. Does this cost anything in performance?

If the *total* rf power used is P_s, half should be used in sending each baseband pulse stream, and hence for each stream

$$\frac{v_0^2}{\sigma^2} = \frac{P_s}{2N_0 B} \qquad (10.7\text{-}2)$$

For the same *total* power, the signal-to-noise ratio in the recovered pulse streams will be less by a factor of 2 than for two-phase modulation, and the error rate will be greater. But, we send twice as many baseband pulses per second. This means that for each single channel, B in (10.7-1) will double, and (10.7-2) will give the same performance as the single channel at the same rate in pulses per second. So, for given energy per bit and noise spectral density, this technique has the same performance in terms of bit error probability as two-phase or single-channel, and occupies half the spectrum. But, the performance of two-channel will degrade more rapidly with phase reference error. This is because the sine channel will cause crosstalk into the cosine channel if there is phase error.

Let us now assume that a $\cos \omega_0 t$ rf pulse and a $\sin \omega_0 t$ rf pulse are sent simultaneously, with amplitudes x and y. Here x and y are $+1$ or -1. We see that there are four combinations of x and y, as shown in Fig. 10.4. Each different combination represents a different phase of the combined rf signal. The result is that in sending two independent binary pulse

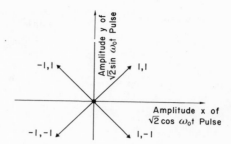

Figure 10.4. Four-phase modulation from two independent binary streams.

streams by means of $\cos \omega_0 t$ and $\sin \omega_0 t$ carriers, we have produced a four-phase phase-modulated rf signal. The four phases are $\pm \pi/4, \pm 3\pi/4$.

Equation (10.7-2) defines the bit error probability. This is the probability that x will be in error, and the probability that y will be in error is the same. We see that four-phase has the same performance as two-phase in terms of bit error probability. This is because it arises also from in-phase and quadrature modulation.

For the received phase to be correct, *both* x and y must be received correctly. The probability of receiving both x and y correctly is

$$(1 - P_e)(1 - P_e) = 1 - 2P_e + P_e^2$$

The probability of an error in received phase or received symbol is 1 minus the probability that both x and y will be correct. This is

$$2P_e(1 - P_e) \qquad (10.7\text{-}3)$$

We call this the symbol error rate. Thus, for four-phase modulation, the symbol error rate is almost twice the bit error rate for small error rates. This is not surprising.

Which is more important, the symbol error rate or the bit error rate? There is no conclusive answer. But let us make the following argument concerning m-phase modulation, or m-level signaling in general.

We first consider the relation between n_2, the number of binary digits necessary to represent some large number N, and the number n_m of bauds, or m-ary digits or symbols, or phase-modulated positions of an m-phase modulated signal, necessary to represent the same number. The relation is

$$N = 2^{n_2} = m^{n_m}$$

$$n_m/n_2 = \ln 2/\ln m \qquad (10.7\text{-}4)$$

Let P_2 be the probability of error in a binary digit and P_m be the probability of symbol error in an m-ary digit. If the probability that the large number N will be received correctly is to be the same in each case (this is one possible criterion for equivalent performance), then

$$(1 - P_2)^{n_2} = (1 - P_m)^{n_m}$$

$$P_2 = 1 - (1 - P_m)^{n_m/n_2} \qquad (10.7\text{-}5)$$

From (10.7-5) and (10.7-4)

$$P_2 = 1 - (1 - P_m)^{\ln 2/\ln m} \qquad (10.7\text{-}6)$$

When $P_m \ll 1$,

$$P_2 = (\ln 2/\ln m)P_m \qquad (10.7\text{-}7)$$

Equations (10.7-6) and (10.7-7) give an *equivalent binary error probability* P_2 for *an m-phase baud* (or *m-ary digit* or *m-symbol*) error probability P_m. We easily see that (10.7-6) agrees with (10.7-3) when $m = 4$.

We have found it easy to compute the ideal error probabilities for two-phase and four-phase phase modulation. Some numerical results are available on symbol error rates for other numbers of phases (References 1 and 5). From these, equivalent bit error rates can be calculated by means of (10.7-6).

The problem of computing the symbol error probability for m-phase modulation ($m > 2$) is illustrated in Fig. 10.5. We have a sinusoidal rf signal of peak amplitude $2^{1/2}v_s$ or power v_s^2, plus a gaussian rf noise with a $2^{1/2} \cos \omega_0 t$ or v_x component n_1 and a $2^{1/2} \sin \omega_0 t$ or v_y component n_2. Here we have choosen the coordinates $2^{1/2} \cos \omega_0 t$ and $2^{1/2} \sin \omega_0 t$ to be of length squared or power 1. In this coordinate system, the length of the signal vector is v_s. The gaussian noises n_1 and n_2 have common variance

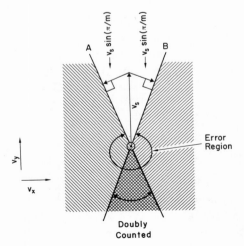

Figure 10.5. Error probability region for m phases.

$\sigma^2 = N_0 B$. Here B is the equivalent baseband bandwidth, which is half the *symbol* rate.

The probability that the components of the noise voltage will lie in the range $dv_x\, dv_y$ at (v_x, v_y) is

$$\left[\frac{1}{(2\pi)^{1/2}\sigma} \exp\left(\frac{-v_x^2}{2\sigma^2}\right) dv_x\right]\left[\frac{1}{(2\pi)^{1/2}\sigma} \exp\left(\frac{-v_y^2}{2\sigma^2}\right) dv_y\right]$$

$$= \frac{1}{2\pi\sigma^2} \exp\left(\frac{-(v_x^2 + v_y^2)}{2\sigma^2}\right) dv_x\, dv_y \quad (10.7\text{-}8)$$

The probability that the sum of the signal voltage v_s and the noise voltage will be incorrectly interpreted is the probability that their vector sum will lie outside of the unshaded area shown in Fig. 10.5 as the upper angle between lines A and B. It is difficult to compute this probability exactly.

However, it is easy to compute the probability that the vector sum will lie to the *left* of the line A or the equal probability that the vector sum will lie to the *right* of the line B. The sum of these two equal probabilities *over*estimates the probability of error because of the small probability that the voltage will be in the doubly hatched lower angular region between lines A and B. This probability is counted twice. When the signal-to-noise ratio is high, this double-counted probability is extremely small and we may take the probability of error as the probability that the voltage lies to the left of A or the right of B.

We get these equal probabilities by rotating the axes along which we measure v_x and v_y so that one is along the line A or the line B. We saw in Section 8.5 that such a gaussian noise distribution is unaffected by rotation. This was basic to our derivation of the gaussian distribution via polar coordinates in equation (8.5-3). We will meet this axis rotation concept again in the next chapter, when we study signal space.

So, if the total probability of error is P_e, we will have

$$\frac{P_e}{2} \doteq \frac{1}{2\pi\sigma^2} \int_{v_x=-\infty}^{\infty} \int_{v_y=v_s \sin(\pi/m)}^{\infty} \exp\left(\frac{-v_x^2}{2\sigma^2}\right) \exp\left(\frac{-v_y^2}{2\sigma^2}\right) dv_x\, dv_y$$

Because the normal density function of variance σ^2 integrates to 1, the v_x integral can be evaluated. The result is

$$\frac{P_e}{2} \doteq \frac{1}{(2\pi)^{1/2}\sigma} \int_{v_y=v_s \sin(\pi/m)}^{\infty} \exp\left(\frac{-v_y^2}{2\sigma^2}\right) dv_y \quad (10.7\text{-}9)$$

We obtain

$$P_e \doteq 2\left[1 - \Phi\left(\frac{v_s \sin(\pi/m)}{\sigma}\right)\right] \quad (10.7\text{-}10)$$

This looks more familiar. Here Φ is the standard gaussian distribution of equation (8.5-10).

As (10.7-10) holds only when

$$\frac{v_s \sin (\pi/m)}{\sigma} \gg 1$$

we may use approximation (8.5-11). Thus, for $m > 2$ and for small probabilities of error, the symbol error probability P_e is approximately

$$P_e \doteq \frac{2}{(2\pi)^{1/2}} \frac{\exp[-v_s^2 \sin^2 (\pi/m)/2\sigma^2]}{v_s \sin (\pi/m)/\sigma}$$
$$= \frac{2}{(2\pi)^{1/2}} \frac{\exp[-(P_s/2N_0 B) \sin^2 (\pi/m)]}{(P_s/N_0 B)^{1/2} \sin (\pi/m)} \quad (10.7\text{-}11)$$

This is true for $m = 3, 4, \ldots$. The formula is the same for the case $m = 2$, except that P_e is only about half as large. The signal-to-noise for every $m \geq 2$ is

$$\frac{S}{N} = \frac{v_s^2}{\sigma^2} = \frac{v_s^2}{N_0 B} = \frac{P_s}{N_0 B} \quad (10.7\text{-}12)$$

What is the spectrum of m-phase modulation around the carrier? For $m = 2$ or $m = 4$, we saw that it must be the same as the sinc^2-shaped power spectrum of baseband rectangular pulses. By simple use of the geometry of complex numbers, we can show that this is also the spectrum for m phases, where m is any positive integer. This must then be the spectrum of continuous-phase modulation if all phase shifts are equally likely, or have a uniform probability density on the phase circle. We can also easily show that the sinc^2 spectral shape holds for m-level amplitude modulation with rectangular pulses, even when m is greater than 2. In Chapter 12 we shall study m-phase again, where we evaluate m-phase communication on the basis of channel capacity.

The next section will show how to obtain the phase reference for demodulating phase-shift keying, whether binary or m-ary.

Problems

1. For m-phase shift keying, the information transmitted per symbol is $\log_2 m$ bits. When the symbol rate is given and the symbol error probability is kept constant at a very small given value, what is the relative energy per bit E_b for $m = 2, 3, 4, 8, 16, 32$? What if the *bit* error probability is kept constant? What other considerations may be important in choosing the number of phases? What number appears best, other considerations aside? What is the gain over two-phase in energy per bit in dB if this number of phases is used?

2. Why is there a difference in the formula for symbol error probability when $m = 2$ and when $m > 2$?

3. Show that the error probability in two-phase shift keying derived by considering it as amplitude modulation by ± 1 is the same as that which would be
derived by considering phase noise. [HINT: equation (4.6-8) shows that the
phase noise spectral density is N_0/P_0, where P_0 is the carrier power and N_0 is
the amplitude noise spectral density. The phase noise can be assumed gaussian
when it is not too strong, for then it is derived by linear operations on gaussian
noise. Finally, the derivation of Section 4.6 really was for noise on the *sine* of
the phase. For small phase angles, this distinction is unimportant, but in
considering the probability of a *large* phase error, it is the *sine* of the phase
error that sets the threshold.]

10.8. Demodulating Phase Shift Keying

In demodulating phase shift keying, it is necessary to have a signal
$\cos \omega_0 t$ as a reference.

We will show how this reference can be obtained for two-phase
modulation, without having to transmit a carrier. This is called *suppressed
carrier* phase modulation. There may be other reasons to transmit a
carrier. We can learn about the propagation medium itself from observing
small phase shifts or other distortions of the carrier. The doppler shift of
the carrier can also be used for radio navigation. We shall not discuss
these uses.

In two-phase modulation the possible signals are

$$\cos \omega_0 t$$

$$\cos (\omega_0 t + \pi) = -\cos \omega_0 t$$

Suppose we square both signals. The squares of the two signals are the
same:

$$\cos^2 \omega_0 t = \cos^2 (\omega_0 t + \pi) = \tfrac{1}{2}(1 + \cos 2\omega_0 t)$$

The square of either input signal has the same double-frequency component $\cos 2\omega_0 t$.

Hence, by squaring the two-phase signal and filtering out dc, we get a
steady $\cos 2\omega_0 t$ signal. If we could divide the frequency of this squared
signal by 2, we would get either

$$\cos \omega_0 t \quad \text{or} \quad \cos (\omega_0 t + \pi) = -\cos \omega_0 t$$

We do not know which we should take for the phase reference in
demodulating data. (This can be true even when we transmit a carrier.) In
suppressed carrier, we cannot tell a data $+1$ and a $\cos \omega_0 t$ reference
from a data -1 and a $-\cos \omega_0 t$ reference. However, we can start with
either phase, and encode 0 as no change in phase between successive
transmitted pulses and 1 as a change in phase between successive pulses.

This is similar to the preprocessing for partial response encoding in Section 10.5. The ambiguity of the phase reference will no longer matter. This approach is called *differential phase shift keying*. Or, we can send a reference data prefix of known value, or use redundancy which is naturally in the data. For example, we can easily tell a TV picture correctly demodulated from one in which blacks and whites are reversed as in a photographic negative. The prefix technique is the one commonly used in deep space communication.

In the next section, we will evaluate the bandwidth occupancy of the most general form of phase shift keying for digital data transmission. This section and the previous one have considered only rectangular pulses.

Problems

1. How can one recover the carrier in four-phase phase shift keying?

2. Suggest a way to encode and transmit a signal unambiguously with four-phase keying despite an ambiguity in the phase of the recovered carrier.

3. What happens to the bit error probability when we use the scheme of encoding 0 as no change in phase and 1 as a change in phase to avoid phase ambiguity? What is the power loss in dB at high signal-to-noise ratio that we must pay for not having the true phase reference?

4. Show that it is not possible to construct a frequency divider without using some form of memory feedback. That is, we cannot build one from a memoryless nonlinear device in series with a filter. (HINT: what is the spectrum of the output when a memoryless nonlinear device is driven by a sine-wave input? Fourier series may help in this.)

5. To a psk signal $\pm\cos \omega_0 t$ is added a pure carrier $\sin \omega_0 t$. Show that this is equivalent to phase modulating a carrier by angles other than $0°$ and $180°$, and find these angles. What fraction of the transmitted power is devoted to the carrier, and what fraction to the data?

10.9. Bandwidth in Phase Shift Keying

One of the advantages of phase modulation as opposed to amplitude modulation is that the transmitter works at a constant amplitude. Some transmitters, such as oscillators, always work at a constant amplitude. Some amplifiers, e.g., traveling wave tube amplifiers or TWTAs, are most efficient only if they operate at a constant amplitude. These amplifiers are used in space microwave systems for their efficiency and reliability.

In Section 9.6 we learned about transmitting in-phase and quadrature components as a way of doubling the data rate within a given bandwidth. If we have data consisting of 0's and 1's, we could use rectangular pulses on each of the two channels. These rectangular pulses phase modulate (or amplitude modulate, for, with biphase modulation, the two are the same) the in-phase and quadrature components. The amplitude is constant, but the bandwidth will be infinite.

We can try to lower the bandwidth by using sinc pulses instead. But here, there is a difference between phase modulation and amplitude modulation. Phase modulation leads naturally to a constant-amplitude system. Amplitude modulation by sinc pulses, unlike phase modulation by sinc pulses, has no frequencies beyond $B = N/2$ above the carrier, where there are N pulses per second. But, the amplitude is no longer constant.

We can make the amplitude more nearly constant by arranging that the sampling times for the in-phase and quadrature channels differ by half a sampling period. The overall power spectrum will be unaltered because the two channels are independent, as in Section 10.7. This approach is called *offset quadrature amplitude modulation*. It is a way of conserving available bandwidth while retaining much of the advantages of the constant amplitude of phase modulation. This may see service in high-rate satellite links where earth terminal complexity is less of a consideration than it has been.

Let us now consider pure phase modulation, in which the rf signal is of the form

$$X(t) = \cos\left[\omega_0 t + \phi(t)\right]$$

Here $\phi(t)$ is a baseband waveform generated from one of two distinct phases N times a second without intersymbol interference. This can be accomplished by making $\phi(t)$ a rectangular wave of amplitude $\pm\pi/2$. But, if we do this, we have seen that the bandwidth of the rf signal is infinite. So, phase modulation with nonrectangular pulses is being used in digital microwave ground links to conserve bandwidth.

How small is it possible to make the bandwidth compared with N? We note that N is the rf bandwidth necessary to transmit N pulses per second without intersymbol interference using amplitude modulation. To attain this am bandwidth, we use sinc pulses for the baseband signal. For this question, we must be more rigorous in our treatment of bandwidth. We have been using the term without a formal definition or formula. We will shortly give several definitions.

Why is frequency or bandwidth the ultimate limitation in the allocation of uses of the electromagnetic spectrum? We shall see in Section 11.2 that, according to Shannon's formula for the information capacity of a gaussian channel, for a specified ratio of signal power to noise power channel capacity is proportional to bandwidth. Shannon's formula is a tool that can be used to analyze all sorts of systems. Here we will merely consider some simple cases involving bandwidth.

There are alternatives to merely dividing the frequency spectrum up among a number of communication channels, each consisting of a transmitter and a receiver that operate in an assigned band of frequencies. We can envision a system in which all users use the same band of frequencies, but different users use it during different short times or time slots (time division multiple access or TDMA). Such systems require active cooperation among users and some sort of central timing.

Or, we can imagine a spread spectrum system (we will discuss this in Section 12.7) in which all users use the same broad band of frequencies, but each must multiply the received signal by an individual synchronized key in order to separate his message from all others (code division multiple access or CDMA).

Allocating a particular band of frequencies to a transmitter–receiver pair seems simpler and more direct. If all transmitters had sharply band-limited signals with flat power spectra we could attain efficient communication without interference simply by making frequency or bandwidth assignments that were contiguous and nonoverlapping. Of course, in real frequency assignments some *guard band* must be left between assigned bands or channels even when transmitters and receivers are nominally band limited, because actual band-limiting filters are imperfect.

But, some signals are by their nature not sharply band limited. In the case of frequency modulation, Carson's formula (4.7-4) for the rf bandwidth B_r in fm,

$$B_r = 2(B + x_p f_d) \qquad (4.7\text{-}4)$$

is merely a useful estimate of the bandwidth of a signal with a very complicated frequency spectrum.

In phase modulation with a constant rf amplitude, the total rf bandwidth required is very close to twice the base bandwidth when the amount of phase modulation is small, but the rf bandwidth becomes greater as the amount of phase modulation is increased, as in fm.

In considering the bandwidth of actual signals, including phase-modulated signals, we face two related considerations. (1) How broad must we make the bandwidth of the actual receiver in order to receive enough of the transmitted signal so that we can demodulate it without serious error (as, with acceptable low intersymbol interference)? (2) How much interfering power will be received from transmitters in adjacent frequency bands? Indeed, how far apart in frequency must rf channels be centered to allow satisfactory operation for a given baseband frequency? It is perhaps too much to expect that one simple definition of bandwidth can resolve such complex issues.

Equation (7.1-15) gave one definition of bandwidth which was useful in determining a particular optimal pulse shape for a given bandwidth. It was the spectral peak definition:

$$B_{pk} = \int_0^\infty [P(f)]^{1/2} \, df/(P_{max})^{1/2} \qquad (7.1\text{-}15)$$

Here P_{max} is the peak of the spectrum $P(f)$.

A closely related definition is the square-root definition:

$$B_{sqrt} = \left(\int_0^\infty [P(f)]^{1/2} \, df \right)^2 \bigg/ \int_0^\infty P(f) \, df \qquad (10.9\text{-}1)$$

This definition, like (7.1-15), gives bandwidth 0 if the spectrum is concentrated at a finite number of frequencies, and bandwidth $2B$ if the spectrum is flat from $f_0 - B$ to $f_0 + B$ and 0 elsewhere (here $B < f_0$). As always, $f_0 = 2\pi\omega_0$, where ω_0 is the carrier frequency. The formula gives infinite bandwidth if

$$\int_0^\infty [P(f)]^{1/2} \, df$$

is infinite, as it is for rectangular pulses. We see that the bandwidth will be infinite whenever the power spectrum $P(f)$ does not drop very rapidly with increasing frequency.

The definition of bandwidth as the length of the smallest frequency interval which contains all energy or power is sometimes too restrictive. Most waveforms or noises have energy or power at very high frequencies, although for practical purposes their energy is confined to a finite region. By this definition, every time-limited pulse has infinite bandwidth. This is not very useful.

Here we will define the *root mean square* (rms) bandwidth B_{rms} and then use the definition to determine the bandwidth in phase shift keying using sinc pulses. The rms definition in this case leads to a simple answer, whereas definition (10.9-1) or (7.1-15) leads to mathematical difficulties. Also, the rms definition leads to a simple bound for interfering power at frequencies far from the carrier frequency. We shall not derive this bound.

We can imagine the spectrum $P(f)$ to be a probability density by normalizing the energy or power to 1. We take $2(3)^{1/2}$ times the square root of the variance of that probability distribution to be the rms bandwidth B_{rms}:

$$B_{rms} = 2(3)^{1/2} \left[\frac{\int_0^\infty f^2 P(f) \, df}{\int_0^\infty P(f) \, df} - \left(\frac{\int_0^\infty f P(f) \, df}{\int_0^\infty P(f) \, df} \right)^2 \right]^{1/2} \qquad (10.9\text{-}2)$$

Subtracting the mean frequency in (10.9-2) tends to center the spectrum around the carrier. For two-sided spectra, we could use the same definition, with integration from $-\infty$ to ∞. But, then we would not be centering around the carrier ω_0, and the rms bandwidth would be f_0^2 or greater. This is not useful except for baseband spectra.

Why is there a $2(3)^{1/2}$ in equation (10.9-2)? This is because we want the rms bandwidth of the spectrum which is constant from $f_0 - B$ to $f_0 + B$ to be equal to $2B$. For baseband spectra where the integration is from $-\infty$ to ∞, the factor of 2 is missing, and the mean frequency is 0.

As an example of the use of the rms definition, the rms bandwidth of a stationary baseband random signal or noise with autocorrelation function $C(\tau)$ can be found, using Fourier transforms of derivatives. The result is

$$B_{\text{rms}} = \frac{3^{1/2}}{2\pi} \left(\frac{-C''(0)}{C(0)} \right)^{1/2} \qquad (10.9\text{-}3a)$$

We shall not derive this. If we have a band-pass rf signal, we want the bandwidth around the carrier ω_0. The one-sided definition is appropriate, and we must subtract f_0^2 and multiply by 2 to obtain the rf bandwidth:

$$B_{\text{rms}} = \frac{3^{1/2}}{\pi} \left(\frac{-C''(0)}{C(0)} - f_0^2 \right)^{1/2} \qquad (10.9\text{-}3b)$$

We can apply this result to the phase modulation signal $X(t)$:

$$X(t) = \cos\left[\omega_0 t + \phi(t)\right] \qquad (10.9\text{-}4)$$

Here $\phi(t)$ is a random pulse process with N pulses per second. The pulse waveform shape is $\Omega(t)$, multiplied by ± 1 data which is random and independent. So

$$\phi(t) = \sum_{n=-\infty}^{\infty} \varepsilon_n \Omega\left(t - \frac{n}{N}\right) \qquad (10.9\text{-}5)$$

Here $\varepsilon_n = \pm 1$ randomly and independently. We shall not discuss the method of recovering the data ε_n. Unless $\Omega(t)$ is small, which would mean that most of the transmitted power is in the carrier rather than in the data, the recovery process can be very complex and nonlinear. We shall not derive it.

We can apply (10.9-3b) and the results of Section 8.3 to this process to find its rms bandwidth. The result is

$$B_{\text{rms}} = \frac{(3N)^{1/2}}{\pi} \left(\int_{-\infty}^{\infty} [\Omega'(t)]^2 \, dt \right)^{1/2} \qquad (10.9\text{-}6)$$

We shall not derive this. Equation (10.9-6) is also the formula for the rms bandwidth if the symmetric pulse $\Omega(t)$ has energy $1/N$ and is used with *amplitude* modulation. The error probabilities in the two cases are not the same, and the spectral shapes are not the same, unless the pulse is rectangular of height $\pi/2$. We shall not go into these matters, and the theory is not yet complete. Reference 6 can be consulted for more on the bandwidth and spectrum of constant-amplitude modulation schemes.

We can use the above results to find the bandwidth of phase modulation by sinc pulses of peak amplitude h. We know that the rms rf bandwidth when sinc pulses are used in amplitude modulation is $2B = N$. We also stated that if the pulses had energy $1/N$ (or peak 1), the rms bandwidth in amplitude modulation is the same as the rms bandwidth in phase modulation. If we multiply the sinc pulses of peak 1 by h, the rms bandwidth of phase modulation goes up by h, according to (10.9-6). So, the rms rf bandwidth of sinc pulses of peak h used in phase modulation is

$$B_{rms} = hN \qquad (10.9\text{-}7)$$

We note that (10.9-7) says something clearly false for allocating channels—that the bandwidth of phase modulation by sinc pulses goes to zero as the modulation index or peak amplitude h goes to zero. This is because all the power is in the carrier. In spite of this, the receiver for phase modulation will still need substantial bandwidth, which must exceed the pulse rate independent of the modulation index. We can make this more quantitative by reasoning analogous to that used in deriving Carson's approximation (4.7-4) for the bandwidth required in fm receivers. But, the rms definition of bandwidth does not measure required receiver bandwidth.

It appears that a better definition than rms bandwidth might be the unnormalized bandwidth or "spectrum pollution index." For this, we multiply (10.9-2) by the square root of the total power. And, we insist that systems of the same pulse error probabillity be compared. If we do this, small-index phase modulation needs extremely high power, and the unnormalized rms bandwidth will not go to zero. We shall not explore this.

With this section, we complete the study of data transmission where the data is sent independently, or with weak dependencies for spectral shaping. In the next chapter, we begin the study of the ultimate limits of data transmission, based on Shannon's information theory.

Problems

1. Give a heuristic explanation of why offset quadrature amplitude modulation tends to level out the fluctuations of amplitude. At which instants of time would you expect the fluctuations in amplitude to be the greatest? Why?

2. Show that the rms bandwidth of a sinc pulse of ordinary baseband bandwidth B has rms bandwidth B. Show that the rms bandwidth is $2B$ if the pulse is used to amplitude modulate a carrier.

3. Show that the rms bandwidth of rectangular pulses is infinite. HINT: use (10.9-6) and delta functions.

4. Show that the rms bandwidth of binary am using the symmetric pulse $\Omega(t)$ of energy $1/N$ is equal to the rms bandwidth given by (10.9-6) when the same pulse is used in phase modulation. [HINT: use the energy theorem and the formula for the Fourier transform of a derivative. Let $X(f)$ be the voltage spectrum of $\Omega(t)$.]

5. Derive a crude expression, analogous to Carson's formula (4.7-4), for the approximate channel separation B, required in phase modulation using N sinc pulses per second of peak height h.

Limits to Error Performance: Information Theory

We have seen in Section 5.1 that a continuous signal of bandwidth less than B can be represented exactly by and reconstructed exactly from $2B$ sample amplitudes per second. We have seen in Section 5.6 that, allowing a certain error in the representation and recovery of the sample amplitudes, these amplitudes can be quantized and represented by a number of discrete amplitudes or levels. Further, if we wish we can designate these amplitudes or levels by binary numbers. We have seen in other parts of Chapters 5, 8, and 10 how signals with a discrete number of levels, including two-level or binary signals, can be transmitted by amplitude or phase modulation, subject to certain errors caused by noise.

Thus we have come to consider transmission systems in which a continuous signal can be represented to some desired degree of accuracy by a sequence of binary digits, and these can be transmitted, perhaps as + or − pulses, with an acceptably low error rate. In transmitting positive or negative baseband sinc or rectangular pulses, a signal power-to-noise power ratio of 31.6, or 15 dB, assures that the chance of an error in interpreting the sign of the pulse will be only about one in 10^8.

We are inevitably led to ask, how best can we transmit a sequence of discrete numbers? How fast can such transmission be, and are errors in transmission indeed unavoidable?

The answers to these questions, and to many other questions as well, are to be found in C. E. Shannon's work on information theory (References 1 and 2).

Information theory by itself could be the subject of several taxing courses. Here we will relate knowledge gained through Shannon's work (Reference 3, esp. Chapter X) to problems raised in this book. Our approach will not be to proceed in the logical order of Shannon's exposition, but rather to abstract material from his work which addresses most directly questions raised in this book.

We shall be interested in the gaussian channel and the photon channel involved in optical communication. We will find the capacity of the gaussian channel in this chapter, but delay finding the capacity of the photon channel until the next chapter. Reference 4 in particular discusses in depth the problem of modeling the channels used in optical communication.

11.1. Signal Space and Noise Samples

In Section 10.7, we studied m-phase shift keying. In determining the symbol error probability, we rotated the coordinates to a more convenient pair. But, the joint probability distribution of the noise in the new directions was the same as in the old. In Section 7.4, we found that the signal-to-noise ratio at the sampling times in optimally filtered pulses did not depend on the pulse shape at all, but only on the pulse energy. These phenomena can best be understood through the concept of *signal space*. We shall show how signal space is defined, and how it is used to determine the maximum possible rate of nearly error-free transmission over a gaussian channel.

The sampling theorem tells us that any band-limited signal can be represented by samples taken at Nyquist intervals, that is, by $2B$ sample amplitudes per second. A single sample can be represented in one dimension—as distance from the origin along one axis. Two samples can be represented in two orthogonal dimensions—the x coordinate can represent one and the y coordinate another. Similarly, n successive samples can be represented by the coordinates of a point in an n-dimensional orthogonal space called *signal space*. We shall carefully construct signal space in what follows.

The total energy associated with n samples is the sum of the squares of their amplitudes. Thus the total energy is the sum of the squares of coordinate values which represent the samples in n-dimensional space. This sum of squares is just the square of the distance from the origin to the point which represents the n samples of the signal.

If the energy of the signal samples has some probability distribution, the probability distribution of the energy of many independent samples will be more peaked. With high probability, as n is made larger and larger the total energy of n samples lies closer and closer to n times the average energy per sample. This follows from the law of large numbers.

Shannon makes use of the concept of signal space and the peaking of the energy probability distribution in finding the capacity of a continuous band-limited channel for a given average signal power and an added white gaussian noise. In our terms, this limiting channel capacity C, measured in bits per second, is

$$C = B \log_2 (1 + P_s/P_n) \qquad (11.1\text{-}1)$$

We should note that Shannon's proof holds for a signal of *average* power P_s and an additive gaussian noise of average power P_n. Transmitters may be limited in *peak* power rather than or in addition to average power. Finding the capacity of a peak-power-limited gaussian channel of finite bandwidth is more difficult and we shall not do it. We will see later that as the bandwidth widens for given average power, peak power constraints become less important.

What does it mean to have capacity C? We can transmit bits over the channel by proper selection of one of 2^m waveforms, each representing m input bits. We find that if we wish to transmit *fewer* than C bits per second, we can by proper design obtain decoded bits of arbitrarily low bit or word error probability. This design includes proper selection of the waveforms and the decoding procedure. Here a word represents m bits, or is an m-bit symbol. To obtain the channel capacity, we must let m become large. This means complex equipment at the receiving end, and a delay in outputting data.

We cannot transmit at rates *above* the capacity C. For large m, the word error probability approaches 1 instead of 0 at rates above capacity. The *bit* error probability remains above a positive value greater than 0 but less than 1. This least bit error rate is given by rate distortion theory, which we will study in Section 13.3.

These same considerations about channel capacity apply to any channel. We shall give the proof only for the band-limited gaussian channel. We will study the capacity for general channels in Section 12.3.

Before we explore the reasoning behind Shannon's theorem, we will rigorously define signal space and the noise distribution in it. This will explain how the channel noise affects the different components of a signal. From this, we will prove Shannon's theorem in the next section. There we will explore some consequences of his fundamental result for efficient communication.

In the next section, we shall be interested in having M $(=2^m)$ signals, waveforms, or codewords, one for each of the 2^m binary numbers, or m-tuples of 0's and 1's, that we wish to transmit. If we transmit at a rate R bits per second, and if the M waveforms are nonzero over the same time interval from 0 to T, we will have

$$m = RT, \qquad M = 2^{RT} \qquad (11.1\text{-}2)$$

We are interested in a *coding* in which large *blocks* of data are represented by one codeword, and so T will be large. For a finite bandwidth B Chapter 5 shows that we have $2TB = N$ sinc pulses or waveforms,

$$\phi_j(t) = (2B)^{1/2} \operatorname{sinc} 2B \frac{t - j}{2B}, \qquad 1 \le j \le N \qquad (11.1\text{-}3)$$

Here the $\phi_j(t)$ have unit energy. And, because T is large, these are almost orthonormal over the interval from 0 to T. Why do we say almost? By this we mean that, for the overwhelming majority of pairs i, j, those for which the peaks are not near the ends of the time interval, we will have

$$\int_0^T \phi_i(t)\phi_j(t)\, dt \doteq \delta_{ij} = \begin{cases} 0 & \text{if } i \ne j \\ 1 & \text{if } i = j \end{cases} \qquad (11.1\text{-}4)$$

This means the integral is nearly 0 if $i \ne j$ and nearly 1 if $i = j$.

Furthermore, Chapter 5 suggests (this can be made precise) that there is no way to obtain substantially more than N orthonormal functions of bandwidth B which have almost all of their energy between 0 and T. So, this particular choice gives as many signal dimensions as possible within the given bandwidth. The M codewords $y_i(t)$ will lie in the N-dimensional space spanned by the $\phi_j(t)$. We see that the number of dimensions $N = 2BT$ rises much more slowly with increasing T than the number of codewords $M = 2^{RT}$.

Suppose $x(t)$ is a function of finite energy and of bandwidth less than B with its energy nearly entirely confined to time 0 to T. Such an $x(t)$ can be represented by a vector x in signal space:

$$x = (x_1, x_2, \ldots, x_N)$$

Here

$$x_j = \int_0^T x(t)\phi_j(t)\, dt$$

This is the component of x along the ϕ_j direction, or the *correlation* of $x(t)$ with $\phi_j(t)$. We will note that the length squared of x is the energy of $x(t)$:

$$\sum_{j=1}^N x_j^2 = \int_0^T x^2(t)\, dt$$

This is because we can write

$$x(t) = \sum_{j=1}^N x_j\phi_j(t)$$

since x is mostly confined to the range 0 to T. Since the ϕ_j are orthonormal, the result follows.

We now define the noise distribution in signal space. We cannot sample white gaussian noise, because it has only infinite values. But we can integrate or cross-correlate it to find its component along the $\phi_j(t)$ direction. The result is a random variable n_j:

$$n_j = \int_0^T n(t)\phi_j(t)\, dt$$

What is the distribution of this random variable? We can replace integration from 0 to T by the integral over all time. This does not change the value much when T is large, because $\phi_j(t)$ is nearly 0 outside this interval. So

$$n_j \doteq \int_{-\infty}^{\infty} n(t)\phi_j(t)\, dt, \qquad 1 \le j \le N \tag{11.1-5}$$

In (11.1-5), the $\phi_j(t)$ are low-pass waveforms whose energy is concentrated within frequencies $-B$ to B. We can if we wish think of (11.1-5) in the frequency domain using the multiplication theorem of Section 3.6. If we first pass the white gaussian noise through the same ideal low-pass filter, the value of (11.1-5) must be unchanged. This is because two ideal low-pass filters in series have the same transfer function as a single low-pass filter. So, in (11.1-5) we may regard $n(t)$ as gaussian noise whose two-sided spectral density is equal to $N_0/2$ between frequencies $-B$ and B and is equal to 0 for higher frequencies.

Since $n(t)$ is now band limited to the frequency interval $-B$ to B, the sampling theorem and (11.1-3) show that n_j in (11.1-5) also represents $(1/2B)^{1/2}$ times the sample of $n(t)$ at time $j/2B$. And, by Section 9.4, these samples must be jointly gaussian random variables. The autocorrelation of the band-limited noise $n(t)$ is, from the Fourier transform table, Table 3.1,

$$C(\tau) = N_0 B \operatorname{sinc} 2B\tau \tag{11.1-6}$$

Equation (11.1-6) allows the covariance of the random variables n_j to be easily found:

$$E(n_j n_k) = \frac{1}{2B} C\!\left(\frac{j-k}{2B}\right); \qquad E(n_j^2) = \frac{N_0}{2} \tag{11.1-7}$$

The random variables n_j have variance $N_0/2$ equal to the two-sided noise spectral density or half of the one-sided density. If $j \ne k$, we see from (11.1-6) that

$$C\!\left(\frac{j-k}{2B}\right) = 0$$

This implies from (11.1-7)

$$E(n_j n_k) = 0, \qquad j \neq k \qquad \qquad (11.1\text{-}8)$$

The random variables n_j and n_k are uncorrelated if $j \neq k$. Since they are jointly gaussian random variables, n_j and n_k are independent if $j \neq k$.

Knowing the joint distribution of the random variables n_j in principle allows word error probabilities to be determined. We did something like this in Section 10.7. More importantly, the joint distribution shows that there is no loss in using only the N samples j with $1 \leq j \leq N$ to make the decision as to which word or waveform was sent. For, we can augment the N sinc functions, the ones with $1 \leq j \leq N$, by all the rest of them, the ones for $j \leq 0$ and for $j > N$. Any band-limited function can be represented by all its sample values, $-\infty < j < \infty$. The collection of these samples uniquely determines what was received.

The samples for j between 1 and N have a signal component and a noise component; the ones for the other j are pure noise. Since the samples for the other j are independent of the samples for j between 1 and N there can be no advantage to using the other samples in deciding which codeword was sent. Thus the best possible decision can be made on the basis of only the samples taken between times 0 and T. This is not surprising.

The channel we have arrived at thus transmits samples [or rather, $(1/2B)^{1/2}$ time samples], and each sample received is the sample transmitted plus an independent gaussian random variable of variance $N_0/2$. The next section derives Shannon's theorem, which tells us how to use this channel most effectively. We do this by careful choice of the $M = 2^m$ waveforms $y_i(t)$ used for signaling.

Problems

1. Which is a more appropriate performance criterion, word error probability or bit error probability, in the following two situations? (a) a common carrier system transmitting voice digitally with many users multiplexed onto the same serial bit stream; (b) a dedicated digital channel for transmitting financial transactions between clearing house banks.

2. Show that the white gaussian noise $n(t)$ in (11.1-5) can be low-pass filtered to contain no frequencies above B without changing the value of n_j. [HINT: use the multiplication theorem of Section 3.6 or equation (6.3-5) with $\tau = 0$.] The same kind of argument shows that the output of filters with distinct pass bands are uncorrelated, hence independent.

3. Equation (11.1-1) gives the channel capacity of a white gaussian channel of bandwidth B with signal power P_s and noise power $P_n = N_0 B$, where N_0 is the one-sided noise spectral density, a constant. The signal power and bandwidth are both increased by a factor K, but N_0 is unaltered. Show that the capacity

also increases by a factor of K. Explain heuristically why this must be so, using frequency division multiplexing (FDM), and using time division multiplexing (TDM). (FDM stacks adjacent frequency bands and uses each as an independent channel. TDM uses adjacent time slots for each of the independent channels.)

11.2. Shannon's Theorem

Equation (11.1-1) gave the capacity of the gaussian channel of bandwidth B. In order to prove that this value is the capacity, we must do two things. We must show that if we try to communicate above this rate, the word error probability cannot approach zero. This is sometimes called the *converse* of the channel coding theorem. We must also show the *direct* part of the coding theorem. This is the harder part. It says that for T large there are almost 2^{CT} codewords of average power no more than P_s which have a word error probability as small as we like. These must all also be contained in an orthogonal space of no more than about $2BT$ dimensions. The smaller the word error probability we desire, the larger we must make T.

We show the converse first, because it is easier. Suppose T is large and that we have M signals $y_i(t)$. Suppose also that the word error probability is small. We have signal space of $N = 2BT$ dimensions. The square of the length L of the noise vector

$$n = (n_1, n_2, \ldots, n_N)$$

is

$$L^2 = \sum_{j=1}^{N} n_j^2 \tag{11.2-1}$$

Here L^2 is a random variable. As T is large, $N = 2BT$ also becomes large. Equation (11.2-1) then expresses L^2 as a sum of a large number of independent identically distributed random variables, which are the n_j^2. By the law of large numbers, the average L^2/N is highly likely to be very close to its expected value. By equation (11.1-7), this expected value is $N_0/2$, the two-sided power spectral density of the additive white gaussian noise $n(t)$.

Thus for large T the noise vector n will almost certainly cause the received signal vector x to be near the surface of a hypersphere of radius $(NN_0/2)^{1/2}$. This hypersphere or noise sphere is centered at the transmitted vector y_k. Here

$$x = y_k + n$$

is the received signal. We can, according to Section 11.1, write

$$x = (x_1, \ldots, x_N) \tag{11.2-2}$$

where

$$x_j = \int_0^T x(t)\phi_j(t)\,dt \tag{11.2-3}$$

Here the $\phi_j(t)$ are sinc functions normalized to be approximately orthonormal over the interval 0 to T. We derived this in Section 11.1. In what follows, we will assume that the $\phi_j(t)$ are exactly orthonormal. This assumption is justified because we will let the codeword duration T approach infinity.

What is the expected length squared of x? The length squared is the sum of the squares of the individual components x_j. This gives

$$E\sum_{j=1}^N x_j^2 = E\sum_{j=1}^N y_{kj}^2 + 2E\sum_{j=1}^N y_{kj}n_j + E\sum_{n=1}^N n_j^2 \tag{11.2-4}$$

The y_{kj}, which are the jth components of the y_k signal vector or codeword, are not random variables. The middle term is then 0, because each $E(n_j)$ is 0. The last term we have seen is $NN_0/2$. And, we observe that

$$\sum_{j=1}^N y_{kj}^2 \le TP_s$$

This expresses the requirement that each y_k have average power over the time T of no more than P_s. The expected squared length of the received signal-plus-noise vector x will then be bounded as follows:

$$E\sum_{j=1}^N x_j^2 \le TP_s + NN_0/2 \tag{11.2-5}$$

Here the N random variables

$$x_j = y_{kj} + n_j$$

are independent. This is because the y_{kj} are given values, while the n_j are random and independent. Thus, by the law of large numbers again, the random variable $\sum_{j=1}^N x_j^2$ will almost certainly lie within, or at least not far outside, a sphere centered on the origin and of radius $[TP_s + (NN_0/2)]^{1/2}$.

The noise spheres of radius not much more than $(NN_0/2)^{1/2}$ about each y_k must fit into the signal sphere of radius $[TP_s + (NN_0/2)]^{1/2}$, or a not much larger sphere. For low error probability, these noise spheres must hardly overlap, because points in the overlap result in errors. This is because we decode a received vector x as the y_k which is closest to it, to make the decoding as likely as possible to be correct (see Fig. 11.1). This is the best possible decoding because the gaussian density in n dimensions decreases in all directions away from its peak. And, we have seen that the probability is concentrated near the surfaces of the noise spheres. Because

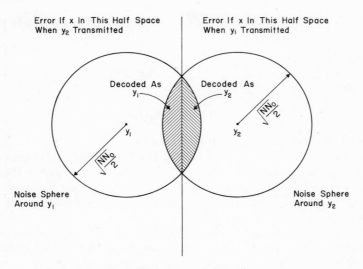

Figure 11.1. Errors in decoding.

the gaussian distribution is spherically symmetric, the probability is proportional to surface area. So a large overlap is the same as a large probability of error.

The volume of a hypersphere of radius r in N dimensions is

$$V_N(r) = c_N r^N \qquad (11.2\text{-}6)$$

where c_N is some constant. The number M of noise spheres that can fit into the received signal sphere with hardly any overlap can at most be the ratio of the volume of the received signal sphere divided by the volume of one noise sphere. Thus from (11.2-6)

$$M \leq \frac{V_N[TP_s + (NN_0/2)]^{1/2}}{V_N(NN_0/2)^{1/2}} = \left(\frac{TP_s + (NN_0/2)}{NN_0/2}\right)^{N/2} \qquad (11.2\text{-}7)$$

If we communicate at rate R, (11.1-2) gives

$$M = 2^{RT}$$

If we are to have low error probability, (11.2-7) must hold. Since the number of signal dimensions N is $2BT$, (11.2-7) becomes

$$2^{RT} \leq (1 + P_s/BN_0)^{BT} \qquad (11.2\text{-}8)$$

Here

$$BN_0 = P_n \qquad (11.2\text{-}9)$$

is the noise power in bandwidth B. We may take the logarithm to the base 2 in (11.2-8) to obtain a bound on the rate R of communication:

$$R \le B \log_2 (1 + P_s/P_n) \qquad (11.2\text{-}10)$$

Capacity is the *largest* rate at which we can communicate with low error probability. So, (11.2-10) shows

$$C \le B \log_2 (1 + P_s/P_n) \qquad (11.1\text{-}1')$$

This is the converse of the coding theorem.

   We now derive the direct part of the coding theorem. No one knows an algorithm for constructing good codes, but we can prove that they must exist. Even when good codes are found, decoding may be very complex. The closer we try to get to the channel capacity, the more complex the decoding will be.

   Our previous discussion leads us to look for a set of almost 2^{CT} codewords y_i (this means 2^{RT} codewords with R close to C) of average power at most P_s, such that the noise spheres of radius $(NN_0/2)^{1/2}$ centered on the codewords are almost nonoverlaping. This is much like packing oranges in a crate. Here the crate is round, and the dimensionality of the crate and of the oranges is high. The higher the dimensionality, the smaller the fraction of the crate the oranges will fill. But, we are interested in the *logarithm* of the number of oranges. Taking the logarithm of the number of oranges divided by the number of dimensions as the measure of fit, it will turn out that for large T the fit of the packing can be made almost perfect.

   How do we choose the code vectors y_i, which correspond to N-tuples of numbers y_{ij}? We will choose them at random and independently, using a technique known as *random coding*. We will show that the error probability averaged over all random codes is small. This is the same as the error probability when both the code and the noise are random, but independent of each other. If this probability is small, there must be some *actual* codes with small error probability, too.

   How do we choose codes at random? The ordinary (hyper-) volume element inside the sphere of radius $(TP_s)^{1/2}$ is used for the probability of selection. We must normalize by dividing by the hypervolume of this sphere to make the total probability equal to 1. The y_i vectors are chosen independently according to this N-dimensional probability. Because each N-tuple is in the sphere of radius $(TP_s)^{1/2}$, each waveform $y_i(t)$ arising from y_i via the expression

$$y_i(t) = \sum_{j=1}^{N} y_{ij}\phi_j(t) \qquad (11.2\text{-}11)$$

will have average power $\le P_s$ in time T.

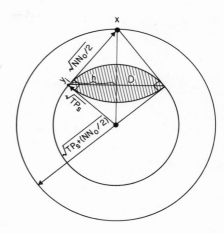

Figure 11.2. The error lens D.

We can see from (11.2-6) or from the law of large numbers that almost all the probability (hypervolume) of the sphere of radius $(TP_s)^{1/2}$ lies close to its surface. We note that this means we can insist that each $y_i(t)$ have average power *exactly* P_s with no loss of capacity, but we shall not pursue this. However, in Fig. 11.2, we have chosen a typical random y_i as if it were on the hypersurface of this sphere, instead of being inside. The received x is as we know very likely to be about $(NN_0/2)^{1/2}$ away from whatever y_i is transmitted. And, we have seen that x will have length about $[TP_s + (NN_0/2)]^{1/2}$.

When do we make an error? There is an error when the random choice of codewords causes some *other* codeword y_k to be in the lens-shaped region D of Fig. 11.2. The region D is the set of possible codewords closer to the received x than the transmitted y_i is. Here we think of the noise vector n as having been chosen first, and then the random code. Because the noise in no way depends on what is transmitted, this makes no difference. The probability of error caused by the choice of y_k for a *given* other k is the normalized volume of this lens D.

Because the codewords are chosen independently of each other, we may assume that y_i is fixed. This will not affect the error probability. Also, the noise vector n is independent of which code $y_1, y_2, \ldots, y_M$ was chosen. So, we can assume the x is of the correct or typical distance from both y_i and the origin. This will be so by the law of large numbers.

Note that .the vector from x to y_i is normal to y_i, because the Pythagorean theorem holds for the three sides of the triangle. We see by using similar triangles that the radius or half-chord h of the lens must be

$$h = \left(\frac{TP_s NN_0/2}{TP_s + (NN_0/2)} \right)^{1/2} \tag{11.2-12}$$

The lens D is contained in a sphere of radius h. Its volume V must be at most the volume of the sphere of radius h. From (11.2-6), this is

$$V_N(h) = c_N h^N \qquad (11.2\text{-}13)$$

The random code was chosen by normalizing the volume of the sphere of radius $(TP_s)^{1/2}$ to 1 so that it was a probability distribution. We did this by dividing by $V_N(TP_s)^{1/2}$. So if the probability of error caused by y_k lying in the lens is $P_{e,k}$, it can be bounded as follows:

$$P_{e,k} \leq \frac{V_N(h)}{V_N(TP_s)^{1/2}} \qquad (11.2\text{-}14)$$

$$P_{e,k} \leq \frac{(TP_s NN_0/2)^{N/2}}{[TP_s + (NN_0/2)]^{N/2}} \bigg/ (TP_s)^{N/2} = \frac{1}{[1 + TP_s/(NN_0/2)]^{N/2}}$$

and

$$P_{e,k} \leq \frac{1}{(1 + P_s/P_n)^{BT}} \qquad (11.2\text{-}15)$$

So, the probability that the lens is *free* of y_k for this one particular k can be bounded by

$$1 - P_{e,k} \geq 1 - \frac{1}{(1 + P_s/P_n)^{BT}} \qquad (11.2\text{-}16)$$

The probability that the lens is free of all $(M - 1)$ of the y_k with $k \neq i$ is the probability $1 - P_e$ of correct reception when the code and the noise are chosen independently. The y_k are independent for all k, so

$$1 - P_e = (1 - P_{e,k})^{M-1} \geq (1 - P_{e,k})^M \geq \left(1 - \frac{1}{(1 + P_s/P_n)^{BT}}\right)^M$$

$$1 - P_e \geq \left(1 - \frac{1}{(1 + P_s/P_n)^{BT}}\right)^M \qquad (11.2\text{-}17)$$

For positive numbers a and M, with a less than 1 and M greater than 1,

$$(1 - a)^M > 1 - Ma \qquad (11.2\text{-}18)$$

This is because the curve $(1 - a)^M$ is concave upward, and $(1 - Ma)$ is tangent to this curve at $a = 0$. We use (11.2-18) with

$$a = \frac{1}{(1 + P_s/P_n)^{BT}}$$

Then (11.2-17) becomes

$$1 - P_e \geq 1 - \frac{M}{(1 + P_s/P_n)^{BT}}$$

$$P_e \leq \frac{M}{(1 + P_s/P_n)^{BT}} \qquad (11.2\text{-}19)$$

We seek to make the number of codewords M that we send in the time T as large as possible, while making the probability of error less than any assigned value, however small. The number of codewords we send in the time T is related to the rate R in bits/second by $M = 2^{RT}$. So, (11.2-19) becomes

$$P_e \leq \left(\frac{2^R}{(1 + P_s/P_n)^B} \right)^T \tag{11.2-20}$$

This approaches 0 as T approaches ∞ if

$$\frac{2^R}{(1 + P_s/P_n)^B} < 1$$
$$2^R < (1 + P_s/P_n)^B$$
$$R < B \log_2 (1 + P_s/P_n) \tag{11.2-21}$$

We said in discussing equation (11.1-1) that the right-hand side of (11.2-21) is the capacity C of a gaussian channel of bandwidth B, average power limitation P_s, and noise power $P_n = N_0 B$, where $N_0/2$ is the two-sided noise spectral density. We have shown in (11.2-21) that we can communicate reliably (this means with word error probability approaching 0) at all rates below this capacity. We showed earlier that this was not possible above capacity. This justifies calling $B \log_2 (1 + P_s/P_n)$ the capacity of the channel. We have proved Shannon's theorem.

Although the ideas used to derive Shannon's theorem appear to rely heavily upon the gaussian nature of the channel, a similar proof is valid in general. We shall not prove it in this book. The way to carry it over to the other channels is through the concept of the mutual information of two random variables. These random variables are the input to and output from a communication channel. We will define mutual information in Section 12.3.

In the next section, we shall use the capacity formula to obtain performance limits for the gaussian channel. The proper comparison is in terms of energy per bit.

Problems

1. Give another reason why we might expect the noise vector n to be almost orthogonal, at least with high probability, to the randomly chosen transmitted signal vector y_i, when the number of dimensions N (or the signal duration T) is large.

2. In Fig. 11.2, we have drawn the error lens D as if it were contained in the sphere of radius h given by (11.2-12) centered at the midpoint of the chord orthogonal to the received signal-plus-noise vector x. What would the noise vector n have had to be for this containment to be *not* true? (This event is of extremely low probability by Problem 1.)

3. Show that we can construct a random code as M independently chosen examples of white gaussian noise waveforms on the interval $0 \leq t \leq T$ low-pass filtered to bandwidth B. [HINT: the orthonormal functions $\phi_i(t)$ are sinc functions.] Find the spectral density N_0' of the noise in terms of the average power P_s. What probability distribution should we use in this case for random coding? (This random code also achieves capacity.)

11.3. Ideal Energy per Bit

Let us rewrite Shannon's formula (11.1-1) for channel capacity, letting

$$P_n = N_0 B$$

We obtain

$$C = B \log_2 \left(1 + \frac{P_s}{P_n}\right)$$

$$2^{C/B} - 1 = \frac{P_s}{P_n} = \frac{P_s}{N_0 B} = \frac{P_s/C}{N_0/(C/B)}$$

$$\frac{P_s}{C} = N_0 \cdot \frac{2^{C/B} - 1}{C/B} \tag{11.3-1}$$

The quantity P_s/C is power, that is, energy per second, divided by bits per second. That is, P_s/C is energy per bit, E_b. We see that for Shannon's ideal performance, energy per bit is given by the power density N_0 times a function of C/B, the ratio of bit rate to bandwidth. This function, which we will call the energy factor Q, is

$$Q = \frac{2^{C/B} - 1}{C/B} \tag{11.3-2}$$

Here the dimensionality of Q is in Hz per bit per second, or cycles per bit. Then (11.3-1) becomes

$$E_b = QN_0 \text{ J/bit} \tag{11.3-3}$$

Here J is joules. When C/B is very small (or B/C is very large),

$$2^{C/B} = e^{(\ln 2)(C/B)} \doteq 1 + (\ln 2)(C/B)$$

$$Q \doteq \ln 2$$

$$E_b \doteq (\ln 2)N_0 = 0.693 N_0 \text{ J/bit} \tag{11.3-4}$$

This is the minimum value of E_b attainable when there is no bandwidth constraint. It is called the Shannon limit. Q is plotted vs. C/B in Fig. 11.3.

The region with C/B small, around 0.02, is where deep space

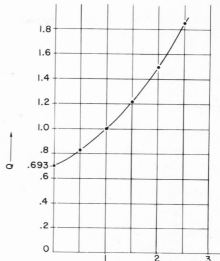

Figure 11.3. Energy factor Q vs. capacity per Hz, C/B.

C/B In Bits Per Sec/Hz Bandwidth⟶

communication channels operate. This is because the low signal-to-noise ratios provide a low capacity compared to the assigned bandwidth of many megahertz. The constraint from deep space is power, not bandwidth. The largest possible capacity for given power is needed, and so B is large. Commerical communication satellites, on the other hand, are more constrained in bandwidth than in power. This is because the communication distances are much shorter, and the potential traffic much higher. On these satellite links, C/B is on the order of 1. Values of C/B greater than 1 are not common on space channels, but are achieved on wire channels. Here signal power may not be a constraint, but the available bandwidth is limited by the physical properties of wire transmission paths. Multilevel signaling is used as in Sections 8.5, 10.5, and 10.7 to achieve data rates greater than the bandwidth.

It is interesting to note that in Section 2.3, in expressions concerning noise-limited free-space transmission, the following quantity occurs:

$$\frac{P_c B}{P_n} = \frac{P_c}{P_n/B} = \frac{P_c}{N_0} \tag{11.3-5}$$

Here P_c is the carrier power, which we may take as P_s. This means that we will convert the entire received signal to carrier to recover the carrier phase. For example, in psk, we can begin this recovery by squaring as in Section 10.8. If P_c/P_n is very small, according to (11.3-4) we can replace P_c/N_0 by $C \ln 2$. This sets a limit to the performance we can attain in carrier-to-noise ratio if we communicate at very low error probability. A lower carrier-to-noise ratio will not support low error probability communication at rate C.

The next section relates the Shannon limit to thermal and quantum noise.

Problems

1. What is P_s/P_n in dB for $C/B = 0.5, 1, 2$?

2. How much power can we potentially save by using coding if we want a bit error probability of at most 10^{-8} on a white gaussian channel of bandwidth B? Assume the uncoded baseline is the case of positive and negative sinc pulses occurring at rate $2B$ pulses per second. Find the potential power saving in dB. HINT: use the asymptotic error probability expression (8.5-12).

3. How *large* a bit error probability would have to be required before the savings in Problem 2 dropped to only 2 dB? (These two problems ignore the fact that we are comparing channels without errors to channels with errors. Section 13.3 shows how to make a better comparison, and allow in dB for the extra benefit of error-free channels.)

4. According to (11.1-1), the capacity C/B of the band-limited gaussian channel *in bits per second per Hz* (or in bits per cycle) depends only on P_s/P_n. Suggest how optimal codes for one gaussian channel with given P_s, N_0, and B can be transformed into optimal codes for another gaussian channel with P_s', B', and the same N_0, if the capacities in bits per second per Hz are equal.

11.4. Shannon's Limit and Thermal and Quantum Noise

Sections 11.1 and 11.2 tell us that the energy per bit used in transmitting messages cannot be less than $(\ln 2)N_0$, where N_0 is the noise power spectral density. Sections 2.1 and 2.9 give us expressions for thermal and quantum noise. What is the relation between these concepts?

For Johnson or thermal noise at low frequencies

$$N_0 = kT \tag{11.4-1}$$

Thus the energy per bit must be at least

$$(\ln 2)kT \tag{11.4-2}$$

For quantum noise alone, Section 2.9 and later 11.9 tell us that if we amplify and convert to baseband or to a low frequency the resulting noise density N_{01} at the output of an ideal amplifier of power gain G is

$$N_{01} = hfG \tag{11.4-3}$$

If we use a high-gain amplifier, this is equivalent to a noise of power density

$$N_0 = hf \qquad (11.4\text{-}4)$$

multiplied by the gain of the amplifier. We shall derive this in Section 11.9. So, if we use an ideal high-gain amplifier and frequency converter in our receiver we can take the quantum noise density to be given by (11.4-4). The energy per bit will be

$$E_b = (\ln 2)hf = 0.693hf \text{ J/bit}$$

But, hf is just the energy of a single quantum in joules per quantum. Hence the energy per bit is less than (0.693 times) the energy per quantum. Or, the number of bits we can transmit per quantum if we amplify the signal with an ideal amplifier is

$$\frac{1}{\ln 2} = 1.443 \text{ bits/quantum} \qquad (11.4\text{-}5)$$

A more effective way of coping with quantum effects in transmitting information will be given in Section 12.5. This will not use an ideal linear amplifier but rather a photon counter.

The next section shows how we may approach Shannon's limit for the gaussian channel by signal design.

Problems

1. Equation (11.4-4) seems to imply that low frequencies are best for transmission of information, because the quantum-induced noise spectral density using ideal linear amplification, $N_0 = hf$, is lower for low frequencies. But, (1.3-1) suggests that for directional antennas of fixed diameters, *high* frequencies give higher data rates. Assume that the total noise spectral density N_0 is a sum of the frequency-dependent quantum term hf and a background thermal noise density N_{02} independent of frequency:

$$N_0 = hf + N_{02}$$

Also assume that antenna efficiencies are perfect, or independent of frequency. For given areas of the transmitting and receiving antennas and given transmitted power, what frequency gives the maximum channel capacity?

2. Suppose in addition the efficiencies of the transmitting and receiving antennas decrease as a function of frequency by the common factor

$$f_0/(f + f_0)$$

What happens to the optimum communication frequency now? Consider the three cases $N_{02} = 0$ (no thermal noise); $N_{02} = kT$, with k Boltzmann's constant 1.380×10^{-23} J/°K and $T = 3$°K (cosmic background radiation); and finally the case $N_{02} \to \infty$ (high thermal noise). (Planck's constant h is 6.62×10^{-34} J sec.) Assume $f_0 = 100$ GHz for this. For what temperature T is $hf_0 = kT$?

11.5. Approaching Shannon's Limit

In Shannon's proof in Section 11.2 of the expression for channel capacity, a signal function or codeword is chosen at random within a sphere of radius $(TP_s)^{1/2}$. We interpret this as follows: $2BT = N$, the number of samples or dimensions in the signal, T is the duration of a codeword, P_s the signal power, and TP_s the energy of a codeword. Physically, this also corresponds very nearly to choosing the samples which make up the signal functions or codewords randomly with a gaussian distribution of amplitudes (see Problem 3 of Section 11.2).

Let us imagine trying to implement such a system. First we choose codewords in this random fashion. Each codeword consists of a block of N samples of different amplitudes. We store these codewords so that we will have them available to use. We use the codewords to encode a message. The signal which is transmitted is made up of a sequence of the codewords. When we receive a block of N samples we compare this block with every allowed or chosen block or codeword, and interpret the received signal as that block or codeword which has the least mean-squared difference from the received block of samples. While not impossible, this form of encoding and decoding is laborious and impractical.

Is it possible to approach Shannon's limiting rate or channel capacity in some simpler and more systematic way?

We obtain some encouragement from Shannon (p. 63 of Reference 1). He showed that white gaussian noise has the peculiar property that it can absorb any other small random signal added to it, with resultant increase in entropy (see Section 12.3) approximately equal to the entropy of a *gaussian* signal of the same power as the small random non-gaussian signal. Thus any weak signal ought to have the same information-carrying capacity as a weak gaussian signal of the same power.

Since we have seen that the optimum signals are approximately gaussian noise waveforms, *any* distribution of weak signal amplitudes ought to be optimum as well. For example, perhaps we can use signal samples of the same amplitudes, when the signal-to-noise ratio is small. This indeed proves to be true, and we shall show it in the next section. When the number of dimensions per second is allowed to be large, the energy per dimension becomes small, and Shannon's concept applies. In this section, we permit two amplitudes, 0 and another which becomes very large as we approach capacity.

We studied continuous-time pulse position modulation or ppm in Section 4.6. Here we study sampled data ppm or its quantized version quantized ppm. This is a simple form of encoding which can approach Shannon's minimum energy per bit as given by equation (11.3-4). We derive this result without appealing to the weak signal absorption property.

In quantized ppm, the signals or codewords consist of sequences N samples long. All the samples but one have zero amplitude. We will call the common amplitude of the nonzero sample v_0. If the nonzero sample is always positive, we have N codewords, one for each of the N time positions the nonzero sample may occupy. If we allow the nonzero sample to be either positive or negative, we get a total of $2N$ codewords of the same energy per codeword. In this case, the average voltage level over all codewords will be zero. We have used N dimensions to send $\log_2(2N)$ bits.

When we receive the signal, we store the block of N samples which constitutes the codeword in a delay line or similar storage device. Then we compare the amplitudes of all samples and choose the sample with the largest magnitude. We see from Section 11.1 that this largest sample, positive or negative, corresponds to the position and sign of the most likely transmitted codeword. This follows because the height of the independent gaussian density function decreases uniformly in all directions with increasing distance from the mean, if the variances in each dimension are the same. We may be in error, for noise may cause the received sample to be the largest in a position in which no energy was transmitted. What is the probability P_c that we will correctly identify the transmitted codeword?

We will assume gaussian noise of variance σ^2. The probability density of the voltage v_n for any sample position for which no power was transmitted is

$$p(v_n) = \frac{1}{(2\pi)^{1/2}\sigma} \exp\left(\frac{-v_n^2}{2\sigma^2}\right) \tag{11.5-1}$$

The probability P_s that $|v_n|$ is not greater in magnitude than some value v_x is

$$P_s = \frac{1}{(2\pi)^{1/2}\sigma} \int_{-v_x}^{v_x} \exp\left(\frac{-v_n^2}{2\sigma^2}\right) dv_n \tag{11.5-2}$$

$$P_s = 1 - 2[1 - \Phi(v_x/\sigma)]$$

$$P_s = 2\Phi(v_x/\sigma) - 1 \tag{11.5-3}$$

Here $\Phi(x)$ is the standard gaussian cumulative probability distribution of (8.5-10).

Let v_x be the received voltage due to both signal and noise in the correct or signal sample position, in which the signal voltage is $+v_0$. The probability density of v_x will be

$$p(v_x) = \frac{1}{(2\pi)^{1/2}\sigma} \exp\left(\frac{-(v_x - v_0)^2}{2\sigma^2}\right) \tag{11.5-4}$$

We will interpret the received signal correctly as resulting from a positive signal in this sample position only if v_x is positive and greater than the magnitude of the voltage received in *any other* sample position.

P_s as given by (11.5-3) is the probability that the magnitude of the voltage in a given *one* other sample position is less than or equal to v_x. The probability that the magnitude of the voltage in *all* of the $N - 1$ other sample positions is less than or equal to v_x is

$$P_s^{N-1} \qquad (11.5\text{-}5)$$

This is because the noise components of different sample positions are independent. The independence arises as in Section 11.1 because the bandwidth B is $1/2T_0$, where T_0 is the time interval between samples.

Considering all possible positive voltages in the signal sample position, the probability that this voltage is positive and larger in magnitude than the voltage in any other sample position, that is, the probability of identifying the correct codeword, is a weighted average of (11.5-5), using the weight function of (11.5-4). From (11.5-3), this becomes

$$P_c = \frac{1}{(2\pi)^{1/2}\sigma} \int_0^\infty \left(2\Phi\left(\frac{v_x}{\sigma}\right) - 1\right)^{N-1} \exp\left(\frac{-(v_x - v_0)^2}{2\sigma^2}\right) dv_x$$

This expression can be made simpler by the following substitution:

$$y = v_x/\sigma$$

$$P_c = \frac{1}{(2\pi)^{1/2}} \int_0^\infty [2\Phi(y) - 1]^{N-1} \exp\left(\frac{-[y - (v_0/\sigma)]^2}{2}\right) dy$$

$$(11.5\text{-}6)$$

Let us now consider the energy per bit. We have assumed ideal (sinc pulse) signaling with an allowable pulse or sample spacing $T_0 = 1/2B$. Thus the energy per pulse E_0 is

$$E_0 = T_0 v_0^2 = v_0^2/2B \qquad (11.5\text{-}7)$$

There is just one pulse per codeword, so E_0 is the energy per codeword. We also have, for a noise density N_0 and a bandwidth B, a mean-squared noise voltage

$$\sigma^2 = N_0 B \qquad (11.5\text{-}8)$$

There are $2N$ codewords, so the number of bits transmittted per codeword is

$$\log_2 2N = \ln 2N/\ln 2 \qquad (11.5\text{-}9)$$

Thus the energy per bit, E_b, will be

$$E_b = \frac{E_0}{\ln 2N/\ln 2} \qquad (11.5\text{-}10)$$

We can see from (11.5-7) through (11.5-10) that

$$v_0/\sigma = (2E_0/N_0)^{1/2} = (2\alpha \ln 2N)^{1/2} \qquad (11.5\text{-}11)$$

Here

$$\alpha = E_0/N_0 \ln 2N = E_b/N_0 \ln 2 \qquad (11.5\text{-}12)$$

From (11.5-12) and (11.5-6), the probability of receiving a codeword correctly, P_c, is

$$P_c = \frac{1}{(2\pi)^{1/2}} \int_0^\infty [2\Phi(y) - 1]^{N-1} \exp\left(\frac{-[y - (2\alpha \ln 2N)^{1/2}]^2}{2}\right) dy \qquad (11.5\text{-}13)$$

We are interested in conditions under which P_c approaches unity, when E_b and N_0 are given. We will certainly have

$$\frac{1}{(2\pi)^{1/2}} \int_0^\infty \exp\left(\frac{-[y - (2\alpha \ln 2N)^{1/2}]^2}{2}\right) dy \doteq 1 \qquad (11.5\text{-}14)$$

very nearly when N is large, because we are picking up almost all the area under the curve

$$\frac{1}{(2\pi)^{1/2}} \exp\left(\frac{-[y - (2\alpha \ln 2N)^{1/2}]^2}{2}\right)$$

from $-\infty$ to ∞. This area is 1, since this is merely a displaced gaussian density, one with a large positive mean.

We want the error probability $1 - P_c$ to be small. In virtue of (11.5-14), when N is large we will have

$$1 - P_c \doteq \frac{1}{(2\pi)^{1/2}} \int_0^\infty \{1 - [2\Phi(y) - 1]^{N-1}\} \exp\frac{-[y - (2\alpha \ln 2N)^{1/2}]^2}{2} dy \qquad (11.5\text{-}15)$$

As in the derivation of (11.5-14), we only need to integrate over an infinite interval of y starting somewhat to the left of $(2\alpha \ln 2N)^{1/2}$, where the left-hand endpoint differs from $(2\alpha \ln 2N)^{1/2}$ by a small positive fraction ε of $(2\alpha \ln 2N)^{1/2}$. This gives almost all the integral when N is large.

In this interval, y is large. Recalling (8.5-11), we can write

$$2\Phi(y) - 1 \sim 1 - \frac{2}{y(2\pi)^{1/2}} \exp\left(\frac{-y^2}{2}\right) \qquad (11.5\text{-}16)$$

$$[2\Phi(y) - 1]^{N-1} \sim \left(1 - \frac{2 \exp(-y^2/2)}{y(2\pi)^{1/2}}\right)^{N-1} \qquad (11.5\text{-}17)$$

As we saw in equation (11.2-18),

$$(1 - z)^{N-1} > 1 - (N - 1)z$$

for z between 0 and 1 and for $N > 2$. Hence we *overestimate* the error probability $1 - P_c$ by using the approximation

$$1 - [2\Phi(y) - 1]^{N-1} \doteq 2(N - 1)\frac{\exp(-y^2/2)}{y(2\pi)^{1/2}} \tag{11.5-18}$$

This gives the upper bound

$$1 - P_c \leq \frac{2}{(2\pi)^{1/2}} \int_L^{\infty} (N - 1)\frac{\exp(-y^2/2)}{y(2\pi)^{1/2}} \exp\left(\frac{-[y - (2\alpha \ln 2N)^{1/2}]^2}{2}\right) dy \tag{11.5-19}$$

In (11.5-19), we must use the lower limit L instead of 0:

$$L = (2\alpha \ln 2N)^{1/2}(1 - \varepsilon) \tag{11.5-20}$$

Here ε is a small positive number. We have observed that this will include almost all the value of the integral. But, y will still be large so that (11.5-18) holds.

The term $[(N - 1)\exp(-y^2/2)]/y(2\pi)^{1/2}$ in (11.5-19) is as *large* as possible when y is as *small* as possible, that is, when y equals L. This is because $\exp(-y^2/2)/y$ decreases for positive y. An upper bound for this factor of the integrand in (11.5-19) is, from (11.5-20),

$$\frac{(N - 1)\exp[-(\alpha \ln 2N)(1 - \varepsilon)^2]}{(2\alpha \ln 2N)^{1/2}(1 - \varepsilon)(2\pi)^{1/2}}$$

$$= \frac{N - 1}{(2N)^{\alpha(1-\varepsilon)^2}(2\alpha \ln 2N)^{1/2}(1 - \varepsilon)(2\pi)^{1/2}} \tag{11.5-21}$$

Since ε is as small as we like, and α is fixed, (11.5-21) is very small if α is greater than 1. For, the exponent of N in the denominator of (11.5-21) will be greater than the exponent 1 of N (or rather of $N - 1$) in the numerator, and N is large. So $[(N - 1)\exp(-y^2/2)]/y(2\pi)^{1/2}$ is small for $y > L$.

Since the remaining factor

$$\frac{1}{(2\pi)^{1/2}}\exp\left(\frac{-[y - (2\alpha \ln 2N)^{1/2}]^2}{2}\right)$$

integrates to 1, the entire right-hand side of (11.5-19) remains small *after* integration. We have shown that the probability of error $1 - P_c$ can be made as small as we please when N gets large. For this, we need only that $\alpha > 1$, or, by (11.5-12), that $E_b > N_0 \ln 2$. We see that if we use even a little more energy per bit than the minimum required by the Shannon limit, the error probability approaches zero.

The result is that quantized ppm approaches Shannon's limit as the number of dimensions N approaches infinity. Here N is $2BT$, where B is the bandwidth of the sinc pulses used, and T is the duration of the codeword. So, we have an explicit coding scheme, one for the infinite-bandwidth gaussian channel. This is because we must allow N to approach infinity to attain the ideal energy per bit given by Shannon's theorem. The ppm coding scheme presented here is the only explicit coding scheme known that achieves channel capacity for a commonly occuring channel.

We also see that quantized ppm has another disadvantage besides its large bandwidth. The peak power becomes very large. We shall see in the next section how to achieve the performance of quantized ppm without having large peak power.

What about the finite-bandwidth gaussian channel? The ppm system has ever-increasing bandwidth as the block length increases, and so cannot be used for a real bandwidth-constrained channel. We have observed that there is no known explicit coding scheme for achieving minimum energy per bit on these real gaussian channels. In Section 12.2, we shall study a good explicit coding technique, convolutional coding, which does not expand the bandwidth very much, but it will not achieve Shannon's limit.

Finding explicit coding schemes that come close to the optimum, but that can be decoded with an affordable investment in decoding equipment, is the central problem of coding theory. We shall look at some examples in the next two sections and in the next chapter.

Problems

1. Show that the capacity of the infinite-bandwidth gaussian channel of average power P_s and noise spectral density N_0 is $P_s/N_0 \ln 2$. Does quantized ppm achieve this capacity?

2. If N dimensions are used for quantized ppm, what is the *peak* power P in terms of the average power P_s?

3. What happens to the E_b in quantized ppm if we send only positive pulses instead of positive and negative? (This shows that capacity is not lowered if we somehow lose our polarity or phase reference.)

11.6. Biorthogonal Coding and Its Performance

In the previous section we thought of the coding as quantized pulse position modulation, that is, modulation in which a positive or negative pulse can appear in any of N successive positions in a block or codeword. There is a geometrical way of looking at this code, or at any block code. As in Shannon's derivation in Section 11.2, we look on a codeword consisting of N samples as a vector in N-dimensional signal space. In the case of the code of Section 11.5, the vector representing a codeword lies

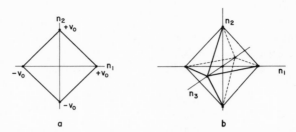

Figure 11.4. Codes as geometric figures: (a) a code as vertices of a geometrical figure in two dimensions, (b) a code in three dimensions.

along one axis of the N-dimensional space only, and has no components along any of the other axes. Each vector has the same length v_0 in N dimensions, and there are $2N$ codewords. Of the two possible vectors for each dimension, one extends in the positive direction and one in the negative direction. The points at the heads of the vectors, which represent the codewords, are the vertices of an N-dimensional figure. Figure 11.4 shows these figures in two and three dimensions. The axes n_1, n_2, n_3 represent the first, second, and third sample positions, or dimensions, along which the voltage for a particular codeword can lie.

We saw in the previous section that Shannon's limiting rate can be approached by using codewords whose samples are represented by the vertices of such a geometrical figure. Suppose, now, that we rotate the figure with respect to the coordinate axes and choose as the samples of the codewords the coordinates (projections on the axes) of the vertices of the *rotated* figure. We saw in Section 11.1 that this does not affect the probability distribution of the noise. This means that the word error probabilities will be the same. And, the energy of each codeword will be unchanged; it will be the sum of the squares of the coordinates of the vertices of the rotated figure. Codewords represented by vertices on different axes of the figure must be orthogonal after the rotation, as before. That is, if x_1, y_1, z_1 and x_2, y_2, z_2 are the coordinates (or samples) of two codewords represented by two vertices, then the dot or inner product of the two vectors is 0 both before and after the rotation. Or

$$x_1 x_2 + y_1 y_2 + z_1 z_2 = 0 \qquad (11.6\text{-}1)$$

Can all the coordinates or samples of the rotated figure have the same magnitude after rotation? We may wish to do this to keep the peak power low, or to obtain a more nearly constant-amplitude system. This is certainly possible in two dimensions, as is shown in Fig. 11.5. Clearly, it is not possible in an odd number of dimensions. A sum formed as in (11.6-1) cannot be zero if the magnitudes of all terms are equal. This is

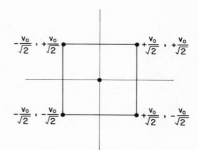

Figure 11.5. Coordinates of rotated two-dimensional figure.

because the sum will be an odd multiple of the magnitude squared of one term. However, if the number N of dimensions is even, it may be possible to rotate the figure in such a way that, in the codewords so produced, all voltages (all coordinates of the codeword points) are equal in magnitude. Thus we will have N orthogonal pairs of codewords of equal energy in which all samples are equal in magnitude. We will illustrate this by listing the signs of the components (of magnitude $v_0/4^{1/2} = v_0/2$) for the samples (or coordinates) of a properly rotated figure in four dimensions:

$$
\begin{aligned}
&+,+,+,+; &&-,-,-,- \\
&+,+,-,-; &&-,-,+,+ \\
&+,-,+,-; &&-,+,-,+ \\
&+,-,-,+; &&-,+,+,-
\end{aligned}
\tag{11.6-2}
$$

We can easily see that each codeword of a pair in one of the four rows is orthogonal to both codewords of any other pair. The samples of codewords of one of the pairs are all of the same sign and *all other codewords* are made up of equal numbers of positive and negative samples.

In N dimensions there are 2^N possible codewords which have equal samples of given magnitude. However, there can be only N orthogonal codewords in an *orthogonal code*. The $2N$ codewords, which include N codewords of an orthogonal code and each of these codewords multiplied by -1, constitute a *biorthogonal code*, or more properly a constant-amplitude biorthogonal code. Such a code is obtained as a linear transformation (a rotation) of the quantized pulse position modulated code, whose geometric representation is as the vertices of a geometrical figure. Each vertex of this figure is a common distance from the origin, as in Fig. 11.4.

The number N of dimensions must be even, if we are to be able to rotate the figure so as to produce a codeword in which all the sample voltages (coordinates) are equal in magnitude as in Fig. 11-5a. One codeword can have all positive voltages (coordinates). The other codeword pairs must then have half positive voltages and half negative voltages.

In order for two codewords which have half positive and half negative voltages to be orthogonal, we must be able to change one codeword into the other by changing half of the positive voltages into negative voltages and half of the negative voltages into positive voltages. We see that this is so by letting v_1, v_2, v_3, v_4, etc. be the voltages of one codeword and u_1, u_2, u_3, u_4, etc. be the voltages of the other codeword orthogonal to it. The magnitudes of all voltages are equal. The codewords are orthogonal if

$$v_1 u_1 + v_2 u_2 + v_3 u_3 + v_4 u_4 + \text{etc.} = 0$$

Half of the u's must have signs opposite to the corresponding v's for the codewords to be orthogonal. But this is just what we get by changing the signs of half of the positive v's and also changing the signs of half of the negative v's. If we change too few or too many, we do not get a word with half positive and half negative voltages.

The rotation we have described can be represented as a linear transformation of the components or samples of the codewords, which represent points in N-dimensional space. This is illustrated in Fig. 11.6. We can take the samples of quantized ppm codewords and convert them to samples of biorthogonal codewords by means of a linear network. We can reconvert the samples of the biorthogonal codewords to samples of quantized ppm codewords by another or inverse linear transformation. Each transformation preserves both energy and orthogonality of codewords. They are *orthogonal transformations.* The transformations can be implemented as matrix multiplications, in which the quantized ppm

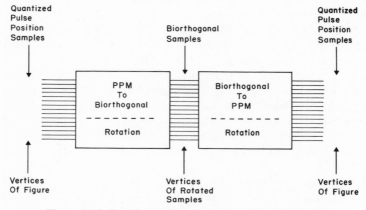

Figure 11.6. Rotation as linear transformation of samples.

codeword samples represent one matrix, the biorthogonal codeword samples represent another matrix, and the transformation is the orthogonal matrix that converts one set of samples into the other.

The conversion matrix from quantized ppm to biorthogonal is easily derived when the biorthogonal code is given. The inverse matrix, which takes the biorthogonal code into quantized ppm, is also easily derived. We shall not go into this further.

In Section 11.5 we analyzed the performance of quantized ppm transmission in the presence of noise. Suppose that we convert to biorthogonal prior to transmission and reconvert to quantized ppm after reception. The conversions or transformations preserve energy, correlation, and probability. So, the random noise power mixed with the signal samples has the same distribution as in the analysis of Section 11.5. For a given signal power and noise density, quantized ppm and biorthogonal coding with constant-amplitude samples give the same error rate. In general, for a white gaussian channel this is true for any two signal sets related by an orthogonal transformation. This follows from the results of Section 9.3 as embodied in the signal space concept of Section 11.1.

Biorthogonal coding with constant-amplitude samples has a practical advantage. The codewords consist of sequences of pulses all of the same energy, rather than of widely separated pulses each with N times as much energy. The samples or pulses of such a biorthogonal code can be sent by phase modulating a transmitter of constant power with rectangular pulses of $\pm 90°$. This is the same as amplitude modulation by ± 1. The spectrum is nearly the same as that of quantized ppm if rectangular pulses are used and N is large. We shall not derive this here. Biorthogonal coding such as this with $N = 32$ was used on the Mariner and Viking spacecraft from 1969 to 1975, when the data rates were still low enough that bandwidth was not much of a consideration. This has given way to convolutional coding, to be studied in Section 12.2. Convolutional coding has better signal-to-noise performance and takes less bandwidth.

The coding of this section uses signals of constant amplitude. We occupy a bandwidth of at least $B = N/2$, where N is the number of samples or dimensions per second. To send n *bits* encoded into such a codeword, we need 2^{n-1} *dimensions* in this binary or two-level signaling scheme. How much is the required bandwidth lowered if we use instead simple multilevel transmission with increased power? The next section examines this.

Problems

1. Are there six orthogonal codewords ($N = 6$), each word consisting of six sample voltages which are $+1$ or -1? Why? What general result does this suggest?

2. List eight orthogonal codewords ($N = 8$), each word consisting of eight sample voltages which are $+1$ or -1. [HINT: start with the orthogonal code of the left half of (11.6-2) and repeat it in the last four positions as the top half of the new code. Repeat it with a sign reversal as the bottom half of the new code.] Explain why this works.

3. Show that for large N, the power spectrum of biorthogonal coding with baseband rectangular pulses and equally likely data is nearly the same as the spectrum of ppm of the same average power. HINT: show that the autocorrelations are nearly the same. It is reasonable to assume that the contents of adjacent positions in the biorthogonal code are approximately uncorrelated for large N.

11.7. Energy per Bit for Multilevel Transmission

We have seen in Sections 11.5 and 11.6 that we can approach Shannon's limiting energy per bit for large ratios of bandwidth to channel capacity by using block coding (quantized ppm or constant-amplitude biorthogonal) with very long blocks. In Section 8.5 we calculated the error rate in multilevel data transmission in which each transmitted pulse represents one or more binary digits. So we may think of multilevel transmission as a form of coding. Suppose that in such transmission we make the signal-to-noise ratio large enough so that the error rate is acceptably low. How does the energy required per bit compare with Shannon's limiting value?

In multilevel pulse transmission we ideally transmit $2B$ pulses or dimensions per second. Let the signal-to-noise power ratio be P_s/P_n. Then

$$P_s/P_n = \text{(energy/bit) (bits/pulse) (pulses/sec) } (1/N_0 B)$$
$$= \text{(energy/bit)(bits/pulse) } (2N_0)$$
$$E_b = \text{energy/bit} = (N_0/2) \, (P_s/P_n)/\text{(bits/pulse)} \qquad (11.7\text{-}1)$$

We can write this as

$$E_b = Q_p N_0 \qquad (11.7\text{-}2)$$

Here

$$Q_p = (1/2) \, (P_s/P_n)/\text{(bits/pulse)} \qquad (11.7\text{-}3)$$

In dB, Q_p is

$$-3.01 + 10 \log_{10} (P_s/P_n) - 10 \log_{10} \text{(bits/pulse)}$$

Here the number of bits per pulse is $\log_2 m$, where m is the number of levels.

Section 8.5 gives a means for calculating approximately the probability of error in multilevel transmission. For two-level transmission, if the

Table 11.1. Power Required for Multilevel Transmission

Number of levels = m	Bits/pulse = $\log_2 m$	Signal power = P_s	$R/B = 2\log_2 m$	Q_p (dB)	Q (dB)
2	1	v_0^2	2	12.2	1.8
4	2	$5\,v_0^2$	4	16.2	5.7
8	3	$21\,v_0^2$	6	20.7	10.2
16	4	$85\,v_0^2$	8	25.5	15.0
32	5	$341\,v_0^2$	10	30.5	20.1

signal-to-noise power ratio is 15 dB, the error rate is approximately one error in 10^8 pulses, or 10^{-8}.

In two-level transmission the signal power is v_0^2, where $2v_0$ is the spacing between levels. The signal powers and bits/pulse are given in Table 11.1 for various cases of multilevel transmission, together with the ratio R/B of bit rate to bandwidth, assuming $2B$ pulses per second. R/B is just twice the number of bits per pulse. Here it is assumed in calculating the power that all transmission digits or levels are equally likely; the signal power is then the expectation of the voltage squared. The spacing between levels is $2v_0$ in all cases, and so the pulse error probabilities are approximately equal for all numbers of levels.

If the signal-to-noise ratio v_0/σ^2 is 15.2 dB for two levels, corresponding to a 10^{-8} error rate, we will have the energy-per-bit performance given in Table 11.1. Here Q is Shannon's minimum attainable energy per bit for the given bit rate to bandwidth ratio, divided by the noise spectral density. This was defined in Section 11.3. It is given by (11.3-2), where the capacity C is the bit rate R.

In Fig. 11.7, we have two scales for the bits per cycle ordinate. The top one is for baseband or single-sideband modulation, and the bottom one is for double-sideband or phase modulation. For phase modulation, we have assumed an rf bandwidth equal to the bit rate, but we know that more rf bandwidth will be needed to attain the performance shown. The values of Q_p given in the table above are shown as crosses. A curve of Q as computed from (11.3-2) is also shown. We take $C/B = R/B$ in the Q curve. This is because Q is the *minimum* attainable energy per bit divided by noise spectral density. So, R must be taken as large as possible, that is, equal to the capacity C. These curves can be read with either scale.

The analogous Q_ϕ curve for phase modulation is also plotted, as the circles in Fig. 11.7. Only the bottom scale is relevant for phase modulation. We see that the curve is about half-way between the two other curves for error probability 10^{-8}. We note from Section 10.9 that the spectrum of m-level baseband rectangular pulses is the same as the

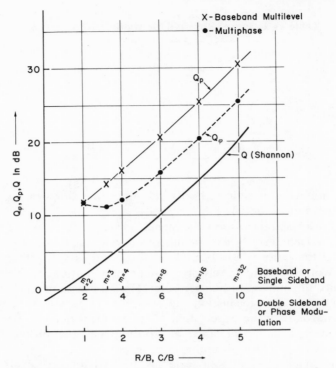

Figure 11.7. Energy per bit for multilevel signaling ($P_e = 10^{-8}$).

spectrum around the carrier of m-phase modulation using rectangular pulses. So, the spectrum of m-level amplitude modulation is the same as the spectrum of m-phase modulation. The Q_p curve for double sideband and the Q_ϕ curve agree for $R/B = 1$, as must be. This corresponds to two-level or one bit per pulse signaling.

We see that for multilevel pulse-by-pulse transmission at the error rate given, the energy required per pulse is consistently about 10 dB higher than Shannon's ideal value, regardless of number of levels. The difference becomes larger at still lower error rates. It becomes arbitrarily large, but rather slowly as the error probability required approaches zero.

We also see that the energy required per pulse increases rapidly with the number of bits per pulse. Going from 1 bit per pulse (two level) to 2 bits per pulse (four level) requires an increase in energy per bit of 4 dB. This is the ratio of $5v_0^2$ to v_0^2 in Table 11.1, divided by the 2 bits/pulse for 4 level.

We can read the tables in another way. If we have a constraint on bandwidth, we can achieve high ratios of R/B, the bit rate divided by bandwidth, by increasing the signal power and using multilevel transmission. The tables tell us what the increase over Shannon's ideal is for the simple multilevel signaling scheme. The power is increased in this way for data transmission over ordinary telephone channels to get more bits per second through the channel without using complicated coding schemes. In this case, bandwidth and coding cost are scarcer resources than received power. If 10^{-8} error probability is desired, we are paying about a 10-dB-power penalty over the ideal for the given bandwidth by using multilevel signaling but no coding. For satellite channels of restricted bandwidth, similar design decisions have been made. For this, multiphase shift keying is used as in Section 10.7. We could also code the multilevel pulses, but we shall not consider this. We note that for narrow bandwidths, the optimum coding must require multilevel signaling.

We have studied the gaussian channel so far in this chapter. The next two sections study the channel in which the limitation is not additive thermal noise but rather the quantum nature of high-frequency electromagnetic radiation. The next section derives the quantum mechanical uncertainty principle which we need to derive quantum noise.

Problems

1. Show that for low pulse error probability P_e, the difference in dB between the upper and lower curves of Fig. 11.7 approaches $10 \log_{10} \left[\frac{2}{3} \ln \left(1/P_e \right) \right]$. This is independent of the number m of bits per pulse. You may assume that the average power of multilevel signaling with $\log_2 m$ bits/pulse (m levels) is $\frac{1}{3}(m^2 - 1)v_0{}^2$, where the levels are spaced $2v_0$ apart. Here $m = 2^k$, where there are k bits/pulse. About what would P_e have to be in order for Shannon coding to save 15 dB over uncoded multilevel transmission?

2. Equation (10.7-11) enables one to calculate the energy per bit required in multiphase modulation as a function of the number of phases for given symbol or baud error probability. Using the results of Problem 1 of Section 10.7, derive the points on the Q_ϕ curve of Fig. 11.7 for error probability 10^{-8}, for $m = 2$-, 3-, 4-, 8-, 16-, and 32-phase moculation. How much better in dB power efficiency is 32 phase over 32 amplitude double sideband when the symbol error rate is 10^{-8}? What is this limiting advantage in dB as m becomes large? (This is independent of error probability P_e for low P_e.) How far above the Shannon Q curve is the Q_ϕ curve for large m and small P_e, in terms of P_e?

3. Suppose that many levels (amplitudes or phases) are used in transmission. In phase modulation these levels lie around a circle in signal space of circumference $2\pi v_p$ where the signal is $v_p(2^{1/2} \cos \omega_0 t)$. The power is $v_p{}^2$. In amplitude modulation the levels lie along a line of length $2v_a$, where v_a is the peak signal voltage. For a given length L of line (circumference or straight), what are the relative powers? In each case, assume that all levels are equally likely. Assume that the number of levels is large. What does this have to do with the previous problem?

11.8. The Quantum Mechanical Uncertainty Principle

Information theory concerns itself with the fundamental limits of communication over a given channel. But, we must know the channel analytically before these fundamental limits can be found. The gaussian channel was characterized in Section 11.1, and its capacity was found in Section 11.2.

We shall be interested here in the fundamental limits of communication by means of electromagnetic waves of very high frequency where quantum effects become important. In this section, we derive the uncertainty relation of quantum mechanics. We use this relation in the next section to explain why there is a noise density hf, as described in Section 2.9, if we try to implement an ideal amplification receiver for this channel.

Suppose we know only that the probability density of the frequency of the field of an electromagnetic wave has a gaussian density

$$V(f) = \frac{1}{(2\pi)^{1/2}\sigma} \exp\left(\frac{-f^2}{2\sigma^2}\right)$$

Here f can be positive or negative. The baseband signal is modulated up to a carrier frequency f_0 to give the actual density of photon frequency in the wave:

$$V(f) = \frac{1}{2(2\pi)^{1/2}\sigma}\left[\exp\left(\frac{-(f-f_0)^2}{2\sigma^2}\right) + \exp\left(\frac{-(f+f_0)^2}{2\sigma^2}\right)\right] \quad (11.8\text{-}1)$$

This is a probability density on frequencies between $-\infty$ and $+\infty$, but it is also a voltage spectral density. Here σ is the root-mean-square deviation around zero frequency of the original baseband frequency spectrum from which we obtained (11.8-1). We note that (11.8-1) is a mixture of two gaussian probability densities with equal probabilities of $1/2$ for each density. One is the original baseband gaussian density moved to be centered around frequency $+f_0$, and the other is centered at $-f_0$.

The probability distribution of photons in energy or frequency (for energy $= hf$) is known from the laws of quantum mechanics to be proportional to $V(f)V^*(f)$, which for real $V(f)$ is $[V(f)]^2$:

$$[V(f)]^2 = \frac{1}{8\pi\sigma^2}\left[\exp\left(\frac{-(f-f_0)^2}{\sigma^2}\right) + \exp\left(\frac{-(f+f_0)^2}{\sigma^2}\right)\right.$$
$$\left. + 2\exp\left(\frac{-(f^2+f_0^2)}{\sigma^2}\right)\right] \quad (11.8\text{-}2)$$

For optical frequencies f_0, the third term above is negligible at all frequencies f.

If we write the distribution of photon energy or frequency in terms of the rms frequency deviation σ_f as an equal mixture of gaussian densities proportinal to

$$\exp\left(\frac{-(f - f_0)^2}{2\sigma_f^2}\right) \quad \text{and} \quad \exp\left(\frac{-(f + f_0)^2}{2\sigma_f^2}\right) \tag{11.8-3}$$

we see that

$$\sigma_f^2 = \sigma^2/2 \tag{11.8-4}$$

Except for a $3^{1/2}$ factor, the variance σ_f^2 is the rms bandwidth, as defined in Section 10.9, of the baseband energy spectrum corresponding to rf energy spectrum (11.8-2).

The time function or classical waveform $v(t)$ of the pulse voltage spectrum $V(f)$ is the Fourier transform of $V(f)$. According to (8.6-7), this is

$$v(t) = \tfrac{1}{2} \exp(-2\pi^2\sigma^2 t^2)[\exp(j2\pi f_0 t) + \exp(-j2\pi f_0 t)]$$
$$v(t) = \exp(-2\pi^2\sigma^2 t^2) \cos 2\pi f_0 t \tag{11.8-5}$$

The laws of quantum mechanics also state that the probability distribution for finding or detecting a photon at a time t is proportional to the square of the classical distribution of field as a function of time, i.e., $[v(t)]^2$, which varies (with the carrier removed, since we must have time resolution coarser than several cycles) as

$$\exp(-4\pi^2\sigma^2 t^2) \tag{11.8-6}$$

If we equate this to

$$\exp(-t^2/2\sigma_t^2) \tag{11.8-7}$$

where σ_t is the rms deviation of the time of detection we obtain

$$4\pi^2\sigma^2 = 8\pi^2\sigma_f^2 = 1/2\sigma_t^2$$
$$\sigma_f \sigma_t = 1/4\pi \tag{11.8-8}$$

Energy is proportional to frequency by Planck's constant, h, so that in terms of the rms uncertainty in energy, σ_e,

$$\sigma_e = h\sigma_f \tag{11.8-9}$$
$$\sigma_e \sigma_t = h/4\pi \tag{11.8-10}$$

This holds if the original $V(f)$ arose from a baseband gaussian frequency pulse.

For any baseband pulse, gaussian or not, we shall show that

$$\sigma_e \sigma_t \geq h/4\pi \tag{11.8-11}$$

Here equality holds *only* for gaussian pulses. This is the uncertainty relation as commonly stated. (Reference 3 of Chapter 3 goes into uncertainty relations in some generality, in its Chapter 4.) It is physically reasonable that there be such a relation; a pulse which has a small σ_t is narrow in time and so must have energy at high frequencies. We can conclude that it must have a large σ_f.

We shall derive (11.8-11). We assume that the carrier has been removed. If $S(f)$ is the baseband frequency pulse, and $s(t)$ is the time pulse which is its Fourier transform, we will have

$$S(f) = \int_{-\infty}^{\infty} s(t) \exp(-j2\pi ft)\, dt \qquad (11.8\text{-}12)$$

The energies of the frequency and time pulses are equal by the energy theorem of Section 3.6. Their energies are 1, because the energies represent probabilities:

$$\int_{-\infty}^{\infty} s^2(t)\, dt = 1 = \int_{-\infty}^{\infty} S(f)S^*(f)\, df \qquad (11.8\text{-}13)$$

Let us center the pulse so that it has mean value 0:

$$\int_{-\infty}^{\infty} ts^2(t)\, dt = 0 \qquad (11.8\text{-}14)$$

We can also assume that we know the value of the spread of the pulse about its mean, which we call σ_t^2. Since the pulse is centered about $t = 0$, we have

$$\sigma_t^2 = \int_{-\infty}^{\infty} t^2 s^2(t)\, dt \qquad (11.8\text{-}15)$$

We define the rms frequency spread σ_f^2:

$$\sigma_f^2 = \int_{-\infty}^{\infty} f^2 S(f)S^*(f)\, df \qquad (11.8\text{-}16)$$

Because $s(t)$ is real, the probability density $S(f)S^*(f)$ is symmetric about zero frequency, so that (11.8-16) measures the mean-square spread of the pulse about the mean or about zero frequency.

We will make use of Schwartz's inequality, which we derived in Section 9.1. The form we use is

$$\left| \int_a^b g_1(t)g_2(t)\, dt \right|^2 \le \int_a^b |g_1(t)|^2\, dt \int_a^b |g_2(t)|^2\, dt \qquad (11.8\text{-}17)$$

We have equality when and only when $g_1(t)$ and $g_2(t)$ are proportional. Equation (11.8-17) follows from (9.1-2) if we regard t as a random variable

T. Here the probability that T is between t and $t + dt$ is $dt/(a - b)$. Then, the integrals with respect to dt become expectations if we use $dt/(a - b)$ instead of dt. But, this only involves dividing both sides of (11.8-17) by $(a - b)^2$. [We could also derive (11.8-17) directly as we did (9.1-2), using integrals in place of expectations.] In any case, (11.8-17) holds for every finite a and b. We conclude that (11.8-17) also holds for infinite a and b:

$$\left| \int_{-\infty}^{\infty} g_1(t)g_2(t)\, dt \right|^2 \leq \int_{-\infty}^{\infty} |g_1(t)|^2\, dt \int_{-\infty}^{\infty} |g_2(t)|^2\, dt \qquad (11.8\text{-}18)$$

Equality holds here when and only when $g_1(t)$ and $g_2(t)$ are proportional.

Let us use (11.8-18) as follows. We take particular g_1 and g_2:

$$g_1(t) = ts(t)$$
$$g_2(t) = \frac{ds}{dt} \qquad (11.8\text{-}19)$$

(11.8-18) becomes

$$\left| \int_{-\infty}^{\infty} ts(t)\frac{ds(t)}{dt}\, dt \right|^2 \leq \int_{-\infty}^{\infty} [ts(t)]^2\, dt \int_{-\infty}^{\infty} \left(\frac{ds}{dt}\right)^2 dt \qquad (11.8\text{-}20)$$

The integral on the left-hand side can be integrated by parts. The result is

$$t\frac{s^2(t)}{2}\bigg|_{-\infty}^{\infty} - \frac{1}{2}\int_{-\infty}^{\infty} s^2(t)\, dt \qquad (11.8\text{-}21)$$

It is reasonable to assume that $ts^2(t)$ is small when t is large (positive or negative), so that $ts^2(t)$ is 0 at $\pm\infty$. Otherwise, σ_f^2 will be infinite, because $s(t)$ will have many sharp large spikes for t large. So, in view of (11.8-13), the left-hand side of (11.8-20) is equal to $(-\frac{1}{2})^2 = \frac{1}{4}$. What about the right-hand side?

The first integral on the right-hand side of (11.8-20) is equal to σ_t^2, by (11.8-15). The second can be found from what we learned in Chapter 3. From Section 3.8, the Fourier transform of ds/dt is $j2\pi f$ times the Fourier transform $S(f)$ of $s(t)$. By the energy theorem of Section 3.6, then, we must have

$$\int_{-\infty}^{\infty} \left(\frac{ds}{dt}\right)^2 dt = \int_{-\infty}^{\infty} [j2\pi fS(f)][j2\pi fS(f)]^*\, df$$

$$= (2\pi)^2\int_{-\infty}^{\infty} f^2 S(f)S^*(f)\, df$$

$$\int_{-\infty}^{\infty} \left(\frac{ds}{dt}\right)^2 dt = (2\pi)^2\sigma_f^2 \qquad (11.8\text{-}22)$$

Inequality (11.8-20) now becomes

$$\frac{1}{4} \leq \sigma_t^2 \cdot (2\pi)^2 \sigma_f^2$$

$$\frac{1}{16\pi^2} \leq \sigma_t^2 \sigma_f^2$$

$$\sigma_f \sigma_t \geq \frac{1}{4\pi} \tag{11.8-23}$$

Schwartz's inequality also tells us that we have equality precisely when

$$\frac{ds}{dt} = kts(t) \tag{11.8-24}$$

Here k is a constant. This differential equation can be solved:

$$s(t) = A \exp\left(kt^2/2\right) \tag{11.8-25}$$

where A is a constant. Since the energy in $s(t)$ is finite, k is negative. Equality holds for gaussian pulses (which we already knew) and only for gaussian pulses. This completes our derivation of the uncertainty relation.

In the next section, we use the uncertainty relation (11.8-23) to show why there is a noise of spectral density hf if we try to use an ideal amplifier to receive high-frequency electromagnetic radiation. We stated this in Section 2.9 without derivation.

In the next chapter on coding we will see how another detection concept for optical communication does better than ideal amplification. It involves photon counting and pulse position modulation. In this way we will find the capacity of optical communication without a restriction to the use of ideal linear amplification in the receiver.

Problems

1. How close does the uncertainty relation (11.8-23) come to being exactly satisfied for a raised cosine pulse? This is the pulse

$$s(t) = \begin{cases} \dfrac{1}{(3\pi)^{1/2}} (1 + \cos t), & -\pi \leq t \leq \pi \\ 0 \text{ otherwise} \end{cases}$$

The $(3\pi)^{1/2}$ makes $s^2(t)$ into a probability density. HINT: use

$$\sigma_f^2 = \frac{1}{4\pi^2} \int_{-\infty}^{\infty} \left(\frac{ds}{dt}\right)^2 dt$$

What is the frequency pulse $S(f)$? HINT: the transform of the $\cos t$ part can be obtained by looking at amplitude modulation of a carrier by a rectangular pulse.

2. How close does the uncertainty relation come to being exactly satisfied for a *rectangular* time pulse? HINT: use the same formula for σ_f^2 as in Problem 1, taking delta functions into account. There is no reason not to assume that the pulse lasts from $t = -\frac{1}{2}$ to $t = \frac{1}{2}$ and has height 1 so that $s^2(t)$ is already a probability density function.

11.9. The Uncertainty Principle and the Ideal Amplifier

In Section 2.9, we stated that an ideal double-detection receiver acted as if there were a noise power density hf_0 at the input, and we also said in equation (2.9-3) that the most ideal amplifier would have a noise power component

$$(G - 1)hf_0B \qquad (11.9\text{-}1)$$

in its output before conversion to baseband. Here G is the power gain of the amplifier, h is Planck's constant, f_0 is frequency, and B is bandwidth.

In this section, we show how this noise is a manifestation of the uncertainty principle of quantum mechanics, as derived in the preceding section. This related the rms frequency spread and the rms time spread by requiring their product to be at least equal to $1/4\pi$.

The uncertainty relation gives the spread in times of arrival or detection of individual photons resulting from the transmission of pulses at precise times. But, we are not counting photons here. Instead, we are amplifying the weak signal so that we have many, many photons. The resulting signal will then be strong.

Why can not we use this strong signal to measure the weak one? It must be because even the best possible amplifier adds noise to the signal and so keeps us from measuring the strong signal exactly. We will assume that this noise is independent of the presence of the signal, and is stationary gaussian noise. This is plausible, and we shall not go into its justification.

We can now evaluate this noise by studying the permissible performance of any convenient scheme of modulation. We shall assume pulse position modulation analog communication in determining the noise density $N_0 = N_0(f_0)$. The same noise density will hold no matter how we signal, as long as we use the same ideal double-detection receiver with linear amplification.

Our strategy is illustrated in Fig. 11.8. We generate a noiseless baseband pulse $v(t)$ of energy E_p centered at $t = 0$. The energy spectrum is $V(f)V^*(f)$, where $V(f)$ is the voltage spectrum of $v(t)$. We multiply this pulse by $2^{1/2} \cos \omega_0 t$ so as to shift all frequencies up to a spectrum centered at $f_0 = \omega_0/2\pi$, where f_0 is very much larger than the bandwidth of the pulse. The frequency-shifted pulse will have an energy E_p and a number of photons E_p/hf_0. We amplify the optical pulse by an ideal

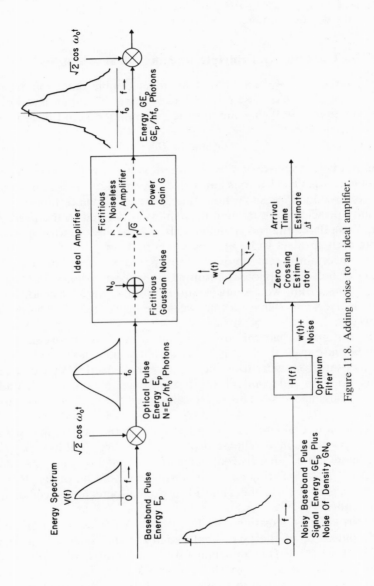

Figure 11.8. Adding noise to an ideal amplifier.

(lowest noise possible) linear amplifier of power gain G to produce an optical signal pulse of energy GE_p, consisting of GE_p/hf_0 photons. We multiply this pulse and any output noise by $2^{1/2}\cos\omega_0 t$ to obtain a baseband signal pulse plus noise. This recovered baseband pulse will have the same energy as the original transmitted pulse. We then make an optimal estimate of the time of arrival of this noisy baseband pulse, and use this time as an estimate of the arrival time of the optical pulse. From this, we infer the ideal amplifier noisiness necessary to satisfy the uncertainty relation. We attribute this noise to a fictitious additive quantum noise of power density N_0 at the input of a noiseless amplifier of power gain G or voltage gain $G^{1/2}$.

The uncertainty relation (11.8-23) tells us the minimum rms deviation in time of arrival of a single photon when the energy spectrum (magnitude squared of the frequency spectrum or wave function) has an rms spread in frequency σ_f. The equality holds when the wave function, and so the energy spectrum, are gaussian in frequency, as we saw in the previous section.

In communicating by means of optical pulses, we will use pulses with much more average energy than hf_0, the energy of a single photon. Measuring the time of arrival of a single pulse of many photons is equivalent to performing the measurement separately for each of the photons in many independent one-photon pulses, and averaging these times. If there are N photons (a pulse energy of Nhf_0), this will lead to a variance σ_{tN}^2, as in Section 8.4:

$$\sigma_{tN}^2 = E\left(\frac{1}{N}\sum_{n=1}^{N}(t_n - \bar{t})^2\right)$$

Here we have

$$E(t_n - \bar{t})^2 = \sigma_t^2 \qquad (11.9\text{-}2)$$

This leads to

$$\sigma_{tN} = \frac{1}{N^{1/2}}\sigma_t \qquad (11.9\text{-}3)$$

Thus, by (11.8-23),

$$\sigma_{tN}\sigma_f \geq \frac{1}{4\pi N^{1/2}} \qquad (11.9\text{-}4)$$

Energy spread σ_e is proportional to frequency spread by a factor h. Thus

$$\sigma_{tN}\sigma_e \geq \frac{h}{4\pi N^{1/2}} \qquad (11.9\text{-}5)$$

Because measuring the arrival time of an N-photon pulse is equivalent to averaging the N independent arrival times of the N individual

photons in the pulse, the probability density of arrival times for the
N-photon pulse will be approximately gaussian when N is large. This is
because of the central limit theorem. So the shape of the N-photon pulse
arrival probability density $v^2(t)$ will be gaussian, and we see that $v(t)$ itself
will be gaussian in shape. The result of the previous section applies, and
we find that for large N, (11.9-4) and (11.9-5) become equalities.

We shall now give an optimum method for finding the arrival time or
peak of such a pulse. We start with a *baseband* pulse which for simplicity
we assume symmetric around $t = 0$. It has known real symmetrical
voltage spectrum $V(f)$. This pulse plus white noise is multiplied by $G^{1/2}$
and passed through a network of transfer function $H(f)$ that is tailored to
give a best estimate of the pulse's peak or center. The output $w(t)$ due to
the pulse is

$$w(t) = \int_{-\infty}^{\infty} G^{1/2}V(f)H(f)e^{j2\pi ft}\, dt \qquad (11.9\text{-}6)$$

We want to estimate the center of the pulse as the time when $w(t)$ passes
through zero. So, we want $w(t)$ to be real and $w(0)$ to be 0. We could also
try to find the location of the peak of the pulse directly. This is more
difficult but leads to the same answer.

We have an N-photon pulse with N large. Since the signal-to-noise
ratio is large, we can assume a linear displacement of the zero crossing of
$w(t)$ by noise. The filtered and slowly varying noise will displace the zero
crossing by Δt, or by an rms amount σ_{tN} equal to the rms noise $(\overline{v_n^2})^{1/2}$
divided by dw/dt at time 0. Figure 11.9 shows this. Here

$$\frac{dw}{dt} = \int_{-\infty}^{\infty} j2\pi f G^{1/2}V(f)H(f)e^{j2\pi ft}\, df \qquad (11.9\text{-}7)$$

We have assumed that $V(f)$ is a real symmetric function, corresponding to
a symmetrical pulse centered at $t = 0$. So, if $jH(f)$ is a real *anti*symmetric
function, $w(t)$ will be real with $w(0) = 0$. Symmetry and the corresponding antisymmetry are not necessary but make the formulas simpler.

We are particularly interested in the value of dw/dt at time zero,
which we will call $(dw/dt)_0$. This is

$$\left(\frac{dw}{dt}\right)_0 = \int_{-\infty}^{\infty} j2\pi f G^{1/2}V(f)H(f)\, df \qquad (11.9\text{-}8)$$

What about the noise of mean-squared voltage $\overline{v_n^2}$ that is mixed in
with the baseband pulse? If N_{01} is the one-sided noise power density in
the *output* of the optical-frequency amplifier, the two-sided noise power
density will be $N_{01}/2$. As in Section 9.6, half of the noise power consists
of sine components which will give no baseband noise when multiplied by
$2^{1/2}\cos\omega_0 t$, and half consists of cosine components whose power will be

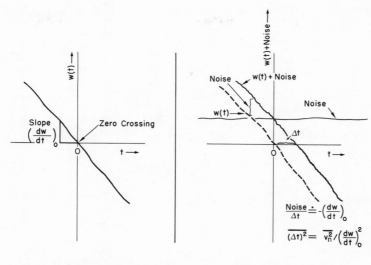

No Noise Situation   **Situation With Noise**

Figure 11.9. Estimate of arrival time Δt.

transferred to baseband by multiplication by $2^{1/2} \cos \omega_0 t$. Hence the baseband mean-square noise voltage $\overline{v_n^2}$ will be

$$\overline{v_n^2} = \frac{N_{01}}{4} \int_{-\infty}^{\infty} [H(f)][H(f)]^* \, df$$

$$\overline{v_n^2} = \frac{N_{01}}{4} \int_{-\infty}^{\infty} [jH(f)]^2 \, df \qquad (11.9\text{-}9)$$

For a given $V(f)$ and a given $\overline{v_n^2}$ we want to choose $H(f)$ to maximize $(dw/dt)_0^2$, because this will lead to the least displacement of the axis crossing by the noise. The unknown function that we have to deal with is $[jH(f)]^2$. The function whose magnitude we want to maximize is

$$\int_{-\infty}^{\infty} 2\pi f G^{1/2} V(f) jH(f) \, df \qquad (11.9\text{-}10)$$

The constraint is (11.9-9), which gives the mean-square noise power. The techniques of Section 7.3 show that

$$jH(f) = \frac{-\pi f V(f)}{K} \qquad (11.9\text{-}11)$$

Here K is a constant. This is matched filtering with the *derivative* of the original pulse. We should not be surprised by this outcome.

From (11.9-11) and (11.9-9)

$$\overline{v_n{}^2} = \frac{N_{01}\pi^2}{4K^2} \int_{-\infty}^{\infty} f^2[V(f)]^2 \, df \qquad (11.9\text{-}12)$$

From (11.9-11) and (11.9-8)

$$\left(\frac{dw}{dt}\right)_0 = -\frac{2\pi^2 G^{1/2}}{K} \int_{-\infty}^{\infty} f^2[V(f)]^2 \, df \qquad (11.9\text{-}13)$$

The energy of the original pulse, E_p, is

$$E_p = \int_{-\infty}^{\infty} [V(f)]^2 \, df \qquad (11.9\text{-}14)$$

The mean-square frequency spread or variance $\sigma_f{}^2$ will be given by

$$\sigma_f{}^2 = \int_{-\infty}^{\infty} f^2[V(f)]^2 \, df \Big/ \int_{-\infty}^{\infty} [V(f)]^2 \, df \qquad (11.9\text{-}15)$$

So we find, from (11.9-12), (11.9-15), and (11.9-14), that

$$\overline{v_n{}^2} = \frac{N_{01}\pi^2}{4K^2} \sigma_f{}^2 E_p \qquad (11.9\text{-}16)$$

Similarly, we also find, using (11.9-13) instead of (11.9-12),

$$\left(\frac{dw}{dt}\right)_0 = -\frac{2\pi^2 G^{1/2}}{K} \sigma_f{}^2 E_p \qquad (11.9\text{-}17)$$

Since the rms time spread $[\overline{(\Delta t)^2}]^{1/2}$ of the N-photon pulse is the rms noise divided by $(dw/dt)_0$, we have found the time spread:

$$\overline{(\Delta t)^2} = \sigma_{tN}^2 = \frac{\overline{v_n{}^2}}{(dw/dt)_0{}^2} = \frac{N_{01}}{16\pi^2 GE_p\sigma_f{}^2} \qquad (11.9\text{-}18)$$

This relation can also be expressed as

$$N_{01} = GE_p(4\pi)^2\sigma_{tN}^2\sigma_f{}^2 \qquad (11.9\text{-}19)$$

Let us note that according to our assumptions

$$N_{01} = N_0 G \qquad (11.9\text{-}20)$$

That is, the baseband noise density is equal to the hypothetical noise density N_0 times the power gain of the amplifying and conversion process, which we assume to be very large so as to avoid any quantum effects in our final observation.

Also, according to our assumptions

$$E_p = Nhf_0 \qquad (11.9\text{-}21)$$

That is, the original baseband pulse energy is equal to the number N of photons in the optical pulse times the energy in each photon.

From (11.9-4) and the fact that the wave function of the N-photon pulse is approximately gaussian when N is large, we find

$$\sigma_{tN}^2\sigma_f^2 = \frac{1}{(4\pi)^2N} \tag{11.9-22}$$

From (11.9-19) through (11.9-22), we obtain

$$N_0 = \frac{Nhf_0G}{G}(4\pi)^2\frac{1}{(4\pi)^2N}$$
$$N_0 = hf_0 \tag{11.9-23}$$

We repeat that N_0 is the fictitious one-sided noise power density which, when added to the input of a fictitious noiseless optical linear amplifier, will assure that the uncertainty relation is satisfied.

Is (11.9-23) the answer we sought? It is indeed the answer to the question involving observation after conversion of the optical pulse back to baseband, at which quantum effects are negligible. A second question we wish to answer in this section, however, is, what is the noise power density N_{01} at the output of an ideal optical amplifier of power gain G, *before* conversion to baseband? Is this noise power density simply Ghf_0?

The answer is no. However large the energy of the optical pulse at the output of the amplifier may be, there must still be a quantum mechanical uncertainty in our observation of the time at which it peaks. If the original optical pulse had N photons, we see from (11.9-22) that the mean-square value of this uncertainty, $(\sigma_{tN}^0)^2$, will be

$$(\sigma_{tN}^0)^2 = \frac{1}{(4\pi)^2\sigma_f^2GN} \tag{11.9-24}$$

The sum of this mean-square uncertainty and the mean-square uncertainty $(\sigma_{tN}^1)^2$ inherent in the baseband output noise power density N_{01} is equal to σ_{tN}^2 for the final baseband pulse. This is because the two sources of uncertainty are independent, so that the uncertainities or variances add. Equation (11.9-18) can be used to give the uncertainty $(\sigma_{tN}^1)^2$ due to the noise density N_{01}, with E_p given by (11.9-21). Equation (11.9-24) gives the uncertainty in observation of the amplified pulse. The sum of these two uncertainties is the total uncertainty σ_{tN}^2 in the baseband pulse:

$$\sigma_{tN}^2 = \frac{1}{16\pi^2\sigma_f^2GN} + \frac{N_{01}}{16\pi^2NGhf_0\sigma_f^2} \tag{11.9-25}$$

Equating the value $1/(4\pi)^2N$ for $\sigma_{tN}^2\sigma_f^2$ obtained from (11.9-22) with the value obtained from (11.9-25), we can solve for N_{01} to obtain

$$N_{01} = (G - 1)hf_0 \tag{11.9-26}$$

This corresponds to (11.9-1) and (2.9-3). We have derived this result using the uncertainty principle.

It appears than that the optical channel can be considered to be a white gaussian channel with noise spectral density hf_0, as in Section 2.9. It would seem that its capacity in bandwidth B is given by (11.1-1), with noise power

$$P_n = hf_0 B \tag{11.9-27}$$

and signal power

$$P_s = Mhf_0 \tag{11.9-28}$$

Here M is the number of signal photons/sec. The capacity of such a channel is

$$C = B \log_2 \left(1 + \frac{M}{B}\right) \text{ bits/sec} \tag{11.9-29}$$

This holds in the absence of thermal noise. For large or infinite B, this becomes, as in Section 11.5, Problem 1,

$$C = \frac{M}{\ln 2} \text{ bits/sec} = \frac{1}{\ln 2} \text{ bits/photon} \tag{11.9-30}$$

We have restricted ourselves to considering only one receiving scheme, linear amplification. This is not well suited to communication using discrete particles such as photons, and the capacity of an optical channel in bits/photon is greater than implied by (11.9-30). We will derive this in the next chapter, where we study the photon channel, in which photons are counted directly.

This chapter has been devoted to modeling communication channels and introducing the notion of channel capacity. The next chapter shows how particular codes may be used to attempt to approach channel capacity. We shall discuss coding for both the gaussian channel and the photon channel. We shall also see how coding can be used to make signals look like noise to the casual observer, without reducing the theoretically attainable communications performance.

Problems

1. Derive the optimum filter (11.9-11) using the techniques of Section 7.3.

2. Give a heuristic explanation why the optimum filter is differentiation followed by matched filtering with the original baseband pulse.

3. An optical channel has capacity 50×10^6 bits/sec given by (11.9-30). What is the *received* signal power P_s if the wavelength is 10 microns (infrared)? What if the wavelength is 5000 Å (green)? Assuming perfect antennas of the same diameters for transmitting and receiving, what is the ratio of the *transmitted* power in the former case to that in the latter (higher frequency) case, if the communication distance is the same? What is the transmitted power in the 5000 Å case if the antennas have diameters of 1 m and the communication distance is the distance from Pluto to the Earth, 4.5×10^9 km? What is the approximate diameter or footprint of the 5000-Å beam on the Earth? What is the equivalent noise temperature of the 5000-Å system? If the capacity of a 30-GHz 20°K microwave system from the same distance with same antenna diameters is also 50 megabits/sec, what are the transmitted and received powers? How large a ground antenna would be needed for the 30-GHz system to make the transmitted powers equal?

Applying Information Theory—
Coding and Randomization

We saw in the last chapter how Shannon's limit may be approached for the infinite-bandwidth white gaussian channel by the use of orthogonal or biorthogonal codes of ever-increasing bandwidth. In this chapter we shall study another coding scheme, convolutional coding. Convolutional codes give good performance at a moderate bandwidth expansion. They are now nearly universally used as the general-purpose coding technique for communication from distant planetary spacecraft to earth We will also generalize the notion of channel capacity, which we have so far defined only for gaussian channels. We will see what the loss in capacity is when we make a gaussian channel into a digital channel. We also will study the capacity of systems that use photon counters, and compare the results with those of Chapters 2 and 11 for ideal amplification.

As an application of combinatorial techniques from the theory of error-correcting codes, we will study ways of making it difficult to interfere with the use of a channel, either by jamming or by snooping. A class of finite-state machine encoders can be used to spread the spectrum of the communicated signal so that it covers a very wide bandwidth when it is transmitted.

12.1. Use of Error-Correcting Codes

We saw from Fig. 11.7 that a multilevel digital pulse transmission system with an error rate of 10^{-8} uses about 10 dB (10 times) more energy per bit than Shannon's limiting criterion. This is because we have tried to reduce errors simply by making the signal power so great that the noise added in transmission very rarely changes a pulse amplitude enough so that the transmitted pulse is misinterpreted at the receiver.

One outcome of Shannon's work has been the devising of a variety of *error-correcting codes* for the binary channel, for the gaussian channel, and for other channels. Suppose that in transmitting a stream of binary digits a transmission link introduces random errors in the binary digits. Shannon's theorem for the binary channel tells us that by using a binary error-correcting code, we can transmit a message consisting of binary digits over such a noisy link with as low an error rate as we desire. This is so even if the initial error rate on the link is considerable. In order to do this we must transmit message digits at a rate lower than the rate at which the transmission link transmits noisy digits. We can make the error rate as low as we like without having to slow down the transmission beyond a certain point. The limiting rate at which we can still get an arbitrarily low error probability is called the *capacity* of the binary channel. Such a definition applies to any channel. We shall go into detail in Section 12.3.

The basic idea of error-correcting codes could have been independent of Shannon's work. Actually, it is implicit in his work, and most error-correcting codes were invented following Shannon's first publication.

The devising of good and practical error-correcting codes has become an extensive field of mathematics. Reference 1, especially Part II, discusses this. Here we will merely note by example that error correction is possible.

Suppose that a system transmits a sequence of binary digits, and makes some errors in doing so. This might be a gaussian channel which detects individual pulses separately and makes a decision on the basis of each pulse alone. We shall study this in Section 12.4. We can correct these digit errors by transmitting *check digits* as well as *message digits.*

As a simple example, we can encode our message digits in blocks of 16 successive digits and add a sequence of check digits after each block which enables us to correct a single error in *any one* of the digits, message digits or check digits. As a particular example, consider the sequence of message digits 1 1 0 1 | 0 0 1 1 | 0 1 0 1 | 1 0 0 0. To find the appropriate check digits, we should write the 0's and 1's constituting the message digits in the 4 by 4 grid shown in Figure 12.1. Associated with each row and each column is a circle. In each circle we put a 0 or a 1, chosen so that there will be an even number of 1's in each column or row (including the circle as well as the squares). The digits in the circles are the check digits. For the particular assortment of message digits used as an example, together with the appropriately chosen check digits, the numbers of 1's in successive columns (left to right) after encoding are 2, 2, 2, 4, all being even numbers, and the numbers of 1's in successive rows (top to bottom) are

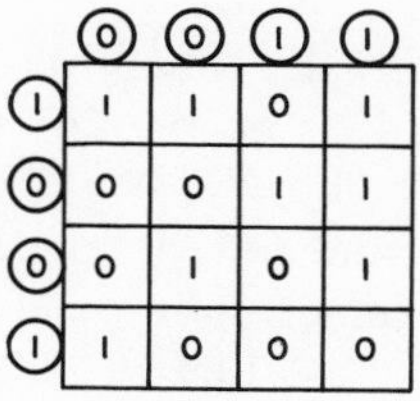

Figure 12.1. The row-column sum parity check code.

4, 2, 2, 2, which are again all even. This is called a *row-column sum parity check code*. The modulo-two sum of the five digits in each row and column is 0.

What happens if a single error is made in the transmission of any *message* digit among the 16? There will be an odd number of ones in some row *and* in some column. This tells us to change the message digit where the row and column intersect.

What happens if a single error is made in a *check* digit? In this case there will be an odd number of ones in some row *or* in some column. We have detected an error, but we see that it was not among the message digits. In that case, a row *and* column parity check would have failed to hold. Since the message digits are all correct, we merely accept them. We could of course correct the error in the check digits by changing the parity bit in the indicated row or column.

We see that this code corrects all single errors. It incorrectly corrects some double errors, confusing them with single errors. And, there are double errors which the code can detect but not uniquely correct. In that circumstance, we might wish to output no word at all, and instead merely indicate that the word was received too unreliably to be corrected. This is called an *erasure* of the word that was transmitted. Erasures are sometimes lumped in with errors in quoting word error probabilities. If two-way communication with retransmission is possible, erasures are not very annoying. We shall not explore this.

The total number of digits transmitted for 16 message digits is 16 + 8 or 24; we have increased the number of digits or expanded the bandwidth needed in the ratio 24/16, or 1.5. Here the code is said to have *rate* 16/24 = 2/3, or *redundancy* 1/3. If we had started out with 400 message digits (a 20 × 20 array), we would have needed 40 check digits and we would have increased the number of digits needed in the ratio of 440/400, or only 1.1. But, we would have been able to correct only one error in 440 digits rather than one error in 24 digits.

Such a coding scheme would not be used nowadays for data or digital voice transmission. These codes are too inefficient compared both to what is theoretically possible according to information theory, and to what is practical. Codes of this simplicity find some use when extremely low word error probability is essential, if the channel already has a fairly low bit error probability (this is called the *input* bit error probability) and if the highest possible transmission rates are not needed. An example is the intermittent transmission of a low volume of data, in which expensive and remote equipment must be sent vital instructions with almost no chance of misinterpretation. On this account, row-column parity check codes were sometimes used in the past to send commands to missiles and spacecraft to execute maneuvers or change mode of operation.

The error-correcting code described above is a *block code*. A block of message digits is encoded as a larger block of signal digits by adding redundancy. The biorthogonal codes of Section 11.6 are also block codes, but they were used on the gaussian channel. There are two problems in block coding. One is to make the encoding and, what is more difficult, the *decoding*, tractable. The other is to make block coding efficient in the sense of approaching Shannon's channel capacity.

Practical algorithms for block coding exist. However, it is difficult to approach Shannon's channel capacity closely using block codes. These matters are discussed in Reference 2. The next section discusses an alternative to block coding, called convolutional coding.

Problems

1. Show that when using the 4×4 row-column sum parity check code of Fig. 12.1 for single error correction on a binary channel of digit error probability p with independent errors, the *word* error probability not counting erasures, or the *undetected* word error probability, is approximately $48p^2$ for small p. Are the output bit errors independent for small p? You may use the fact that the probabilty of a double error dominates the probability of all higher types of multiple errors when p is small. First find out which double errors are confused with single errors, and which are detectable but not correctable. NOTE: we usually want to use codes for moderate p, not small p.

2. Suppose a 25th digit is added to the code by putting the overall modulo-two sum of the 16 message digits in the upper left corner of Fig. 12.1. This gives a code of rate 0.64. Show that all double errors are detected, but none corrected. Conclude that the undetected word error probability behaves like a constant times p^3 for small p.

3. What is the minimum number of ones in any codeword of each of the above two codes, not counting the all-zero word?

12.2. Convolutional Coding

So far we have examples of *block coding*, where batches of message bits or information bits are encoded and transmitted as a unit. In this section, we will investigate *convolutional coding*, which is a continuous process rather than a batch process. Bits go into the encoder one at a time and emerge from the encoder at a rate which is an integer times the rate at which they go in. (The ratio can be a fraction, but we shall not consider this.) Here the concept of word error probability cannot readily be defined. Instead, bit error probability alone is used to specify the performance of convolutionally coded systems.

Convolutional coding has been judged to be somewhat simpler and more effective than block coding and is more widely used (Reference 2). It has the following three advantages:

1. The codes apply to channels of *fixed* bandwidth.
2. Better performance is attained than by block coding for an equivalent complexity of hardware.
3. The decoding algorithm can be adapted to take advantage of the data source statistics in the decoding operation. In decoding, we do not need to assume that each input data sequence was equally likely to have been sent. We may take advantage of the nonrandomness of the input, especially the short-term dependencies. Or, we can remove intersymbol interference.

A disadvantage of convolutional coding is that the existence of good codes with bit error probabilities approaching zero depends on a random coding argument as in Section 11.2, rather than on a specific construction as in the case of orthogonal codes. This has not slowed the adoption of convolutional coding.

We shall examine convolutional coding for the case in which the channel has already been made into a binary channel by making a *hard decision* (0 or 1) on each received binary pulse before decoding. We will call the binary pulses *symbols* to distinguish them from the information bits that were input to the encoder. Convolutional coding is also used on the gaussian channel. In doing this, the actual voltage of each received, filtered, and sampled pulse must be used in decoding, rather than only the sign of this voltage. We call this *soft decision* decoding. The decoder principles and functional design remain about the same.

The theoretical penalty for making a gaussian channel into a *binary symmetric channel* by making hard decisions in the receiver will be considered in Section 12.4. A binary symmetric channel is one which is time invariant or stationary and in which errors in different symbols are independent. It is also required that the two error types $0 \rightarrow 1$ and $1 \rightarrow 0$ be equally likely.

We shall consider only one kind of decoding for convolutional codes. This is called *maximum-likelihood decoding* or more usually *Viterbi decoding* after the inventor, A. J. Viterbi (Reference 3). Other methods

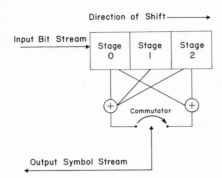

Figure 12.2. Rate 1/2 convolutional encoder of constraint length 3.

exist with their own ranges of validity, but Viterbi decoding is replacing them, and indeed convolutional coding with Viterbi decoding is replacing other coding schemes, or being used where coding was never used before.

Let us consider the convolutional encoder of Fig. 12.2. Serial input bits go into a three-stage shift register without either feedback or feed forward connections. Here three is the *constraint length*. (Practical constraint lengths for Viterbi decoding are currently limited to 10 or less.) The encoder starts in any state, usually all zeros. This initial state need not be known to the decoder, and so Viterbi decoding is self-synchronizing.

The output stream in Fig. 12.2 has twice as many output symbols per unit of time as the input stream has, so we say the rate is 1/2. Rates in use are rarely less than 1/4. A commutator alternately picks up each of two *output* symbols in one *input* bit time. The first output is the modulo-two sum of the present and the previous two input bits. The second is the modulo-two sum of the present and the next to last input bit, i.e., the one immediately previous is omitted from the second sum. These two sums are sampled by the commutator and transmitted. Here we have a linear code, and the block code of the previous section was linear as well. This means that modulo-two addition of two input bit streams corresponds to modulo-two addition of the corresponding output symbol streams.

Suppose that initially there were 0's in the two leftmost stages of the shift register. As an example of convolutional encoding let us run the input stream $011010\cdots$ through the encoder. We can visualize the process as shifting the digits to the right in the shift register of Fig. 12.2. The successive outputs obtained from each of the modulo-two-adders (outputs are underlined) are shown in Table 12.1. Converting this into a single-channel output stream by a raster scan (left to right, top to bottom) of the adder outputs in Table 12.1 gives the output serial symbol stream $00110101010010\cdots$.

We have described the convolutional *encoder*. How does the *decoder*

Table 12.1. Outputs from Modulo 2 Adders

Stage 0 content	Left adder $(y_n = x_n \oplus x_{n-1} \oplus x_{n-2})$	Right adder $(z_n = x_n \oplus x_{n-2})$
0	0	0
1	1	1
1	0	1
0	0	1
1	0	0
0	1	0

for Viterbi decoding work? To understand this, we view the encoder as a finite-state machine. There are four states $\underline{a}, \underline{b}, \underline{c}, \underline{d}$, corresponding to the four possible combinations of leftmost two bits in the encoder of Fig. 12.2. The states are:

State:	$\underline{a}$	$\underline{b}$	$\underline{c}$	$\underline{d}$
Contents:	00	10	01	11

The state diagram is shown in Fig. 12.3.

When a new bit enters, the new contents of the shift register are completely determined by the previous contents of the leftmost two bits (the state) and by the new bit. The two output symbols are also determined by the state and the new bit.

Let us assume that *node sync* is known. This means that we know which pairs of symbols in the received symbol stream correspond to information bits. There are two possible phases for node sync and we assume that we know which one is correct. This is not essential, because node sync can be derived from the received data stream itself. We shall not explore this here.

In the *trellis diagram* of Fig. 12.4, we view the encoder dynamically. The nodes correspond to states at a given time, measured left to right. Nodes along a vertical line correspond to the same time. Assume the register to be originally in state $\underline{a} = 00$. Two states, $\underline{a}$ and $\underline{b}$, are possible after one bit enters the encoder. After that, all four states are possible. Transitions caused by a 0 entering the register are shown by solid lines, and transitions caused by a 1 as dotted lines. On each line or link between nodes, the corresponding output symbol pair is shown. Every input sequence corresponds to a unique path through the trellis. From this path, the output or encoded sequence can then be read.

In this book we have studied only memoryless channels, in which errors or noise fluctuations are independent at different bits or at different times. The most important of these are the binary symmetric channel and

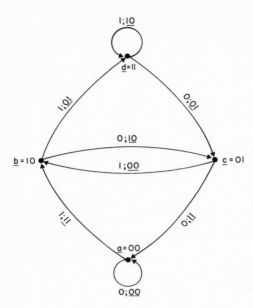

Figure 12.3. State diagram of encoder. In the triple of numbers x; uv on each link, x is input and uv is output.

the white gaussian channel. For such channels, the probability of any path which goes through a given node depends only on the probability of the past of the path, which is the part of the path that leads *into* the node. In computing future probabilities, we do not need to know in detail what the path leading *to* the node was, but only its probability.

So, in order to determine the most likely message, the decoder need only remember two things. The first is the most likely path that goes into each of the four nodes. These are called the *survivors* corresponding to the nodes. The decoder also needs to remember the *probability* of the most likely path into each node. With this small amount of memory, the

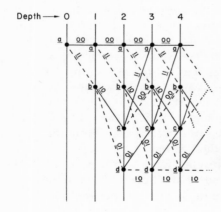

Figure 12.4. Trellis diagram. The solid lines are transitions caused by 0, the dashed lines transitions caused by 1. The output symbol pair is given above each link. The nodes are the states $\underline{a} = 00$, $\underline{b} = 10$, $\underline{c} = 01$, and $\underline{d} = 11$.

decoder can still make the best possible decision, or the decision most likely to be correct. From these probabilities, the most likely path in the decoder up to the present time is known at all times. This most likely path determines the most likely original message sequence.

But, how do we ever reach a decoder decision? There are various methods. The easiest is to tolerate some particular decoding delay, for example, five times the constraint length. The penalty in performance for tolerating a moderate delay in decoding can be shown to be small compared with the alternative of delaying a decision for a very long time. A longer delay also means that the decoder needs to be built with more memory.

In our case, the constraint length is 3, so the decoder delay is 15. After a delay of 15 information bit times, we will output one *bit* from the decoder for every two *symbols* that enter the decoder. This output bit is the information bit at time or depth 15 in the past corresponding to the most likely path into the currently most likely node. We note here that a tie-breaking rule will be necessary to resolve situations in which several nodes or paths have equal likelihood or probability.

How do we compute the probabilities or likelihoods required by the decoder? The decoder does not need to remember probabilities, but only some *metric*. This metric allows, in principle, probabilities to be uniquely reconstructed, including most likely nodes and survivors. For the binary symmetric channel, the metric for any path up to a given node is simply the number of disagreements in received *symbols* between the transmitted sequence corresponding to the path and the actual received sequence. The survivor into a given node is then the path that has the fewest number of disagreements with the received sequence. This is because the probability of a given error pattern depends only on the number of errors, and not on their location.

The logical design of a Viterbi decoder is easily accomplished from the above description. We shall not describe this design, but shall decode a specific example. It will be assumed that there is synchronism between the encoder and decoder, so that the decoder knows when transmission begins and also knows that the encoder was initially in the $\underline{a} = 00$ state.

We assume for simplicity that the information sequence was all 0's. The transmitted symbol sequence will be all 0's, but at twice the input bit rate. Let us suppose that we accept a decoding delay of only 3 (instead of 15). Let the received sequence be $\widehat{01}\ \widehat{01}\ \widehat{10}\ \widehat{00}$, with leftmost bit received first. These are the eight received symbols corresponding to the four information bits. The appearance of three 1's or three errors in eight symbols implies a rather noisy channel.

Let us agree that the decoder always chooses the higher or highest node in the event of a tie in lowest metric. For a given node, if there is

Table 12.2. Decoding Table

Depth	a = 00 Start		b = 10		c = 01		d = 11	
	Metric	Survivor	Metric	Survivor	Metric	Survivor	Metric	Survivor
1	1	0	1	1	—	—	—	—
2	2	00	2	01	3	10	1	11
3	3	000	3	001	2	010	1	111
4	3	0000	2	0101	2	1110	2	1111

more than one way to enter the node which results in the same metric, pick as the new survivor the link coming in from as high as possible.

Table 12.2, a decoding table, shows the metric at each node at each time, together with the most likely paths as a sequence of information bits into each node. After receipt of the seventh and eighth symbols, corresponding to the fourth input bit, the decoder declares that the first bit was a 0 (correct). This is because the bit three in the past (at depth 1) is a 0 for the survivor into node $\underline{b}$ at depth 4 . This is the most likely node, if we invoke the tie-breaking rule. The path taken to this decision is shown in Table 12.2, with → denoting a 0 transition and - - → denoting a 1. The corresponding encoder output symbol pairs are shown encircled.

Let us see how the table was filled out in the case of the column corresponding to state $\underline{b}$ = 10. The metric is 1 at depth 1 because the first received pair was $0\underline{1}$. This differs in one position from the $\underline{11}$, which should have been transmitted to get the encoder from the initial state 00 into the state 10. This is apparent from the state diagram of Fig. 12.3, or the trellis diagram of Fig. 12.4. The survivor is the information bit 1 because the $\underline{11}$ output from state $\underline{a}$ can only be produced by a 1 input.

At the next stage, depth 2, node $\underline{b}$ is reached only from node $\underline{a}$, through transmission of $\underline{11}$. The metric at node $\underline{a}$ was 1, although we have not described the sequence of events there. The $0\underline{1}$ that was received as the second pair of symbols differs from the assumed transmission $\underline{11}$ in one digit, so we must add 1 to the metric at $\underline{a}$, which had value 1. This gives a new metric of 2 at node $\underline{b}$. The survivor at $\underline{a}$ at depth 1 was 0. The transition to state $\underline{b}$ at depth 2 can only be caused by inputting a 1. So, the survivor at $\underline{b}$ at depth 2 is the information bit sequence 01.

Depth 3 for the first time presents two possibilities: state $\underline{b}$ can be reached from either state $\underline{a}$ or state $\underline{c}$. We have to compute candidate metrics in two ways and take the smaller. The pair $\underline{10}$ was received. If the decoder were in state $\underline{a}$, 1 would be added to the metric from state $\underline{a}$ at depth 2. This metric was 2, so we would have metric 3 in state $\underline{b}$ at depth 3. But, if the decoder were in state $\underline{c}$, the metric would be the metric in state $\underline{c}$ at depth 2, which is shown as 3, plus 1 more. This 1 is added to the metric because the transition from state $\underline{c}$ to state $\underline{b}$ produces a $\underline{00}$ output, which differs from the received $\underline{10}$ in one position. So, from state $\underline{c}$ the metric would be 4, which is larger than 3. We therefore assume we reached state $\underline{b}$ from state $\underline{a}$, and the new metric is 3. The survivor is the old survivor 00 from state $\underline{a}$ at depth 2, followed by the 1 which causes the transition from state $\underline{a}$ to state $\underline{b}$. Thus the new survivor is 001.

At depth 4, $\underline{00}$ is received. If we reach $\underline{b}$ from $\underline{c}$, a transition caused by a 1, we do not increase the metric from its value 2 at $\underline{c}$. This metric 2 is lower than the metric 5 obtained by coming from state $\underline{a}$, which has metric 3. This 5 is because the transition $\underline{a}$ into $\underline{b}$ is associated with the transmission of $\underline{11}$. Thus we assume the $\underline{c}$ into $\underline{b}$ transition, a transition caused by an input 1. The $\underline{c}$ survivor at depth 3 was 010, so the survivor at node $\underline{b}$ at depth 4 is 0101.

We continue in this way for each of the four states, using previously defined rows to determine the next row, both metric and survivor.

Suppose we have to make a decision after eight symbols are received as to what the first information bit is. The tie-breaking rule chooses node $\underline{b}$ with metric 2, even though nodes $\underline{c}$ and $\underline{d}$ also have metric 2. The survivor is 0101, whose first or leftmost bit is 0. So we declare the first bit to be zero, and we do not make an error. (With a different tie-breaking rule, we would have made an error.)

The path through the decoding table is also shown. It tells us how we reached node $\underline{b}$ at depth 4, or which states we went through to get to node $\underline{b}$. Note that the *second* bit would be in error (a 1) if we had to declare it at this time. But, we will not declare the second bit until depth 5, where we will be at another node. That other node might not be reached from this node, so the second bit could be decoded as a 0 instead of as a 1. We cannot tell at this stage of decoding.

In a way, Viterbi decoding is much like exhaustive search decoding for block codes. In such decoding on the binary symmetric channel, we compare the received block with every possible transmitted block or codeword, and choose as the most likely transmitted codeword the one that has the fewest disagreements with the block that was received. The difference is that with Viterbi decoding of convolution codes, the block is not fixed but shifts or varies with time. We count the disagreements incrementally as we receive the symbols corresponding to a new information bit.

Viterbi decoding on the gaussian channel with *soft* decisions proceeds in the same way, with the same logical design. The only difference is in the metric, which is derivable from the gaussian density in signal space as in Section 11.1. We shall not go into detail on this.

It is possible to show that for the particular encoder of Fig. 12.2 used with soft-decision decoding on the gaussian channel, with energy per bit E_b and noise spectral density N_0, the output bit error probability P_e behaves as

$$\ln \frac{1}{P_e} \doteq 3E_b/N_0 \qquad (11.2\text{-}1)$$

This is for high signal-to-noise ratios or low bit error probabilities and arbitrarily long decoding delay. A biorthogonal code (see Section 11.6) of length 16 which encodes 5 bits at a time has lower rate (5/16) and so requires more bandwidth than the convolutional code. The encoder and decoder have approximately the same implementation complexity as those for the convolutional code. The bit error probability expression for the biorthogonal code is

$$\ln \frac{1}{P_e} \doteq \frac{5}{2} \frac{E_b}{N_0} \qquad (11.2\text{-}2)$$

We see that the biorthogonal code has about 0.8 dB (3 ÷ 5/2) less performance at high signal-to-noise ratios. This is in spite of the fact that the biorthogonal code occupies bandwidth 1.6 [(16/5) ÷ 2] times the bandwidth that the convolutional code does. At moderate signal-to-noise ratios, where these codes may be used, more exact calculation must be made. When this is done, the conclusion is similar: convolutional codes with Viterbi decoding are superior to biorthogonal codes in both error probability and bandwidth, if the complexity of both systems is about the same.

We have considered several examples of coding systems in the last two sections. In the next section, we will define channel capacity for more general channels. In particular, this will give us Shannon's limit for the binary symmetric channel. By using this, we will be able to find the theoretical minimum penalty for making hard decisions. It is about 2 dB.

Problems

1. Continue the decoding of the example table assuming eight more zeros are received. That is, assume that the sequence 0 1 0 1 1 0 0 0 0 0 0 0 0 0 0 0 ⋯ is received. Decode with decoding delays of 3, 4, and 5.

2. What happens in Viterbi decoding if we are ignorant of the initial state of the encoder? How does the ability to recover with minimum errors depend on decoding delay?

3. In a convolutional code, the decoder may not have node sync, that is, it may assume the wrong phase which tells which pairs of received symbols correspond to original bits. Suggest a method of using the metrics computed by the decoder to detect an out-of-sync condition when random data are being sent.

4. How many dB below Shannon's limit for the infinite-bandwidth white gaussian channel is the constraint-length 3 rate 1/2 convolutional code of the example if we can tolerate a bit error probability P_e of 10^{-8}? How does it compare with two-level uncoded signaling when the same very low error probability is desired in each case?

5. What happens in Problem 4 if we compare the convolutional code to Shannon's limit for the gaussian channel which accommodates baseband signaling at a rate of 2 symbols per sec per Hz?

6. For the following convolutional code of constraint length 3 and rate 1/3, draw the state diagram and the trellis:

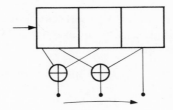

12.3. Shannon's Limit and Error Correction

The use of error correction does not give us something for nothing. At best, there is a limit to the rate at which we can send error-free digits over a noisy binary channel, just as there is over the gaussian channel.

Shannon treats the following case: We have a transmitter that transmits a sequence of symbols chosen from some discrete set of symbols. A channel involving finitely many symbols for input and output is called a *discrete channel*. The set might be $+1$, -1, or $+3$, $+1$, -1, -3, or the letters of the alphabet, or any other finite set of discrete symbols. Each symbol in the sequence of transmitted symbols is chosen randomly and independently. The probability that the symbol x is chosen and transmitted is $p(x)$. We say that $p(x)$ is the probability that the random variable X is equal to x. The probability distribution $p(x)$ defines the source statistics or simply the *source* which we may identify with the random variable X.

When the symbol x is transmitted, a symbol y is received. The probability $p(y \mid x)$ defines the *channel*. This is the conditional probability that y is received when x is transmitted, or that the random variable Y is equal to y, given that $X = x$. We assume that $p(y \mid x)$ depends only on x

and y and does not change with time. The communication channel we have described is called a *discrete stationary memoryless* channel. The white gaussian channel is a continuous channel but it is also stationary and memoryless.

The probability $p(y)$ that a received symbol will be y, or the probability that $Y = y$, is

$$p(y) = \sum_x p(y \mid x)p(x) \tag{12.3-1}$$

In terms of these quantities, Shannon defines the *entropy* $H(Y)$ and also $H(Y \mid X)$, the *conditional entropy* of Y when X is given. The entropy is

$$H(Y) = -\sum p(y) \log_2 p(y)$$
$$= -E \log_2 p(Y) \tag{12.3-2}$$

Here E denotes expectation. $H(Y)$ represents the number of bits necessary to specify Y. We shall see this in another light in Chapter 13.

The conditional entropy is

$$H(Y \mid X) = -\sum_x \sum_y p(x)p(y \mid x) \log_2 p(y \mid x)$$
$$= -E_X E_{Y \mid X} \log_2 p(Y \mid X = x)$$

Here E_X denotes the expectation with respect to X and $E_{Y \mid X}$ denotes the expectation with respect to Y given that X equals x. This can also be written

$$H(Y \mid X) = -E \log_2 p(Y \mid X) \tag{12.3-3}$$

The conditional entropy represents the average number of bits necessary to specify Y when X is known, averaged over both X and Y.

The quantity

$$I(X; Y) = H(Y) - H(Y \mid X) \tag{12.3-4}$$

is defined by Shannon as the *mutual information* between X and Y. It represents the *decrease* in the number of bits needed to specify Y resulting from knowledge of X. Or, it is what X tells us about Y. Mutual information is easily shown to be symmetric:

$$I(X; Y) = I(Y; X) \tag{12.3-5}$$

We also have the following inequality:

$$I(X; Y) \geq 0 \tag{12.3-6}$$

And, $I(X; Y) = 0$ is an *exact* condition that X and Y be independent.

These last two facts are slightly more difficult than (12.3-5) and we shall not derive them. We conclude that mutual information is a measure of independence more powerful than correlation is (see Section 9.1). Reference 1 exploits the concept of mutual information in the first half of the book.

Shannon proves by a random coding argument that the maximum rate at which error-free binary digits or bits can be sent over such a discrete memoryless channel when a given source X is input to the channel is the rate R given by

$$R = I(X; Y) \tag{12.3-7}$$

Here the random variable Y is related to the channel input X by (12.3-1), so that Y is the channel output when X is the input. To approach this rate we may have to use a long, complicated error-correcting code, because the proof involves long, complicated random codes.

For some particular distribution of probabilities $p(x)$ among the symbols to be transmitted, the rate R will be a maximum. This maximum rate is called the *channel capacity* C. It is the largest rate at which the channel can be used virtually error free, or with error probabilities which can be made as small as we please.

A channel in which we can send and receive only pulses of amplitudes $+v_0$ and $-v_0$ is a *binary* channel because there are two possible symbols or characters per pulse, in reception as well as transmission. It is a *symmetric* channel if we are as likely to receive a $+v_0$ when a $-v_0$ was sent as we are to receive a $-v_0$ when a $+v_0$ was sent. The probability of either sort of bit error will be called P_e.

Let us calculate the channel capacity for such a channel. Because the channel is symmetric, the rate is seen to be greatest when, at the transmitter, we choose $+v_0$ and $-v_0$ with equal probabilities, so that

$$p(+v_0) = p(-v_0) = \tfrac{1}{2}$$

This means that to achieve the capacity of this channel, we should use a source X for which the two source outputs are equally likely. Because the channel is symmetric, when the probabilities $p(x)$ at the transmitter are each $1/2$, they will also be the probabilities $p(y)$ at the receiver.

In terms of the error probability P_e, we have the transition probabilities

$$p(+v_0 \mid -v_0) = p(-v_0 \mid +v_0) = P_e$$
$$p(+v_0 \mid +v_0) = p(-v_0 \mid -v_0) = 1 - P_e$$

Thus,

$$H(Y) = -(\tfrac{1}{2}\log_2 \tfrac{1}{2} + \tfrac{1}{2}\log_2 \tfrac{1}{2})$$

$$= 1 \tag{12.3-8}$$

$$H(Y \mid X) = -[\tfrac{1}{2}P_e \log_2 P_e + \tfrac{1}{2}(1 - P_e)\log_2(1 - P_e) + \tfrac{1}{2}P_e \log_2 P_e \\ + \tfrac{1}{2}(1 - P_e)\log_2(1 - P_e)]$$

and so

$$C = 1 + P_e \log_2 P_e + (1 - P_e)\log_2(1 - P_e) \tag{12.3-9}$$

This expression is in units of bits per symbol or bits per pulse, and capacity means error-free bits per pulse.

The positive quantity

$$H(P_e) = -P_e \log_2 P_e - (1 - P_e)\log_2(1 - P_e) \tag{12.3-10}$$

is called the *entropy* function of the probability P_e. In terms of this function, (12.3-9) becomes

$$C = 1 - H(P_e) \tag{12.3-11}$$

When $P_e = 0$, a noiseless channel, the capacity C is 1 bit per pulse, as we expect. When $P_e = \tfrac{1}{2}$, a useless channel, $H(P_e) = 1$ and the capacity is, as we also expect, 0.

How can we relate these concepts to the gaussian channel of Chapter 11? We will not define $H(Y)$ and $H(Y \mid X)$ for continuous random variables X and Y with a joint density function $p(x, y)$ and individual densities $p_1(x)$ and $p_2(y)$. If we did define them, these entropies would have to be infinite. Instead, we define the *differential entropy* $h(Y)$ and the *conditional differential entropy* $h(Y \mid X)$ as follows:

$$h(Y) = -E(\log_2 p_2(Y)) = -\int_{-\infty}^{\infty} p_2(y)\log_2 p_2(y)\,dy \tag{12.3-12}$$

$$h(Y \mid X) = -E(\log_2 p(Y \mid X)) = -E\left(\log_2 \frac{p(X, Y)}{p_1(X)}\right)$$

$$= -\int_{-\infty}^{\infty}\int_{-\infty}^{\infty} p(x, y)\left(\log_2 \frac{p(x, y)}{p_1(x)}\right)dx\,dy \tag{12.3-13}$$

Here $p(x, y)/p_1(x)$ is the conditional probability density $p(y \mid x)$ of Y given X.

Shannon showed that the proper definition of mutual information for these continuous random variables is

$$I(X; Y) = h(Y) - h(Y \mid X) \tag{12.3-14}$$

This has all the properties we described in the discrete case.

Shannon's random coding proof of the coding theorem for the gaussian channel given in Section 11.2 did not use mutual information. Nevertheless, a proof using mutual information that applies to both the discrete and continuous cases can be given. The random variable X that maximizes $I(X; Y)$ in (12.3-14) for the gaussian channel will also be a gaussian random variable. Using this, the capacity can be computed, and the answer of (11.1-1) obtained. We saw in Section 11.2 that we could have used a gaussian random variable X for the random coding probability distribution. Here we see that this random variable is the one that maximizes (12.3-14) subject to a constraint on its average power or on its variance. We shall not pursue this further.

In the next section, we will compare the capacity of the gaussian channel with the capacity of all the binary symmetric channels that may be created from it at all possible pulse rates. This will determine the power penalty for hard limiting.

Problems

1. Plot channel capacity vs. P_e as given in (12.3-9) for the range $P_e = 0$ to $P_e = 1$. Why is the curve symmetric about $P_e = \frac{1}{2}$? Derive the symmetry from considerations of communication rather than in terms of algebra.

2. Find the entropy of the distribution which takes values $-1, 0, +1$ with probabilities $\frac{1}{4}, \frac{1}{2}, \frac{1}{4}$. What is the *maximum* entropy for a distribution taking three values? Check this result by comparing the entropy you have found with the maximum.

3. Find the capacity of the discrete memoryless channel with two inputs ± 1 and three outputs $-1, 0, 1$, where

$$p(0\,|+1) = p(0\,|-1) = p \qquad \text{(erasure probability)}$$
$$p(+1\,|+1) = p(-1\,|-1) = q$$

Here $p + q = 1$. This is called the *binary symmetric erasure channel*, with 0 corresponding to an erasure. It can arise from a gaussian channel if a *null zone* detector is used. This outputs a 0 when the detected voltage for a given pulse is too close to 0, so that an error is likely. (You may assume that the capacity is achieved for equiprobable inputs.)

4. Show that the gaussian distribution of mean 0 and variance σ^2 has the greatest differential entropy among all distributions with continuous density which have mean 0 and variance σ^2. (HINT: use the technique of Section 7.3.) What is that differential entropy? (The entropy power mentioned in Section 11.4 is the variance of the gaussian distribution which has the same differential entropy as the given distribution. This result shows entropy power ≤ power.)

5. Why is the *ordinary* entropy $H(X)$ of a random variable X with a continuous density infinite?

12.4. Channel Capacity for Binary Data Transmission

We can send $N = 2B$ baseband binary $\pm v$ pulses per second over a bandwidth B, but in the presence of additive gaussian noise we will make errors in interpreting them at the receiver. Suppose that we hard limit and take a positive received voltage to imply that a $+v$ was sent and a negative voltage to imply that a $-v$ was sent. Under these circumstances, what is the least possible energy per bit of information transmitted if we use ideal binary error correction?

For such binary transmission the channel capacity C in bits per second is given by expression 12.3-9 (in bits per pulse) times the $2B$ pulses we transmit per second:

$$C = 2B[1 + P_e \log_2 P_e + (1 - P_e) \log_2 (1 - P_e)]$$
$$C = (2B/\ln 2)[\ln 2 + P_e \ln P_e + (1 - P_e) \ln (1 - P_e)] \quad (12.4\text{-}1)$$

This is zero for $P_e = 1/2$, as must be.

It is reasonable that the greatest channel capacity in bits per second for a given power constraint will be attained at very low signal-to-noise ratios, for which the probability of error P_e is only a little less than 1/2. This corresponds to large bandwidth B. We shall not derive this result here but shall merely assume it.

We can get the channel capacity in this range of P_e most easily by differentiating C with respect to P_e:

$$\frac{\partial C}{\partial P_e} = \frac{2B}{\ln 2} [\ln P_e - \ln (1 - P_e)] \quad (12.4\text{-}2)$$

We see that at $P_e = 1/2$, $\partial C/\partial P_e = 0$. Thus the channel capacity in the vicinity of $P_e = 1/2$ will depend on $\partial^2 C/\partial P_e^2$:

$$\frac{\partial^2 C}{\partial P_e^2} = \frac{2B}{\ln 2} \left(\frac{1}{P_e} + \frac{1}{1 - P_e} \right) \quad (12.4\text{-}3)$$

At $P_e = 1/2$,

$$\left. \frac{\partial^2 C}{\partial P_e^2} \right|_{P_e = 1/2} = \frac{8B}{\ln 2} \quad (12.4\text{-}4)$$

Now let

$$P_e = \tfrac{1}{2} - \Delta P_e \quad (12.4\text{-}5)$$

Here ΔP_e is a small positive number for large bandwidths. Then, from

(12.4-4) and the Taylor series for C,

$$C \doteq \frac{1}{2} \frac{\partial^2 C}{\partial P_e^2}\Bigg|_{P_e=1/2} (\Delta P_e)_2$$

$$C \doteq \frac{4B}{\ln 2} (\Delta P_e)^2 \qquad (12.4\text{-}6)$$

It is easy to obtain ΔP_e for a noise voltage with a gaussian probability density

$$p(n) = \frac{1}{(2\pi)^{1/2}\sigma} \exp\left(\frac{-n^2}{2\sigma^2}\right) \qquad (12.4\text{-}7)$$

When we add a signal voltage v small compared with σ, we shift the overall distribution (for signal plus noise) to the right of the original center by the amount v. So, the amount ΔP_e by which the probability of error is reduced from $1/2$ is v times the value of (12.4-7) for $n = 0$. Or

$$\Delta P_e = \frac{v}{(2\pi)^{1/2}\sigma} \qquad (12.4\text{-}8)$$

From (12.4-6) and (12.4-8)

$$C = \frac{2Bv^2}{\pi \ln 2\sigma^2} \qquad (12.4\text{-}9)$$

Signal power P_s and noise power density N_0 are related to σ^2 and v by

$$P_s = v^2 \qquad (12.4\text{-}10)$$

$$\sigma^2 = N_0 B \qquad (12.4\text{-}11)$$

Thus

$$C = \frac{2}{\pi \ln 2} \frac{P_s}{N_0} \qquad \text{(two-level hard limiting)} \qquad (12.4\text{-}12)$$

$$E_b = \frac{P_s}{C} = \frac{\pi \ln 2}{2} N_0 = 1.09 N_0 \text{ J/bit} \qquad (12.4\text{-}13)$$

This is the least energy per bit E_b with which binary digits can be transmitted over a hard-limited binary channel with additive gaussian noise. It is $\pi/2 = 1.57$ times greater (1.96 dB greater) than Shannon's minimum, which was given in Section 11.3. So, we lose 2 dB by hard limiting a gaussian channel of infinite bandwidth.

We have derived this result for baseband communication, but the same result will be true for biphase modulation. Here we make a hard decision as to which of the two phases was sent. The equivalence follows because biphase modulation by $\pm 90°$ is equivalent to baseband signaling with a $\pm\cos \omega_0 t$ pulse which lasts for one bit period. We saw this in Section 10.7.

In practical situations, the loss due to hard limiting the output will be greater than the 2 dB indicated above. This is because we cannot build a communication system in which bits of error probability near 1/2 arrive at a very rapid rate. But, that is what the above capacity calculation provides. The code will be a low rate code, but it will be hard to implement.

Thus, applying perfect error correction to the binary output of a noisy channel which transmits $+, -$ pulses does not attain Shannon's ideal gaussian channel capacity in terms of signal power and noise density. In interpreting received pulses as $+$ or $-$ we have not used all of the information available in the received signal. If the received signal is very nearly zero the probability that a $+$ pulse was transmitted is almost the same as the probability that a $-$ pulse was transmitted. If the received signal is very positive, it is much more likely that a positive pulse was transmitted than that a negative pulse was transmitted. If the received signal is very negative, it is more likely that a negative pulse was transmitted than that a positive pulse was transmitted. Not taking this into account has cost us at least 2 dB. We have seen that, in practice, the loss will be greater, because error correction for very unreliable bits is extremely complex.

Nonetheless, it is simplest to say that a $+$ signal represents a $+$ pulse and a $-$ signal represents a $-$ pulse. For this reason, hard-limiting detection is still used in common carrier transmission systems, and especially in systems made up of many cascaded digital links, each of which must be cheap and simple. In such a system, error correction may be used from end to end, but it would be too costly to use coding in each link. Hard limiting is used when power is relatively cheap, so that the error probability in detecting individual pulses can be made low. Overall error correction can then be added with no change to the individual binary links.

Systems of transmission that approach Shannon's limit more closely have been used in space communication systems, especially from deep space. These include biorthogonal coding (Section 11.6) and more recently convolutional coding using Viterbi decoding with soft decisions (Section 12.2). The tradeoffs have changed because of the revolution in digital components. Hard limiting is slowly giving way for near-earth satellite links as well.

For a time, it was thought that hard limiting with subsequent binary error correction might be used even where complexity could be tolerated.

The rationale was that binary error correction was simpler than the soft decision decoders for the gaussian channel. But, the power improvement realized through soft decoding has been found to more than make up for the extra equipment complexity, which is not large.

We can also consider using three-phase modulation on the gaussian channel. We compared three-phase with two-phase at high signal-to-noise ratios in Sections 10.7 and 11.7. Hard decisions were made in both cases. We saw that if we compare three-phase with two-phase at the same low bit or pulse error probability, three-phase is better than two-phase by a factor of

$$\frac{3}{4}\frac{\ln 3}{\ln 2} = 1.189 = 0.75\,\text{dB}$$

Here we will compare three-phase and two-phase on the basis of channel capacity. As in two-phase, it will be true that the capacity is maximized when the bandwidth is very large and the signal-to-noise ratio per pulse is very low. The pulse error probability for three-phase will then be only slightly less than 2/3. Using the same kind of derivation as we did for two-phase to derive (12.4-12), the capacity using three-phase with hard decisions is

$$C = \frac{27}{16\pi \ln 2}\frac{P_s}{N_0} \qquad \text{(three-phase hard decisions)} \qquad (12.4\text{-}14)$$

We shall not derive this here. We see that in this case three-phase is *worse* than two-phase by a factor of

$$\tfrac{32}{27} = 1.185 = 0.74\,\text{dB}$$

But, we would not consider using two-phase with hard decisions on channels which were being designed from the beginning. Rather, soft decisions would be used, and the same is true for three-phase modulation. So, the above result may not be of much practical importance. Here we will discuss an advantage of three-phase modulation with soft decisions that may be of importance.

In using two-phase modulation on the gaussian channel with soft-decision decoding, we are not restricting the output, but are restricting the *input* to two levels or phases. For the infinite-bandwidth gaussian channel, we have seen in Section 11.6 that Shannon's ideal energy per bit can be attained using biorthogonal codes of ever-increasing bandwidth. These codes can be chosen to have only two levels. So, for the infinite-bandwidth gaussian channel, there is no theoretical penalty for restricting the inputs to only two levels.

But, for a finite-bandwidth gaussian channel, there will be a theoretical penalty for restricting the inputs to only two levels. This penalty increases as the capacity in bits/cycle increases. We can see this by observing that a two-level coding scheme can never get more than 2 bits/cycle over a channel even in the absence of noise. In this way, we see that three-phase signaling with soft-decision decoding will be theoretically superior to two-phase with soft-decision decoding, as bandwidth constraints become more important. We shall not pursue this further here.

This completes our discussion of coding for gaussian channels. In the next two sections, we make a careful information-theoretic study of the photon channel of optical communication, where we are not restricted to detection by ideal linear amplification.

Problems

1. Graph the capacity formula of (12.4-1) for various values of $x = P_s/P_n$ approaching 0. Show both x and P_e, the bit error probability of the resulting binary symmetric channel, on the horizontal axis Use Table 8.1 of the gaussian distribution. [Here the capacity should be normalized by dividing by P_s/N_0, and $B = (P_s/N_0)/x$.]

2. Find the loss in theoretically required energy per bit due to hard limiting the output of a gaussian channel, if the bit error probability out of the hard limiter is 0.01 and 0.1. HINT: $\Phi(1.282) = 0.9$; $\Phi(2.326) = 0.99$, where $\Phi(x)$ is the cumulative gaussian probability distribution of mean 0 and variance 1.

12.5. Capacity of a System Using a Photon Counter

In Section 2.9 we considered the signal-to-noise ratio of a system using an ideal amplifier of high gain. In Sections 11.4 and 11.9 we derived the capacity per photon of such a system. Could we do better simply by counting photons? The answer is an emphatic YES.

It turns out that when quantum effects are involved, in some cases we pay a high price if we amplify the signal ideally as discussed in the above sections. We preserve both amplitude and phase information, but at the cost of introducing a noise hfB in the final baseband output signal. Here f is the frequency used for communication. This noise is what we would get by amplifying a noise of a spectral density hf, but the output noise is not amplified input noise; it is a noise associated with an inherent uncertainty in trying to observe very small signals.

Let us now consider a system in which we pay attention only to the amplitude or power of the signal. We consider a transmitter which is a coherent light source such as a laser. When the transmitter is turned *on* we receive on the average N photons per second. We turn the transmitter on or off for successive periods of duration T. For *off*, 0 photons are received in the absence of thermal noise. For *on*, NT photons are received on the average, but it may be that no photons are received. We need to know the probability that no photons are received during an *on*.

To obtain this probability, we divide the interval T into m parts, each T/m sec long. As m is made large, the probability that more than one photon will be received in time T/m becomes vanishingly small compared with the probability that one photon will be received. The probability that one photon will be received becomes NT/m (because on the average NT photons are received in the time T) and the probability that no photons will be received becomes $1 - NT/m$. The probability $p(0 \mid 1)$ that *no* photon will be received (the zero represents this) when the transmitter is on (the 1 represents this) is the product of $(1 - NT/m)$ taken m times, because the receipt of a photon in one short time interval in no way affects the reception in any other time interval. Or

$$p(0 \mid 1) = (1 - NT/m)^m$$

as m approaches infinity. This gives for the limit

$$p(0 \mid 1) = e^{-NT} \tag{12.5-1}$$

The probability $p(1 \mid 1)$ that one *or more* photons will be received in the time interval T when the transmitter is *on* must then be

$$p(1 \mid 1) = 1 - e^{-NT} \tag{12.5-2}$$

When the transmitter is *off* no photons can be received. Hence,

$$p(0 \mid 0) = 1 \tag{12.5-3}$$

$$p(1 \mid 0) = 0 \tag{12.5-4}$$

The last four numbered equations determine the transition probabilities for the photon-counting channel, where we record only the presence or absence of photons.

If the transmitter is *off* we will receive no photons. When the transmitter is *on* the probability of receiving one or more photons can be made as close to unity as we wish by making NT, the number of photons in time T, large enough.

Knowing this, let us now consider a modulation scheme in which we have M different codewords. Each codeword consists of M successive and adjacent time intervals of duration T, so that the duration of each codeword is MT seconds. In each codeword the transmitter is *on* during exactly one particular time interval and *off* during all other time intervals in the codeword. This is quantized pulse position modulation much as in Section 11.5.

The information rate R of the transmitter will be

$$R = \log_2 M \text{ bits/codeword} \tag{12.5-5}$$

We can make the bits per codeword as large as we desire by making M large. This, of course, increases the duration of each codeword.

The average number of photons per codeword is S, the expected number of photons in the time interval T:

$$S = NT \qquad (12.5\text{-}6)$$

The average transmitted bits per photon, R/S, is

$$R/S = (\log_2 M)/NT \text{ bits/photon} \qquad (12.5\text{-}7)$$

In the absence of an interfering signal or noise, a photon is never received in a wrong time interval, because the transmitter is turned *off*. If we make NT large enough we can make the probability $p(1 \mid 1)$ that one or more photons will be received in the proper time interval as close to unity as we wish. Thus the word error probability is as small as we like. But, for any arbitrary value of NT we can make R/S, the bits per photon, as great as we wish by increasing M, the number of codewords. So, in the absence of interfering noise, there is no limit to the reliable bits per photon if we use a photon counter as a receiver and if there are no noise photons or no counts in the absence of signal photons (i.e., no dark current).

Let us summarize what we have done. One would expect that in the absence of external noise, the capacity of the photon channel would approach infinity as the average signal power (number N of mono-chromatic photons per second) approaches infinity. This is because the self-noise due to quantum mechanical uncertainties becomes a smaller fraction of the signal strength as the signal power is increased. What is surprising is that the capacity *per photon* approaches infinity.

The capacity of the noiseless photon channel *in bits per second* is also infinite, even if there is an average power limitation. This is because we can use as large a counting time T as we need, without affecting the rate in bits per photon. The number of photons in one counting time will be the same, but the number of photons per second can be anything we like. Hence the capacity in bits per second can be made very large. So, the capacity of the average-power-limited photon channel is infinite in the absence of noise. We should note that the above scheme has high peak power, as ppm did in Section 11.5 for the gaussian channel.

What about *peak*-power-limited photon channels? In ppm, the pulses occur so rarely that the average power does not grow, but the peak power is very high. Reference 5 shows that the capacity is *finite* in this case, as we may expect. We shall not discuss this further.

Let us consider photon channels for simple or uncoded binary data transmission. Here we shall assume equal probabilities for transmitting 0 and 1. There may be practical reasons such as a peak power limitation for requiring the transmitter to be on half the time. And, the complexity of coding may be too high for some applications, so that we must send data uncoded. We shall compare uncoded photon counting with ideal amplification. For photon counting, the average bit error probability P_e is, from (12.5-1),

$$P_e = \tfrac{1}{2}p(0 \mid 1)$$

$$P_e = \tfrac{1}{2}e^{-NT} \qquad \text{(photon counting)} \qquad (12.5\text{-}8)$$

Suppose average bit error probability P_e is desired, with P_e small. For ideal amplification, we have a gaussian channel. From (8.5-12),

$$P_e \doteq \frac{1}{(2\pi)^{1/2}(P_s/P_n)^{1/2}} \exp\left(\frac{-P_s}{2P_n}\right) \qquad (12.5\text{-}9)$$

Here

$$P_n = N_0 B = \frac{N_0}{2T} = \frac{hf}{2T}$$

when the frequency f is used for communication. Since $P_s = Nhf$,

$$\frac{P_s}{P_n} = 2NT \qquad (12.5\text{-}10)$$

$$P_e \doteq \frac{1}{2(\pi NT)^{1/2}} e^{-NT} \qquad \text{(ideal amplification)} \qquad (12.5\text{-}11)$$

We see that (12.5-11) is smaller by a factor $1/(\pi NT)^{1/2}$ than the P_e for photon counting given by (12.5-8), but this is not equivalent to any fixed rate increase (smaller T) or power gain (smaller N). The result is that in simple binary data transmission, we do slightly better with an ideal amplifier than with an ideal photon counter. Another advantage is that by heterodyning, we can perform filtering operations at an intermediate frequency instead of at the carrier frequency in controlling bandwidth and noise. If we use a photon counter, we have to exclude noise photons or undesired frequencies by means of narrow optical filters, which may be difficult or impossible to realize.

This completes our discussion of the photon channel without noise. With noise, the situation is more difficult. We shall study this in the next section.

Problems

1. Show that the probability that *exactly* one photon is received when the transmitter is on is NTe^{-NT}. Generalize to t photons, $t \geq 1$. We say that the number of photons received has a *Poisson distribution* with parameter $\lambda = NT$.

HINT: the number of ways t time slots can be chosen out of m is the so-called *binomial coefficient*

$$\binom{m}{t} = \frac{m(m-1)(m-2)\cdots(m-t+1)}{t!}$$

Here $t!$ (t factorial) is $t(t-1)(t-2)\cdots 3\cdot 2\cdot 1$. When $t = 0$, $\binom{m}{0}$ is 1.

2. If the number of photons per bit in binary transmission is large enough to give an error rate of 10^{-8} using an ideal amplifier, what is the error rate using an ideal counter?

12.6. Systems Using a Photon Counter with Noise

In optical communication, we have seen that ideal amplification of the received signal leads to a limiting signaling rate of $(1/\ln 2)$ bits per photon. This much inferior to the optimum limit of $(hf/kT)(1/\ln 2)$ bits per photon, which we will show we can theoretically approach by counting photons. Practically, the rates we can attain by photon counting may be limited by how elaborate the codes are that we can instrument rather than by thermal photons. We shall see how this arises. Reference 4 gives more detail concerning this phenomenon. Reference 8 of Chapter 2 is a good reference on optical receivers in general, including ideal amplification and direct detection or photon counting.

Why should we think of using optical frequencies for signaling? From equation (1.3-1), for a wavelength λ, a distance L and transmitting antennas of effective areas A_T and A_R the ratio of received power P_R to transmitted power P_T is

$$P_R/P_T = A_T A_R/\lambda^2 L^2 \tag{12.6-1}$$

This suggests the use of a short wavelength. Because we can avoid quantum noise with photon counting, we may be able to realize the advantages of (12.6-1). But, going to optical wavelengths requires very smooth antenna surfaces and very precise pointing. Here we disregard antenna and pointing problems and consider only how quantum effects will limit a communication system.

In receiving a signal of optical frequency f mixed with Johnson noise, we have the option of amplifying the signal with an ideal amplifier and shifting frequency to baseband or to a low-frequency range at which quantum effects are so small that they can be disregarded. The overall power gain is G and the bandwidth is B. When we do this, according to Section 11.9, the gaussian output noise at the *baseband* or low frequency will consist of two independent pieces. The first is Johnson noise power at the original optical frequency f and bandwidth B amplified by the power gain G. The second contribution is a noise $GhfB$ of quantum origin.

Because the two noises are independent the powers add:

$$P_n = \frac{hf}{e^{hf/kT} - 1} GB + GhfB \qquad (12.6-2)$$

Here T is the temperature of the thermal noise source. When $hf \ll kT$, the noise power density at the input of the amplifier is nearly kT. According to Section 11.2, when noise of this power density is added to a signal, the limiting information rate C in bits per joule of transmitted power, which is attained as B approaches infinity, is

$$C = 1/(kT \ln 2) \text{ bits/J} \qquad (12.6-3)$$

We saw this already in (11.4-2).

When $hf \gg kT$, the second term on the right of (12.6-2) dominates. Following Shannon, the limiting information rate C in bits per joule of transmitted power will be

$$C = 1/(hf \ln 2) \text{ bits/J} \qquad (12.6-4)$$

The energy per photon is hf. Thus the limiting rate given by (12.6-4) is $(1/\ln 2)$ bits per photon. This limit holds only if we amplify the received signal with an ideal amplifier. As we saw in the previous section, it does not hold for photon counting nor for other forms of transmission and reception described by Helstrom in Reference 4. Indeed, we saw that if no noise is added to a signal of given average power, the number of bits per second or per photon we can transmit is unbounded. This is as we expect no longer true in the presence of noise. We shall examine this phenomenon.

Oddly enough, the theoretical limit for *high* frequencies turns out to be the classical limit given by (12.6-3) for *low* frequencies, that is, $kT \ln 2$ J/bit. We shall show this later in this section.

But, let us first ask how many bits per photon do we need in order to attain the limiting $kT \ln 2$ J/bit?

A photon has an energy hf. If we require $kT \ln 2$ J/bit, the number Q of bits per photon must be

$$Q = hf/(kT \ln 2) = 6.92 \times 10^{-11}(f/T) \text{ bits/photon} \qquad (12.6-5)$$

How big will Q be for optical frequencies and attainable temperatures? Let us assume a wavelength of 5000 Å, corresponding to a frequency of 6×10^{14} Hz, and a temperature of 400°K, corresponding roughly to the average temperature of deep space at visible frequencies. For these values,

$$Q = 104 \text{ bits/photon}$$

If we tried to attain this by means of quantized pulse position modulation, the number M of possible positions would have to be at least

$$M = 2^{104} = 10^{31}$$

This, of course, is ridiculous.

The practical conclusion is that in optical signaling at moderate temperatures we will encounter insuperable problems of encoding and decoding long before we approach the theoretical limit of $kT \ln 2$ J/bit. With codes of any reasonable length and elaborateness, we will fall short of $kT \ln 2$ J/bit, so far short that we can afford to ignore thermal photons. For optical frequencies and moderate temperatures, the rate at which we can signal on the photon channel, measured in bits per photon, may be limited by our ability to implement codes, not by thermal photons. But, photon counting will be an improvement over ideal amplification even with this limitation.

This difficulty seems plausible from another point of view. At what rate do we receive thermal photons? The first term in the right-hand side of (12.6-2) gives the thermal photon power in one mode of propagation if we set $G = 1$. The rate n_0' at which we receive thermal photons in a coherent one-mode system is thus

$$n_0' = (e^{hf/kT} - 1)^{-1} B \text{ photons/sec} \qquad (12.6\text{-}6)$$

Let us set $B = 1$ Hz, which allows counting time intervals of around half a second. For $f = 6 \times 10^{14}$ Hz and $T = 400°$K,

$$n_0' = 2^{-104} = 10^{31} \text{ photons/sec}$$

There are almost no noise photons counted in 1 sec, or, to be precise, 10^{-31} of them on the average. Thermal photons will not be important until our codewords last for 10^{31} seconds or so, which is more than 3×10^{15} billion years. We shall exploit this fact later in this section when we determine the capacity in bits/photon.

So, we will explore the theoretical limitations of a photon communication system in which there is a *noise* source of photons with an average of n_0 photons per counting time interval τ, and a *signal* source of photons with an average of n_s photons per time interval τ when the transmitter is on. This is when the shutter in Fig. 12.5 is open. Here, a light source

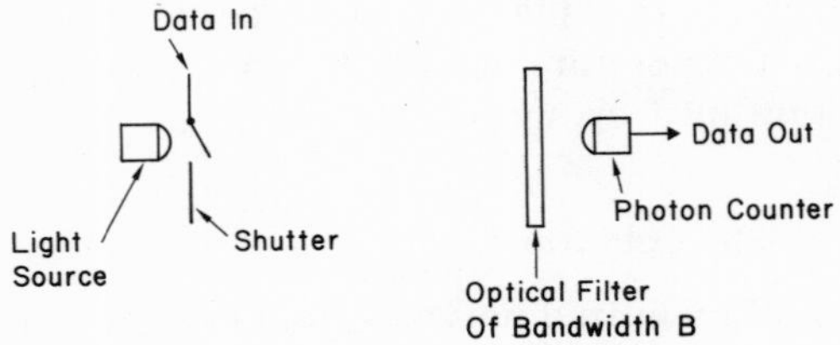

Figure 12.5. A photon channel.

transmits photons during intervals when the shutter is open. Received signal and thermal photons pass through an optical filter intended to eliminate out-of-band thermal photons. The bandwidth B is really a sort of mean bandwidth of the optical filter. This must be made wide enough so as not to lose many signal photons when the shutter is open but narrow enough not to let in too many noise photons. This means that B should be around unity (if the shutter is open for a half second), but probably somewhat larger. We will assume

$$B = \frac{1}{2\tau}$$

in what follows. In a real system, the light source and shutter could be replaced by a pulsed laser.

Each photon that passes the optical filter is counted by a photon counter. Photons are detected by this photon counter with a probabilty of detection depending on the energy E in the electric field. For a given electric field energy, the distribution of the number of photons detected over a given time τ is a *Poisson distribution*

$$pr(j) = \frac{1}{j!} n^j e^{-n}, \qquad j = 0, 1, 2, \dots \qquad (12.6\text{-}7)$$

We saw this in Problem 1 of Section 12.5. Here $n = n(E)$ is the mean number of photons detected in time τ. It is equal to the expected field energy in time τ, divided by hf. Note that $0! = 1$ so that

$$pr(0) = e^{-n} \qquad (12.6\text{-}8)$$

is the probability that *no* photons are detected. The number n *is* called the *Poisson parameter*.

Now let us assume that

$$hf \gg kT \qquad (12.6\text{-}9)$$

This takes us into the range in which quantum effects are very strong. Then, accurately enough, (12.6-6) becomes

$$n_0' = B e^{-hf/kT} \qquad (12.6\text{-}10)$$

Since $B = 1/2\tau$,

$$n_0 = n_0'\tau = \tfrac{1}{2} e^{-hf/kT} \qquad (12.6\text{-}11)$$

This does not depend on the counting time τ.

We note that the capacity of the photon channel in bits per photon does not depend on τ but only on n_0 and n_s. Here n_s is the number of signal photons counted in time τ when the shutter is open and noise is absent. This independence of τ arises because the probability of receiving various numbers of signal photons and noise photons in time τ depends only on n_0 and n_s.

We can also find the capacity in bits per second. For photon counting at visible frequencies this is not simply related to the capacity in bits per photon in the way that it is for amplification at microwave frequencies. The capacity in bits per second for an average-power-limited photon channel will be slightly less than the capacity in bits per photon times the number of photons per second. To the accuracy of the approximation being used in this section, however, we may ignore this difference. Reference 5 gives details concerning this.

The case in which n_0 is very small corresponds to a low temperature. But, n_0 will be very small for visible frequencies even for temperatures of thousands of degrees Kelvin. This is because hf is equivalent to a temperature of tens of thousands of degrees when f is in the visible range of frequencies.

When n_0 is small and signal is absent, the probability of one noise photon is about equal to the expected number of noise photons, because the probability of more than one photon is small compared to the probability of one photon. This is true for a constant field strength by (12.6-7), and it is also true if we average with respect to the noise field distribution. We shall not go into detail on the noise field distribution, but merely assume this to be true. The result is that the probability of receiving a noise photon in the absence of signal, $p(1 \mid 0)$, is

$$p(1 \mid 0) = n_0 \qquad (12.6\text{-}12)$$

Here we assume a binary channel with inputs and outputs 0 and 1. We assume also that the number of signal photons n_s in time τ is small. The true capacity in bits/photon must at least be equal to the value that we get with this extra restriction. And, it turns out that the capacity we obtain will be very close in ratio to the true capacity at low noise temperature. We shall assume this, and so n_0 and n_s will both be small.

We expect that, as in the noiseless case, the capacity is achieved when the fraction of time that the shutter is open is very small. This corresponds to pulse position modulation. This can be proved but we shall merely assume it.

If we take a block of M counting times τ and use pulse position modulation, we will find how large M can be so that the probability of error will still be low. We can have an error if there is a noise photon during one of the $M - 1$ time intervals when the shutter is closed. Or, there may be no noise photons in these intervals, but we may fail to receive a signal photon when the shutter is open. This corresponds to an *erasure*, in which no photons at all are received. In Section 12.3, Problem 3, we saw that in an erasure channel, the erasures lower the capacity in bits per second, but only by the probability of an erasure. So, it is reasonable that we can ignore the probability of erasure in computing the capacity *in bits per photon*. This is true and we assume it.

The capacity in bits per photon is then

$$C = \log_2 M \text{ bits/photon} \qquad (12.6\text{-}13)$$

Here M is the largest number of intervals for which the probability of receiving a noise photon in some interval during which the shutter was closed is small. We note that when there is no erasure, it is overwhelmingly more likely that one photon was received than that more than one was received. So (12.6-13) really does represent the capacity in bits/photon.

If

$$M = (1/n_0) + 1$$

the expected number of noise photons in $M - 1$ intervals during which signal is absent is

$$(M - 1)n_0 = 1$$

We can make M a small fraction of $1/n_0$:

$$M = (\beta/n_0) + 1 \qquad (12.6\text{-}14)$$

The *expected* number of noise photons in the $M - 1$ intervals during which signal is absent will then be small:

$$(M - 1) \cdot n_0 = \beta \ll 1$$

This also means that the probability of a noise photon in *any* of the $M - 1$ intervals will also be small, and the error probability will be close to 0.

From (12.6-13), the capacity will be

$$C = \log_2 \left(\frac{\beta}{n_0} + 1 \right)$$

$$C \doteq \log_2 \beta + \log_2 \frac{1}{n_0} \qquad (12.6\text{-}15)$$

We note that β must be chosen small, but it can be chosen very large compared with n_0. If this is so, then we will have

$$\log_2 (1/\beta) \ll \log_2 (1/n_0) \qquad (12.6\text{-}16)$$

In view of (12.6-16), (12.6-15) becomes

$$C \doteq \log_2 \frac{1}{n_0}$$

$$C = \frac{-\ln n_0}{\ln 2} \text{ bits/photon} \qquad (12.6\text{-}17)$$

This can also be derived by more rigorous analysis.

From (12.6-17) and (12.6-11) the capacity C in bits/photon satisfies

$$C = \frac{hf}{kT \ln 2} + 1 \text{ bits/photon} \tag{12.6-18}$$

Because of assumption (12.6-9), the first term on the right of (12.6-18) is much larger than unity. So we can disregard the second term in (12.6-18):

$$C = \frac{hf}{kT \ln 2} \text{ bits/photon} \tag{12.6-19}$$

How many bits per joule is this? Because there are hf joules/photon, we see that the maximum number of bits per joule is

$$\frac{1}{kT \ln 2} \text{ bits/J} \tag{12.6-20}$$

Or, the minimum number of joules needed to transmit one bit is

$$E_b = kT \ln 2 \text{ J/bit} \tag{12.6-21}$$

This is the same as the classical low-frequency result, derived in Section 11.4.

In the next section, we will study the application of information theory and coding to the transmission of information between cooperating transmitter and receiver, in the presence of an unauthorized transmitter engaged in jamming our transmission, or receiver engaged in snooping on our transmission. Jamming and snooping can be prevented by using guided waves instead of radio. But, in many cases we cannot use guided waves. The next section describes a particular way of coping with jamming and snooping on radio channels called *spread spectrum communication*.

Problems

1. How many joules are needed to communicate one bit over a photon channel of wavelength 5000 Å with thermal photons of temperature $T = 400°K$ if the number M of possible pulse positions is limited to at most 2^{20} because of current technological limits? What is the loss in dB due to this limitation?

2. How small is n_0 if $T_0 = 1000°K$ and $f = 6 \times 10^{14} \text{ Hz}$ (a wavelength of 5000 Å)? By how large a factor is the capacity in bits/photon increased if photon counting is used instead of ideal amplification? What is this factor in dB? If technology limits M to 2^{20}, how much better is photon counting than ideal amplification?

3. Use (12.6-7) to show that the probability of two or more photons with a constant electric field is small compared with the probability of one photon when the expected number n of photons is small. The ratio is about $n/2$ if the expected number of photons is n. HINT: for small n, $e^{-n} \doteq 1 - n + n^2/2$.

4. Photon counters sometimes have a *dark current* of counts even when there are no signal or noise photons impinging on them. Suppose there are a small number n_d' of such dark current counts per second. Why does this not lower the capacity in bits/photon?

5. We could vary the counting time in Section 12.5 to get the capacity in bits/sec. Show that we can here as well.

6. A photon-counting system uses a wavelength of 5000 Å and has a counting interval τ of 10^{-6} sec. For reasons of technology, the bandwidth is limited to being no smaller than 0.1 Å. The noise temperature is 400°K.
   (a) What is the smallest bandwidth B in Hz rather than in Å?
   (b) How much larger is this bandwidth than the overall minimum possible B_{min} for a photon-counting system with 1 μsec counting time?
   (c) According to (12.6-10), how many more noise photons per second n_0' or per counting interval n_0 are there with this system than with the overall minimum-bandwidth system?
   (d) According to (12.6-17), by how much is the capacity in bits/photon reduced in the 0.1 Å system? What is this in dB?
   (e) How many positions would be needed to approach this capacity by means of quantized pulse position modulation?
   (The point of this problem is that coding complexity constraints are more important in limiting the potential benefits of photon counting than our ability to implement narrowband optical filters is.)

12.7. Jamming and Snooping

Military radio communication systems are subject to jamming. The jammer transmits high-power unwanted signals which completely mask the desired signal. There are various ways to deal with this.

If one can, one uses frequencies high enough so that they do not travel beyond the horizon, and keeps his receivers out of the line of sight from earth-based locations convenient to hostile jamming transmitters. Microwaves are pretty much confined to line of sight, but they can be scattered beyond the horizon by atmospheric inhomogeneities. Frequencies in the UHF range from 100 to 1000 MHz are scattered more than microwaves.

Figure 1.8 shows the location of water vapor and oxygen absorptions, at frequencies around 22 and 40 GHz. There is a 60-GHz absorption for oxygen off the scale. If we use frequencies in one of these ranges, we can design a radio system in which atmospheric attenuation, not path loss, is the major transmission loss. As we have seen in Section 1.8, doubling the distance will then double the path loss *in dB*, rather than increasing it by only 6 dB as when space loss is predominant. Doubling of the loss in dB will preclude jamming from a distance.

Suppose, however, that a ground or satellite receiver is in line of sight of a hostile or inadvertent jammer, and at about the same distance as our own transmitter. In this case, receiving antenna or array directivity, or a nulling on the interfering source may help us. Another, or additional, approach is to use a *spread-spectrum* signal. We will show how to do this in the next section. In this section, we will examine the benefits of spread spectrum.

We have already noted in Section 11.2 that the way to get by with the least energy per bit in the presence of white gaussian noise is to use a broad bandwidth signal and so a low signal-to-noise ratio. We have seen in Chapter 11 that ppm and biorthogonal coding can be used in attempting to approach the limiting energy per bit when there is no bandwidth constraint.

In these considerations we assumed the noise power to be proportional to the bandwidth used. This is so for Johnson (thermal) noise. Suppose, however, that the total average noise power P_n is available to an enemy jammer, and is independent of the bandwidth we use to communicate. A jammer will be limited in his available power, no matter how he distributes it in frequency. Let us suppose that P_n is much larger than the thermal noise, even for the widest bandwidth that could be used. If our average signal power is P_s, Shannon's formula from Chapter 11 tells us that the attainable channel capacity C is

$$C = B \log_2 (1 + P_s/P_n) \qquad (11.1\text{-}1)$$

Shannon's formula holds only for white gaussian noise. However, we saw in Section 12.3 that gaussian noise is the worst kind of noise, because it has the greatest differential entropy for given variance or noise power. So, a jammer should try to generate gaussian noise, and it should be white to make the noise samples independent. The channel capacity is greater than that given by (11.1-1) for any other kind of noise.

Let us consider the case in which the jamming power P_n is much greater than our signal power P_s. This could be the case if a large ground antenna were used to jam transmission between aircraft. Then, very nearly

$$C = \frac{BP_s}{P_n \ln 2} = 1.443 BP_s/P_n \qquad (12.7\text{-}1)$$

For example, if the jamming power P_n is 1000 times our signal power, (12.7-1) tells us that in principle a signal bandwidth of 1.7 MHz will give us a reliable 2450-bit/sec channel. Or, we have taken a 2450-bit/sec channel and spread its spectrum to 1.7 MHz. Of course, practical limitations on coding mean that we cannot get this high a bit rate at low error probability, but it is still worth doing. We see that spread spectrum is an excellent way to counter jamming. Reference 6 summarizes much of the present state of this technology.

We note that we have assumed a signal-to-noise ratio of 1/1000. If a system is designed to operate at such a low signal-to-noise ratio *against thermal noise*, it will be very hard for a hostile snooper to find out that the wideband signal is even being transmitted. Thus spread spectrum is also useful in countering snooping.

Before closing, we should note that jumping unpredictably from one frequency to another or from one communication channel to another can be very effective in countering jamming and snooping. Such frequency hopping is really a form of spread spectrum. In the next section, we will describe the most common method of spreading the spectrum. It involves the use of pseudonoise sequences.

Problems

1. A communication system using 60-GHz millimeter waves undergoes an atmospheric attenuation of 4 dB/km. Such a system is set up over a 15-km path using parabolic antennas, receivers, coders, and decoders, so that it attains an acceptable bit rate and error rate with some margin. A jammer operating from 25 km away can be effective in this case if he can cause the receiver to see jamming power 10 dB more than the signal power. The jammer can generate 25 times as much power at the 60-GHz frequency as the communication system being jammed can. How many meters in diameter must the jamming antenna be, if the communication system itself uses a 2-m transmitting antenna? (Path loss must be considered, as well as attenuation.)

2. How should the capacity equation (12.7-1) be modified to take into account thermal noise N_0B as well as jamming noise P_n? Assume P_n large compared with P_s. What is the capacity penalty from unintentional jamming (radio frequency interference or RFI) where N_0B is much smaller than P_n, as it may be for deep space microwave communication systems? What is this capacity loss in dB if the spacecraft is 400×10^0 km from earth, the interferer satellite is at synchronous altitude (36,500 km from earth), the satellite antenna and earth station antenna are pointed so that the loss is 40 dB over ideal pointing, the spacecraft and satellite transmitter powers and antenna diameters are the same, the satellite spectrum covers the 10-MHz-wide receiving bandwidth of the earth station, and the original unjammed capacity was 1 megabit/sec?

3. Suppose that the "jamming" is caused by interference from the simultaneous operation of $M - 1$ other transmitters using the same system as the transmitter–receiver pair being jammed, each of the same power P_s and bandwidth B but with uncorrelated or independent signaling. According to Shannon's formula, what is the channel capacity per transmitter–receiver pair in terms of P_s, B, and M? Assume no noise power except that of the interfering transmitters. What form does this take when M is very large?

12.8. Pseudonoise and Spectrum Spreading

How can we widen the bandwidth of a signal, especially when it is inherently low rate? And, how can we do this so that even if an

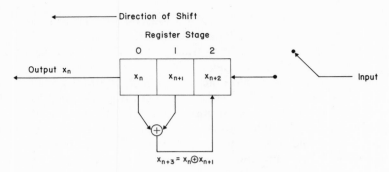

Figure 12.6. Three-stage linear feedback shift register.

unfriendly transmitter knows the system we are using, his jamming or snooping job is made as difficult as possible? One way, the most common, is by using *pseudonoise sequences*. There are other uses of these sequences in communication, and we will explore these as well.

Pseudonoise sequences are instances of shift register sequences (Reference 4 of Chapter 10). We have seen shift registers before, in Section 10.6 for randomizing the source and in Section 12.2 for convolutional coding. In the present case, we use the shift register as a block coder.

Consider the length 3 feedback shift register of Fig. 12.6. At any time, the three register stages are x_{n+2}, x_{n+1}, x_n. Shifting to the left, x_n is outputted while x_{n+1} shifts into register stage 0, the stage which x_n occupied. x_{n+2} shifts into register stage 1, which x_{n+1} occupied. Register stage 2 is loaded with the modulo-two sum $x_n \oplus x_{n+1}$, or, the modulo-two sum is fed back to stage 2. Because the operation $x_n \oplus x_{n+1}$ on x_n and x_{n+1} is linear, this is a linear feedback shift register.

Initially, the *output* is disabled, and the contents of the register are irrelevant. The register is loaded with three input bits: x_0 in position 0, x_1 in position 1, and x_2 in position 2. The *input* is then disabled, the output is enabled, and the register becomes a free-running finite-state machine, or one without further input.

Successive outputs are transmitted in this case until seven symbols have been outputted. We will see that the output sequences do not repeat unless the input was 000. The register is now ready for another input load of 3 bits. Or, it can run forever, producing a sequence of period 7 whose phase is determined by the initial nonzero input. We will explain these phenomena.

The output of the free-running register satisfies the *recursion*

$$x_{n+3} = x_n \oplus x_{n+1} \tag{12.8-1}$$

The recursion *defines* the register. If we load the register initially with

001, the infinite output sequence will be

$$x_0 \quad x_1 \quad x_2 \quad x_3 \quad x_4 \quad x_5 \quad x_6 \quad x_7 \quad x_8 \quad x_9 \quad \cdots$$

$$\boxed{0 \quad 0 \quad 1} \quad 0 \quad 1 \quad 1 \quad 1 \quad \boxed{0 \quad 0 \quad 1} \quad \cdots$$

(12.8-2)

Note that the "window" $x_7 x_8 x_9$ contains the same bits 001 as the window $x_0 \, x_1 \, x_2$. So, the sequence starting from x_7 is a repeat of $x_0 \cdots x_6$. This is because the contents of the three stages or window at any instant of time determine the output for all future and past time.

Could the period have been greater than 7 for any register of length 3? No, because if 000 ever occurs, then the sequence is identically 0. But there are $2^3 - 1 = 7$ possible window contents or windows other than 000. There are *eight* windows starting with each of the eight positions $x_0, \ldots, x_6, x_7$. Hence some two windows must be the same, and the sequence has period shorter than 8. Seven is the maximum length possible, so (12.8-2) defines a *maximum-length shift-register sequence*, or a *pseudonoise sequence*, or a *PN sequence*. Every one of the seven nonzero windows occurs in the sequence in one of the seven possible positions or phases, if we wrap the sequence around a circle. This is because there are seven possible positions and seven possible windows, and the period is seven as well. The same conclusion holds for a PN sequence corresponding to a register of any length.

Algebraic theory shows that there exists a maximum-length linear feedback shift register sequence of every length $2^n - 1$. The connections for feedback are not obvious, but they do exist. We shall not go into this.

Let the single period $x_0 \, x_1 \cdots x_{2^n - 2}$ of a maximum-length linear shift register sequence be wrapped around a circle. Every phase shift is another maximum-length shift register sequence, an output from the same shift register but with a different initial window. What happens if we modulo-two add two such shifts?

Because the recursion relation (12.8-1) is linear, such a sum will also satisfy the recursion. The sum will be 0 in each position, if the two shifts are by an identical amount. Otherwise, the sum is another cyclic shift. This is because it is not the 0 sequence, and satisfies the recursion. We have derived the *cycle-and-add* property of maximum-length linear shift register sequences: the modulo-two sum of two shifts of such a sequence is another shift.

We will use the cycle-and-add property to derive a correlation property of PN sequences, to explain their use in spread spectrum and in other areas of communication. It will also explain why they are called pseudonoise sequences.

A single period of length $2^n - 1$ of a PN sequence contains 2^{n-1} 1's and $2^{n-1} - 1$ 0's. This is because looking at the sequence cyclically, every

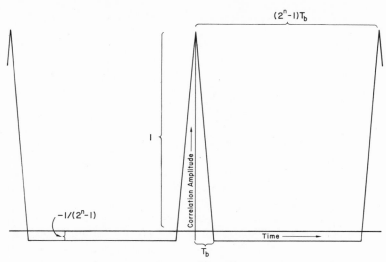

Figure 12.7. Autocorrelation of a PN sequence.

one of the 2^n windows of n 0's and 1's must occur exactly once, except the all-0 window. Because of the cycle-and-add property, two different cyclic shifts of a PN sequence disagree in exactly 2^{n-1} of their $2^n - 1$ positions. The reason is that 1's in the sum correspond to disagreements in the sequences being added. We note that the number of disagreements is half the length or period, rounded up.

In the PN sequence, suppose we replace 0's by 1's and 1's by -1's. We suppose that each ± 1 is of bit duration T_b, so we have a constant-power periodic waveform of period $(2^n - 1)T_b$. The autocorrelation of Section 6.6 is easy to derive, knowing the number of disagreements between any two different shifts of the PN sequence. This is shown in Fig. 12.7. The peak autocorrelation amplitude is $(2^n - 1)$ times the steady negative autocorrelation amplitude away from the peak. The correlation is periodic with the same period as the original sequence. The area of each spike or triangle is equal to T_b. (This counts the negative part as well.)

Thus the autocorrelation of a ± 1 version of a PN sequence is a periodic stream of near delta functions of strength about T_b. For this reason, the power spectrum will be nearly flat. There is a dc spike in the spectrum owing to the negative offset, and the spectrum consists of spectral lines because of the periodicity. Otherwise, the spectrum looks very much like that of white noise. This accounts for the term "pseudonoise sequence."

In one form of spread-spectrum communication, individual data bits or pulses of duration T_d, which we assume here are rectangular, are multiplied or chopped by a ± 1 PN sequence whose total *period* is the data bit *duration*. The relation between T_d and T_b is then

$$T_d = (2^n - 1)T_b \qquad (12.8\text{-}3)$$

We assume that a PN bit starts at the same time a data bit starts, so that PN chopping of the data stream is equivalent to using the ± 1 PN sequence itself as the signaling pulse shape. The chopping spreads the data spectrum, because the chopped sequence has as we expect nearly the same power spectrum as the PN sequence has. For random data, the data power spectrum was originally a sinc^2 spectrum with power mostly confined to a frequency interval $-1/2T_d$ to $1/2T_d$. With the PN chopping, the spectrum is spread out to be about $(2^n - 1)$ times as wide, and $1/(2^n - 1)$ times as high because the average power is the same. More exact calculations of the spectrum of the chopped sequence can be carried out, but the above description illustrates all the essential features.

We will show how the original data bits may be recovered with no loss, as long as the PN sequence, including its phase, is known. We only need to multiply the received signal by the PN in its correct phase, to remove the chopping exactly, and then detect the resulting waveform as if there never were any PN sequence. We can see that this is matched filtering for PN-chopped pulses, as in Section 7.4. So, this is the best possible detection scheme. And, if a snooper knows the PN except for the phase, the output of the multiplier or dechopper will look much like white noise if the snooper is off even by as little as the small pseudonoise bit time T_b. But, for receivers which do have the correct phase, the signal-to-noise ratio will be the same as if the spectrum were not spread.

PN sequences can also be used for cooperative communication to provide a multiple-access or multiple-channel capability analogous to frequency-division multiple-access or FDMA. We will call this code-division multiple access or CDMA. Here different users of a channel use the same PN for their chopping sequence, but in a different phase. Signals from unwanted or incorrect user transmitters will merely add a small amount of noise after multiplication by the correct PN phase at the receiver. The small noise is due to the fact that the autocorrelation of a PN sequence is not exactly 0 away from the peak. This is very similar to in-phase and quadrature modulation which we studied in Section 9.6.

This discussion has assumed baseband multiplicative chopping or amplitude modulation. PN sequences are also used with other forms of modulation for both spread-spectrum and multiple-access communication.

There are still more uses of PN sequences in communication. The delta function autocorrelation makes them useful for synchronization. They can be used instead of orthogonal or biorthogonal codes on a gaussian channel. Geometrically, the phase shifts of a PN sequence constitute a *simplex code*. This gives very slightly better performance than the biorthogonal codes of Section 11.6. But, decoding is more complex than the best orthogonal or biorthogonal decoders. There are also uses of PN sequences in cryptography.

PN sequences are good random number generators. In communication, they have been used to generate controlled synthetic gaussian noise for channel simulation in certain convolutionally coded communication systems, where the performance down to the last tenth of a dB may be hard to determine analytically. The autocorrelation property also makes them useful in distance-measuring or ranging systems. They are used for this in radar location and in radar mapping, where they generate what is called a range gate. Here different round-trip distances correspond to different delays or phases of the PN sequence. The radar return from undesired ranges is rejected much as are the other users of the multiple-access communication system described earlier in this section. This is the basis of earth-based radar mapping of the planets. Doppler shift provides the second dimension to the maps.

Finally, about 60 years ago a sequence of length 32, much like a PN sequence with the missing 0 inserted, was used to position the print wheel in teletype. This was done long before any theory of these sequences was developed. There are 32 characters on a teletype print wheel. Each of the 32 wheel positions corresponded to a unique window of 5 zeros and ones, and there was a 0 or 1 physically located below each teletype character. The print wheel was made to rotate to the unique position whose window matched the window corresponding to the character to be printed.

In the last two chapters, we have examined the *channel* of the communication system. In the next chapter, which is the last, we examine the *source*, the other component of communication, and the one usually associated with the users who pay for communication services.

Problems

1. What are all the possible output sequences and their periods for the shift register of three stages with feedback $x_{n+3} = x_n \oplus x_{n+1} \oplus x_{n+2}$? Are any of these maximal-length sequences?

2. Explain the steady negative correlation in Fig. 12.7.

3. If a PN sequence is used to generate a simplex code of ± 1's, what is the Euclidean distance between any two different codewords? What is the cosine of the angle between them? Is the angle acute or obtuse?

Sources, Source Encoding, and Source Characterization

It is reasonable to expect that any behavior that we model mathematically will differ a little from the mathematical model. The picture of gaussian noise in a multidimensional signal space we drew in Section 11.1 is a good mathematical model. Thermal noise and certain other kinds of noise are quite accurately gaussian. When we take more and more samples, the sum of the squares of the samples is more and more nearly the number of samples n times the variance σ^2. More and more nearly, the signal represented by the n samples will lie within, and, indeed, almost on the surface of an n-dimensional sphere of radius $n^{1/2}\sigma$.

Thus, in principle at least, we really should be able to transmit $B \log_2 (1 + P_s/P_n)$ independent binary digits (or bits) per second over a channel of bandwidth B and noise power P_n by using an average signal power P_s.

This is a very important result in itself. But, in considering or using it, we should always remember that neither the numbers we wish to transmit as data, nor the letters we wish to transmit as text, nor samples of audio or video signals are independent of one another. Most actual signals have a good deal of order. Columns of numbers on an adding machine often have a string of zeros in the leftmost places, and frequently end in .00 cents. In text, e occurs much more frequently than z, and q is almost always followed by u. In speech, words are more common than meaningless sounds. Television pictures are not random snow. They have areas of gradually changing shade and color, and boundaries or edges between regions.

It must be easier to send an orderly signal than a chaotic one. It would be easy to transmit the utterances of an afflicted person who spoke only one phrase or sentence; one would merely have to send the phrase or sentence once and then send a simple signal or pulse each each time he repeated it.

Shannon showed that an entropy can be assigned to an idealized message source, and this entropy gives us the least channel capacity that will transmit without error messages generated by the source. If we define the source entropy in such a way that it takes the required accuracy of reproduction into account, we shall see in this chapter that the entropy also gives the least channel capacity that will transmit messages from the source with some fixed or allowable error.

The idea of the entropy in connection with a real message source is an insightful and useful one, but it can degenerate into nonsense if we press it too far. What it does suggest is that by using some sort of encoding we can reduce the channel capacity needed to transmit messages from a given message source, or a given sort of message source. We will find that this is indeed true. We shall be especially interested in the ultimate mechanism by which messages are generated. We call this the message behind the message.

13.1. The Huffman Code

Shannon coped with the problem of order in signals from various sources, such as a writer, a speaker, or a TV camera, by assuming the signals to be ergodic, as in Section 9.4. This implies that any particular signal from the signal source is stationary, that is, that its statistical properties do not change with time. More importantly, the ergodic property also means that an ensemble average over all signals produced by the signal source is the same as a time average over any one signal.

Shannon then defined an entropy or entropy rate of such a signal source. This is the number of bits of informaton per outcome or per symbol produced by the source. In this section, we shall see why entropy is a useful definition for studying sources of data.

If the source X produces letters or symbols independently of one another, and if the symbol x is produced with a probability $p(x)$, then the entropy $H(X)$ of the source X is, as in Section 12.3,

$$H(X) = -\sum_x p(x) \log_2 p(x) \text{ bits/symbol} \qquad (13.1\text{-}1)$$

Here the sum is over all possible symbols x that X can take on as values.

Suppose the symbols are not produced independently. Then we divide long messages from the source into blocks of N symbols. Suppose that there are m possible different blocks $x^{(N)}$ of N symbols, and that $p_N(x^{(N)})$ is the probability that the xth *block* will occur. Here m increases as N increases. We call this source $X^{(N)}$, or the *Nth extension* of the original source. Let us consider the quantity

$$H_N(X) = \frac{1}{N} H(X^{(N)}) = -\frac{1}{N} \sum p_N(x^{(N)}) \log_2 p_N(x^{(N)}) \text{ bits/symbol}$$

$$(13.1\text{-}2)$$

As N approaches infinity, that is, as we make the blocks longer and longer, $H_N(X)$ approaches a limiting value which we call the entropy of the source X in bits per symbol. If successive outcomes are independent, this value will be equal to the original source entropy $H(X)$. This is not hard to show, and we shall assume it.

Shannon showed that in principle a signal of entropy H bits per second can be transmitted over a channel if the channel capacity C is greater than H by even a vanishingly small amount. To do this, we need to encode the signal properly. If C is less than H, no encoding will work.

For a truly ergodic source, this makes good if impractical sense. We can see the impracticality by observing that for a message written using capital English letters and the space only, without punctuation, there are

$$27^N$$

different stretches of text of length N. Long before N approaches infinity, the number of stretches of text we must examine in order to find the probability of any one stretch becomes unmanageably large.

For real text, or any other real signal, (13.1-2) degenerates into nonsense for large values of N. It is not only that we cannot examine all stretches of text 1000 letters long; many possible and plausible stretches have never been written. We do not have a good model for generating text; none can exist which relies only on symbol probabilities. The intent of the writer becomes of major importance.

Nevertheless, there is a simple variable-length coding scheme, called the Huffman code, that can be useful in efficiently encoding symbols or blocks of symbols when the probabilities of occurrence of the symbols or blocks of symbols are known. The encoding is efficient because short codewords are assigned to frequently occurring symbols and longer codewords to less frequently occurring symbols. Morse utilized this principle in designing his telegraph code. We shall explain Huffman's scheme, but we will not derive the number of bits per symbol it requires on the average. This number is as low as possible, so the Huffman code is optimum.

We have seen in equation (13.1-2) that if we divide the message into a block of letters or words and treat each possible block itself as a new kind of symbol, we can compute the entropy per block by the same

**Table 13.1. Probabilities
of Words**

Word	Probability
the	0.50
man	0.15
to	0.12
runs	0.10
house	0.04
likes	0.04
horse	0.03
sells	0.02

formula we used for independent symbols and get as close as we like to the source entropy merely by making the blocks very long.

Thus the problem is to find out how to use binary digits to efficiently encode a sequence of symbols chosen from a group of symbols, each of which has a certain probability of being chosen. Shannon and Fano both showed ways of doing this, but Huffman found the best way. We shall consider his encoding method here.

Let us for convenience list all the symbols vertically in order of decreasing probability. Suppose the symbols are the eight words *the, man, to, runs, house, likes, horse, sells*, which occur independently. The probabilities of their being chosen are as listed in Table 13.1.

We can compute the entropy per word by means of equation (13.1-1); it is 2.25 bits per word. However, if we merely assigned one of the eight 3-digit binary numbers to each word, we would still need 3 bits to transmit each word. How can we encode the words more efficiently?

Figure 13.1 shows how to construct the most efficient code for encoding such a message word by word. The words are listed to the left, and the probabilities are shown in parentheses. In constructing the code, we first find the two lowest probabilities, 0.02 (*sells*) and 0.03 (*horse*), and draw lines to the point marked 0.05, the probability that a word is either *horse* or *sells*. We then disregard the 0.02 and 0.03 and look for the next two lowest probabilities, which are 0.04 (*likes*) and 0.04 (*house*). We draw

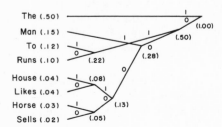

Figure 13.1. Constructing the most efficient code.

Table 13.2. The Huffman Code

Words	Probability p	Codeword	Number of digits N in codeword	Np
the	0.50	1	1	0.50
man	0.15	001	3	0.45
to	0.12	011	3	0.36
runs	0.10	010	3	0.30
house	0.04	00011	5	0.20
likes	0.04	00010	5	0.20
horse	0.03	00001	5	0.15
sells	0.02	00000	5	0.10

lines to the right to a point marked 0.08, which is the sum of 0.04 and 0.04. The two lowest remaining probabilities are now 0.05 and 0.08, so we draw a line to the right connecting them, to give a point marked 0.13.

We proceed in this way until paths run from each word to a common point to the right, the point marked 1.00. We now traverse this tree backwards. We label each upper path going to the left from a point with a 1, and each lower path with a 0. The codeword for a given word is the sequence of digits encountered going left to the word from the common point 1.00. The codewords are listed in Table 13.2. We see that the Huffman code is generally a variable-length code, as opposed to the error-correcting block codes of Chapters 11 and 12, which are of constant length.

In Table 13.2 we have shown not only each word and its encoded version but also the probability of each codeword and the number of bits in each codeword. The choice of upper and lower paths from each point will depend on how the paths are drawn, but the lengths of the Huffman codewords are independent of this. The probability of a word times the number of bits in the word gives the average number of bits per word in a long message due to the use of that particular word. If we add the product of the probabilities times the numbers of bits for all the words, we get the average number of bits per word, which is 2.26 in this case. This is only a little larger than the entropy per word, which was 2.25 bits per word. And, it is substantially smaller than the 3 bits per word we would have used if we had merely assigned a different 3-bit codeword to each word.

It can be demonstrated that the Huffman code is the most efficient code for encoding a set of symbols having different probabilities. It can also be demonstrated that Huffman coding always calls for less than one binary bit per symbol (or word) more than the entropy. (In the above case, it calls for only 0.01 extra bits per symbol.) Huffman coding must also require a number of bits per symbol at least equal to the entropy. And, if the largest probability of any symbol is small, Huffman coding will

Table 13.3. Probabilities of English Letters

Value of j	Letter referred to	Probability of letter, $p(j)$
1	E	0.13101
2	T	0.10465
3	A	0.08149
4	O	0.07993
5	N	0.07096
6	R	0.06880
7	I	0.06344
8	S	0.06099
9	H	0.05257
10	D	0.03787
11	L	0.03388
12	F	0.02923
13	C	0.02757
14	M	0.02535
15	U	0.02458
16	G	0.01993
17	Y	0.01982
18	P	0.01981
19	W	0.01538
20	B	0.01439
21	V	0.00919
22	K	0.00420
23	X	0.00166
24	J	0.00132
25	Q	0.00121
26	Z	0.00077

be very close to the entropy of the source. We see that the entropy will be attained by Huffman coding of long blocks of independently occurring symbols.

We can now see that the entropy of a collection of words or symbols represents the minimum average number of binary digits or bits necessary to encode the symbols, taking their probability of occurrence into account. This is why the entropy occurs in information theory.

Table 13.3 is a famous table. It gives the probabilities of occurrence of the letters of the English language. The first 12 most likely letters are

ETAONRISHDLF

Such a table is useful in elementary cryptanalysis in deciphering simple alphabetic substitution ciphers. A Huffman code can of course be constructed for it too.

Let us now turn to an entirely different matter, involving the Huffman code given in Figs. 13.1 and Table 13.2. When we encode a message by using this code and get an uninterrupted string of symbols, how do we tell whether we should take a particular 1 in the string of symbols as indicating the word *the* or as part of the codeword for some other word?

We should note that of the codewords in Table 13.2, none forms the first part of another. This is called the *prefix property*. It has important and, indeed, astonishing consequences, which are easily illustrated. Suppose, for instance, that we receive the message of 38 bits

$$10010000010001101110011000010100111001 \qquad (13.1\text{-}3)$$

We can decipher this message uniquely, even without markers for the beginnings and endings of codewords. For, only one codeword, 1, starts with 1. It is the codeword for *the*. We then have the sequence $0010000 \cdots \cdots$. The codeword 001 is recognized as the codeword for *man*, and is stripped off. Unambiguous decoding can proceed, since each time we are faced with the remainder of the sequence, the prefix property guarantees that in the absence of errors there is exactly one codeword that can be stripped off. This property is called *unique decipherability*, *self-acquirability*, or *self-synchronizability*.

What happens if we lose the first few bits of such a message? For a general Huffman code, we are usually able to recover. But, for example, if there are four words each of probability 1/4, the Huffman code is 00, 01, 10, 11. If we lose the first bit of such a message we will forever be "out of sync," and are not able to detect this fact if successive words are equally likely. Unless all the words are equally likely, simple reasoning shows that we can recover sync. If there are also bit errors, the situation is very bad. We note that in a Huffman code, every unending string of 0's and 1's can be decoded. We shall not discuss this.

Huffman codes always have the prefix property. The reason is simple. If a given n-tuple is a codeword, the Huffman path leads to it in n steps. If the n-tuple were also a prefix of another codeword of more than n steps, the Huffman path for the longer codeword would reach the word encoded by the shorter codeword before all its steps were traversed. Huffman's construction does not allow this. So, we always have the prefix property with Huffman coding.

In the next section, we will investigate some additional methods for encoding sources, ways which take advantage of particular kinds of redundancies or nonindependencies in consecutive pieces of data.

Problems

1. Use the definition (13.1-2) to compute the entropy of a binary source in which the first output is 0 or 1 with probability 1/2 and in which the probability that the next symbol differs from the present symbol is p. HINT: encode no change as 0 and a change as 1.

2. Complete the decoding of the message (13.1-3).

3. Suppose the first 6 bits of the message (13.1-3) are lost in transmission, but this fact is unknown at the receiver. Decode the message, and compare with the correct decoding.

4. Can the Huffman code corresponding to a probability distribution on a set of two words have an average word length close to 1 more than the entropy of the distribution?

5. The probabilities of various letters occurring in English text are given in Table 13.3. Disregarding the fact that the probability of occurrence of a letter actually depends on the preceding letters, and disregarding the absence of the space, (a) what is the entropy in bits per letter; (b) what is the Huffman code; (c) how many bits per letter are needed to transmit the letters using the Huffman code?

13.2. Various Source-Encoding Schemes

The best encoding schemes for sources are usually come upon, not mathematically, but by ingenuity and intuition.

For example, black typescript on white paper, or black line drawings on white paper, are mostly white. Suppose that 0 represents white and 1 represents black. As a facsimile scanner or TV camera scans over such material there are long *runs* of consecutive 0's between the few consecutive 1's that represent black. In *run length* encoding, the lengths of runs of zeros and ones are transmitted, rather than the individual 0's and 1's. Usually such run lengths are encoded by fixed-length codewords, but they may be encoded by Huffman coding. We shall examine the performance of run-length coding later in this section.

In endeavoring to save bit rate or bandwidth in transmitting voice or TV signals, we can try to predict the value of a quantized sample from the values of the preceding samples. We can then transmit the error in prediction. This error can, if one wishes, be encoded by a Huffman code.

In these cases, an effort is made to reproduce a quantized signal exactly. Another approach is to allow errors to which the recipient is not sensitive. The commonest example is the chrominance or color signal in color TV. This signal is narrower-band than the brightness signal. Colors are smeared in transmission, but the eye is not sensitive to such smearing.

When the sound or view to be transmitted is not to be reproduced exactly, the received signal or message must meet some *fidelity criterion* which depends on the recipient's detection ability or sensitivity to errors in reproduction. Shannon uses such a fidelity criterion in his treatment of continuous signal sources, which cannot be transmitted and reproduced with perfect fidelity. We shall examine this in the next section.

Let us consider run-length coding. We propose to find the effectiveness of such coding for a particular sort of signal source. The model we shall discuss is best understood in terms of facsimile or TV. Here we consider only one scan line at a time, and so ignore two-dimensional structure. Better methods of transmitting images by source encoding must take the two-dimensional structure into account.

We will assume perfectly hard-clipped digital TV as a model, so that the only values of the brightnesses of picture elements or pixels are 0 and 1. A "Markov" source model is assumed, which means that only the previous pixel affects the probability distribution of the current pixel. Here there are two states, called simply 0 and 1, and a known transition probability p. When the source is in the 0 state, it always outputs a 0 pixel; when it is in the 1 state, it outputs a 1. The probability of a *change* of state at any time is independent of the past history and of the present state and is always equal to p. This model may be applied to single scan lines of a two-level picture.

For facsimile used to transmit pages of typewritten text, there will be two values of this change-of-state probability. The probability of a change of state from the 0 or "black" state would be higher than the probability of a change from the 1 or "white" state, because much more of the page is white than black. But, we shall assume the change of state probability independent of which state we are in.

Because in our simplified model the output sequence of 0's and 1's is a Markov process, the probability distribution of the next symbol depends only on the present symbol, not on past history. The probability that the next symbol is the same as the present symbol is $1 - p$, whereas the probability that it is different is p.

For such Markov sources, Section 13.1 shows that the entropy $H(X)$ is the entropy of the distribution of the next symbol given the present symbol. Since that distribution takes two values, one with probability p and the other with probability $1 - p$, we have

$$H(X) = H(p) = -p \log_2 p - (1 - p) \log_2 (1 - p) \text{ bits/pixel}$$
$$(13.2\text{-}1)$$

In the run-length coding of this source, we will begin by transmitting a 0 or 1, depending upon what the first symbol of a line is. We assume that the line is very long compared with $1/p$. Then the fractional penalty for having to transmit this added bit is small. The *lengths* of successive runs, that is, strings of consecutives 0's and 1's of length as large as possible, are transmitted. We do not need to transmit the *type* of run it is (0's or 1's), since the two types alternate. And, by using Huffman coding, which we know is uniquely decipherable, we do not need any "overhead" bits. In the absence of errors, we can reconstruct the line uniquely.

How close does run-length coding come to the minimum of $H(X)$ bits per outcome given by (13.2-1)? It takes about B bits (less than one more than B bits, certainly) to describe the run length by Huffman coding, with B given by

$$B = -\sum_{n=0}^{\infty} p(1-p)^n \log_2 p(1-p)^n \text{ bits/run} \qquad (13.2\text{-}2)$$

This is the formula for the entropy of the distribution $\{p(1-p)^n\}$, $n = 0, 1, 2, \ldots$. The probability $p(1-p)^n$ is the probability that the run lasts $n + 1$ bits. Here $(1-p)^n$ is the probability that the source stays in the same state for n successive times, and p is the probability of a change in state at the $(n + 1)$st time. Because the line length is long compared with $1/p$, we let the sum in (13.2-2) go from zero to infinity.

Let us use a little algebra of series:

$$\sum_{n=0}^{\infty} (1-p)^n = \frac{1}{p}$$

$$\sum_{n=0}^{\infty} n(1-p)^n = -(1-p)\frac{d}{dp}\left(\sum_{n=1}^{\infty} (1-p)^n\right) = \frac{1-p}{p^2} \qquad (13.2\text{-}3)$$

Equation (13.2-3) coupled with (13.2-2) gives us a simpler expression, because the log terms are independent of n and factor out:

$$B = -\log_2 p - \frac{1-p}{p}\log_2 (1-p) = \frac{H(p)}{p} \text{ bits/run} \qquad (13.2\text{-}4)$$

In a large number n of pixels, there will be N_n different runs, with N_n given approximately by

$$N_n \doteq np \qquad (13.2\text{-}5)$$

This is because, in n pixels, there are very close to np changes of state, by the law of large numbers. Every change of state corresponds to the start of a new run, and a state change has probability p.

So, in n pixels, we have about np runs with about $H(p)/p$ bits/run. Thus there are about

$$np \cdot \frac{H(p)}{p} = nH(p) \text{ bits} \qquad (13.2\text{-}6)$$

to be transmitted to completely describe a TV line of n pixels. We see that run-length coding very nearly achieves $H(p)$ bits per pixel. By (13.2-1), this is the source entropy, and so is as low as any method can achieve.

Run-length coding, with Huffman coding for the run lengths, is optimum for Markov sources. But, real TV is not hard clipped, and is not Markov either. The performance of run-length or other coding for real TV signals must be evaluated experimentally.

In the next section, we will bring source encoding into the framework of Shannon's information theory. We do this via the study of rate distortion theory.

Problems

1. What happens in run-length coding if we miss the first few bits of transmission of a line, but otherwise have no transmission errors? What happens if there is an occasional transmission error? What does this say about the bit error probability required for run-length coding?

2. Although typewritten text does not satisfy the simple assumptions we have made on Markov sources, let us assume that the number of bits per pixel needed for run-length coding of digital facsimile is about the same as coding of a Markov source with $p = 1/10$. If a page of text has 1200 scan lines with 900 pixels/line, how many bits per page are needed? Contrast this with an electronic mail system which directly sends the text as 8 bit characters. Assume 50 lines of text per page for this, with 70 characters per line. What is the fractional or dB saving of electronic mail over digital facsimile with run-length coding? What other advantages and disadvantages might electronic mail have over facsimile?

13.3. Rate Distortion Theory

Where are we at this point in our study of information transmission? We see that the concept of channel capacity is of supreme importance in information and communication theory and practice. This applies to both analog and digital communication, to voice, video, data, handwriting, and anything that will be built.

We have seen that by proper channel coding a channel can *literally* transmit up to C bits per second with an error as small as assigned. And, by assigning a binary code to the channel input symbols or to strings of input symbols, the bits can *literally* come in and go out as binary digits.

We have also seen in Sections 13.1 and 13.2 (Huffman coding and run-length coding) that there are more efficient ways of encoding sources than an arbitrary assignment of codewords to messages from a source. We have even given in (13.1-2) an expression for the entropy of a source with memory, in terms of the probabilities of blocks. This implies that it may be valuable to encode entire blocks.

What is still missing is something like a source entropy when we do not insist on reproducing the messages from the source exactly.

We have already encountered quantization in Section 5.6, in which a continuous, infinite entropy signal such as band-limited white gaussian noise can be converted into a finite-entropy stream of bits at the cost of some error in reproduction. What we need is just a generalization of this with an appropriate cost assigned to errors or differences between the original signal and the reconstructed signal.

Shannon provided this generalization in his original information theory papers (Reference 1 of Chapter 11). The theory was fully developed in his rate distortion theory paper (Reference 1 of this chapter), which appeared about 12 years after his original paper on information theory. One of the main conclusions of rate distortion theory is that the same parameter of a channel, the capacity, is adequate to describe the suitability of a channel for passing information when errors are tolerable as well as when errors are not tolerable.

This section explains the concepts of rate distortion theory. We shall see how much channel capacity is required to transmit sources when errors of a specified amount are tolerable. Knowing the tolerable errors tells how much capacity is needed. Or, knowing the capacity tells us how much error we must accept. This applies to a channel of a given capacity, whether the channel is high rate but of high error probability, or of low rate but makes no errors at all.

Shannon considers both the source probability distribution and the data quality. The source produces random outputs according to some distribution. Here we shall treat independent outputs or memoryless sources only. The extension to sources with memory uses definition (13.1-2). Most real sources, of course, have a great deal of memory, and this memory can be used to greatly simplify the transmission of outcomes from the source. We shall devote the next two sections to this.

To measure data quality or to extend the concept of error to continuous signals, Shannon introduced a *fidelity criterion* and *distortion measure*. Here, we will consider only a *single-letter* distortion measure, $d(u, v)$. This tells how close the output letter v is to the input letter u.

We have a sequence of data k letters or symbols long to transmit over a memoryless channel. The input and output sequences u and v are

$$u = (u_1, \ldots, u_k)$$
$$v = (v_1, \ldots, v_k)$$

Here the u and v are individual letters or symbols, and v_i is received when u_i is transmitted. We define the probability of receiving v_i when u_i was sent to be $p(v_i \mid u_i)$.

The total distortion on the *block*, $d(u, v)$, is defined to be additive:

$$d(u, v) = \sum_{i=1}^{k} d(u_i, v_i) \qquad (13.3\text{-}1)$$

The total distortion $d(u, v)$ is the sum of the k distortions $d(u_i, v_i)$. This is analogous to counting errors in a block by adding a 1 to the count every time an error is made in a symbol of the block.

The distortion per letter or per symbol is the distortion given by (13.3-1), divided by the number of symbols k in the block. If the distortion d counts errors $[d(u_i, v_i) = 1$ if $u_i \neq v_i$; $d(u_i, v_i) = 0$ if $u_i = v_i]$, the per-symbol distortion is the observed or measured symbol error probability. The expected value of the distortion per symbol, or the *fidelity criterion*

$$\text{fidelity criterion} = E\left(\frac{1}{k} d(U, V)\right) \qquad (13.2\text{-}2)$$

depends on the probability distribution of the source U. From this, the probability distribution V is determined via the channel $p(v_i \mid u_i)$ which outputs symbols v_i in response to inputs u_i.

Shannon was interested in the lowest rate, or worst, channel over which the source U could be transmitted with the expected or average distortion, or fidelity criterion, or loss of fidelity at most equal to a given quantity δ. This rate $R(\delta)$ is the *rate distortion function*, which is the minimum mutual information between channel input U and output V we can have when the expected distortion per letter, or fidelity criterion, is not more than δ. We can if we wish express this definition mathematically by the equation

$$R(\delta) = \min\left\{ I(U; V) : E\left(\frac{1}{k} d(U, V)\right) \leq \delta \right\} \qquad (13.3\text{-}3)$$

Here $I(U; V)$, the mutual information from Section 12.3, is the rate of a channel given by probabilities $p(v \mid u)$ with input probabilities those of U. The minimum is taken over all channels $p(v \mid u)$ which hold the average distortion to at most δ.

Just as there are codes for combatting noise, there are codes for combatting redundancy. Codes for combatting redundancy in the source are called *source codes*. The Huffman code is a source code for reproducing an input sequence exactly. Here the concept of redundancy should include the possibility that we have some distortion in the output signal. We should not need to transmit as many bits as if we had to reproduce the input sequence exactly.

A *source code* is a set C of M k-vectors or codewords v_i of length k. The components are from the alphabet of *received* symbols.

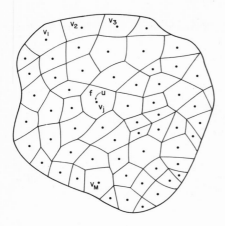

Figure 13.2. Space of source blocks of length k. The encoding function f maps u to v_i if u is in region j.

We write the code as

$$C = \{v_1, v_2, \ldots, v_M\} \qquad (13.3\text{-}4)$$

Here we only consider block codes, or codes of constant length k. There are also convolutional source codes as well. We will have no loss from considering only block codes, if we encode long blocks of outcomes.

There is an *encoding function* $f(u)$ associated with a source code. It takes each source codeword or k-vector u into a v_i closest to u in the distortion measure. This f can be defined by cutting up the space of source blocks into disjoint regions, as in Fig. 13.2. In the jth region, any source block u is mapped by the decoding function f into the same codeword v_j. The jth region basically consists of those u closer to v_j than to any other v_i.

As an example, the values u of U can be voltages from -1 to $+1$ V. The regions of Fig. 13.2 can be 16 intervals of voltages $\frac{1}{8}$ of a volt apart. Here $k = 1$ and $M = 16$. The v_j can be the 16 centers of these intervals. The encoding function will output the voltage v_j, the center of jth interval, if the actual voltage u is in the jth interval. This will be recognized as a 16-level or 4-bit quantizer or analog-to-digital converter. If the distortion $d(u, v)$ is the square error

$$d(u, v) = (u - v)^2$$

then the average distortion is the mean-square error. We found the average distortion for any reasonable distribution of U in Section 5.6, where we studied errors in representing signals approximately by quantization.

The average per symbol distortion $d(C)$ of the source code C is defined to be

$$d(C) = \frac{1}{k} E(d(U, f(U))) \qquad (13.3\text{-}5)$$

In this case U denotes the entire random *vector* $(U_1, \ldots, U_k)$ of the first k source outputs. We have assumed these to occur independently. The expectation is taken over the distribution defining the original single-letter source U.

The average distortion (13.3-5) tells how close we can get to original codewords on the average if we represent or replace a source codeword or k-vector u by its corresponding closest codeword v_j. The *rate* of the source code C is

$$r(C) = \frac{1}{k} \log_2 M \text{ bits/symbol} \tag{13.3-6}$$

Here M is the number of k vectors in the source code, and k is the number of symbols encoded at one time. We see that $r(C)$ is the number of bits per source symbol needed to describe the source codeword that is closest to the actual source outcome, if we use a constant-length code to transmit the occurrences of the codewords v_j.

By using the definition of mutual information, it is not hard to show that a source code C of average per-symbol distortion $d(C) \leq \delta$ requires a transmission rate $r(C)$ no less than the rate distortion function $R(\delta)$. Conversely, we can find source codes of large block length k, of rate $r(C)$ not much more than $R(\delta)$, and of average per-symbol distortion at most δ. Shannon proved this converse by a random coding argument, much as in Section 11.2. The result is that we can get good source codes (this means the distortion is at most δ) of rate about $R(\delta)$, but no lower. Shannon's proof also implies that the M codewords in a good source code are all about equally likely to occur. So, no further rate reduction is possible by Huffman coding.

We can use a source code to "compress" a given source to a binary symmetric source, in which 0 and 1 are independent and have equal probability. There will be $R(\delta)$ bits from this source for each symbol from the original source. The average distortion between an outcome and its encoded version will be at most δ. Every source is equivalent to a binary symmetric source in this way.

How can we take into account the *channel* that we use to transmit the source U? Consider the generalized communication system of Fig. 13.3. The outcomes from the source are encoded in some way, and sent over the channel. The *encoder* can comprise anything done to the source before transmitting it. A lens on a TV camera can be part of the encoder in Shannon's theory. But, the encoder of Fig. 13.3 also includes any channel coding to combat noise. In the decoder, we decode or reconstruct the original source outcomes, and send the result $\hat{U}$ to the destination or sink.

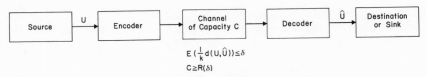

$$E\left(\frac{1}{k}d(U,\hat{U})\right) \le \delta$$

$$C \ge R(\delta)$$

Figure 13.3. General communication system.

Shannon showed the following about this general system. If the decoder outputs symbols with average per-symbol distortion of at most δ,

$$E\left(\frac{1}{k}\,d(U,\,\hat{U})\right) \le \delta \tag{13.3-7}$$

the channel must have capacity C at least $R(\delta)$:

$$C \ge R(\delta) \tag{13.3-8}$$

This is proved using simple properties of mutual information. The same concept of capacity that occurred in Chapter 11 in discussing *error-free* transmission occurs in discussing transmission *with* errors.

Shannon also showed the "direct" part of this result. This states that transmission with average distortion at most δ *is* possible if the channel capacity exceeds $R(\delta)$. These two facts comprise the source–channel coding theorem. We shall derive the direct part, knowing the existence of good source codes.

In Fig. 13.3, we required an encoder as a black box matching the source to the channel. In Fig. 13.4, we have designed such an encoder. The *source encoder* in the black box converts the source into a binary symmetric source of rate $R(\delta)$ bits/symbol. The *channel encoder* makes the channel virtually noiseless, and of capacity C bits per symbol or per channel use. If C exceeds $R(\delta)$, we can send the output of the source encoder over the channel. Since the probability of *word* error is very small, we can ignore the possibility of any error in transmission. So, we can reconstruct the source encoder output exactly.

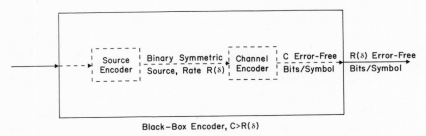

Black–Box Encoder, $C > R(\delta)$

Figure 13.4. Source and channel encoder in series.

The output of the channel decoder is, with extremely high probability, the same as what was input to the channel encoder. This input was the output of the source encoder. So, the receiver reconstructs the output of the source encoder exactly.

The average distortion of this communication system will then be the distortion of the source code alone, as given by (13.3-5). But, the source code has average distortion at most δ. In this way we can indeed reconstruct the source with average distortion at most δ. We note that in particular cases, there may be less complex designs for the encoder than the one of Fig. 13.4.

Rate distortion theory can be used to find the channel capacity needed to transmit random bits when a final average *bit* error probability δ is allowed, instead of a very low or zero word or bit error probability. The source is a binary symmetric source. The distortion measure $d(u, v)$ is

$$d(u, v) = \begin{cases} 0 & \text{if } u = v \\ 1 & \text{if } u \neq v \end{cases} \qquad (13.3\text{-}9)$$

We note that the expectation of $d(U, V)$ is merely the probability that $U \neq V$. If u, a k-vector, is the source sequence and the k-vector v is received, the per-symbol distortion is, from (13.3-1),

$$\frac{1}{k} d(u, v) = \frac{1}{k} \sum_{i=1}^{k} d(u_i, v_i) = \frac{1}{k} (\# \text{ bit errors}) \qquad (13.3\text{-}10)$$

We require that the expected per-letter distortion be no more than δ:

$$\text{average bit error probability} = \frac{\text{expected \# bits in error}}{\text{number of bits transmitted}} \leq \delta$$

$$(13.3\text{-}11)$$

Here δ is the final allowed bit error probability.

What is the rate distortion function of the binary symmetric source with this bit error distortion measure? From (13.3-3), we seek a channel $p(v \mid u)$ such that the mutual information $I(U; V)$ is minimized, subject to the constraint that the expected distortion be at most equal to δ. It is apparent that we can assume that the expected distortion be *exactly* δ in seeking the minimum. This means that we must have

$$E(d(U, V)) = pr(V \neq U) = \delta \qquad (13.3\text{-}12)$$

The term $pr(V \neq U)$ can be expressed as

$$pr(V \neq U) = pr(U = 0)p(1 \mid 0) + pr(U = 1)p(0 \mid 1)$$

$$pr(V \neq U) = \tfrac{1}{2}(p(1 \mid 0) + p(0 \mid 1)) \qquad (13.3\text{-}13)$$

We wish to minimize

$$I(U; V) = H(V) - H(V \mid U)$$
$$= H(pr(V = 0)) - \tfrac{1}{2}H(p(1 \mid 0)) - \tfrac{1}{2}H(p(0 \mid 1)) \quad (13.3\text{-}14)$$

subject to constraint (13.3-12). In (13.3-14), $H(p)$ is the entropy function of the probability p. Here

$$pr(V = 0) = \tfrac{1}{2}pr(0 \mid 0) + \tfrac{1}{2}pr(0 \mid 1) \quad\quad (13.3\text{-}15)$$

We can see from (13.3-15) that (13.3-14) is symmetric in $p(1 \mid 0)$ and $p(0 \mid 1)$. Since constraint (13.3-12) is also symmetric, the minimum occurs when $p(0 \mid 1)$ and $p(1 \mid 0)$ are equal, and we let their common value be p:

$$p(0 \mid 1) = p(1 \mid 0) = p \quad\quad (13.3\text{-}16)$$

We have here a binary symmetric channel, and so

$$p(V = 0) = p(V = 1) = 1/2 \quad\quad (13.3\text{-}17)$$

From (13.3-12) and (13.3-16), the error probability p of this channel satisfies

$$p = \delta \quad\quad (13.3\text{-}18)$$

The rate distortion function $R(\delta)$ given by (13.3-14), (13.3-16), and (13.3-18) is then merely

$$R(\delta) = 1 - H(\delta) \quad\quad (13.3\text{-}19)$$

We see that we only need a channel of capacity infinitesimally greater than $1 - H(\delta)$ to transmit a binary symmetric source with a final average bit error probability of δ. Or, if we wish to communicate a binary symmetric source over a channel which has capacity C at most $R(\delta)$ bits/channel use, we must accept an average bit error probability of at least δ. Here δ is given implicitly in terms of C by solving the equation

$$C = 1 - H(\delta) \quad\quad (13.3\text{-}20)$$

This equation permits us to compare the efficiencies in energy per bit of channels which make errors with channels which do not. We could have used this in Section 11.3.

Source coding is not used much in communication systems, but channel coding is used often. What are the reasons for this difference? In decreasing order of importance, we feel that they are as follows.

(1) The distortion measure is not known analytically or experimentally for the most important sources. These include voice and video. And, the single-letter distortion concept does not apply to voice and video in any simple way.

(2) Probabilistic models of sources are much less realistic than probabilistic models of channels.

(3) There are many types of sources, but fewer types of channels. So, investment in source encoding cannot be distributed among as many types of users.

We shall explore the second reason. The simplifying assumptions that must be made to model a source probabilistically often must ignore the underlying physical structure or mechanism or meaning that determines the outcomes. Taking into account the known structure, or the real meaning, of the source will result in a much larger saving than could ever be attained through an approach which makes much formal and elegant use of information theory, but ignores the underlying structure. One can use both approaches in the same system, but the major savings are to be obtained by getting at the underlying physical structure of the source.

In the next two sections, we shall examine such major reductions in required channel capacity. In the next section, the source is human speech, and the encoder is called the *vocoder*. In the section following, we investigate a possible theory of the characterization of more general data sources.

Problems

1. A source code C for a binary symmetric source is constructed by blocking outcomes into groups of three. We send 0 if the majority of the three bits is a 0 and a 1 if the majority is a 1. At the receiver, a 0 is interpreted as three 0's and a 1 is interpreted as three 1's. If distortion is final decoded bit error probability, what is the average per-letter distortion of this code? What is the average distortion of a system in which this source code is used on a binary symmetric channel with bit error probability p? Check your result by letting $p = 1/2$. Explain why the check works.

2. A binary symmetric source outputs a times per second and is to be sent over a binary symmetric channel of bit error probability $p < 1/2$ with b channel uses per second. If the average bit error probability after decoding is to be δ, which of the following combination of parameters is possible?

   (a) $p = 0.09,$ $\delta = 0.11,$ $a = 1,$ $b = 1$
   (b) $p = 0.1,$ $\delta = 0.02,$ $a = 2,$ $b = 3$
   (c) $p = 0.02,$ $\delta = 0.001,$ $a = 5,$ $b = 6$

3. A binary symmetric source is to be sent over a binary symmetric channel with bit error probability 0.01. The desired distortion is also 0.01. Figure 13.4 suggests that we should first convert the source into a binary symmetric source of rate $R(0.01)$, and then build a channel encoder which slows down the channel so as to make it noiseless. Suggest a much better way of meeting the requirements of the encoder in this case.

4. The human eye is especially sensitive to sharp edge transitions, but brightness over large regions is still important. Regarding each entire picture as a single "letter," suggest a reasonable distortion measure for still video such as is used in deep space transmission of pictures of planets.

5. A source produces white gaussian noise $x(t)$ (in volts) as a signal, limited to the band 0 to B_1. The two-sided spectral density of the source is $N_1/2$. The mean-square distortion $d(x(t), y(t))$ is

$$d(x(t), y(t)) = \frac{1}{T} \int_0^T (x(t) - y(t))^2 \, dt \; W$$

Here T is a given length of time interval and $y(t)$ is the waveform which is output from the system when $x(t)$ is input. It is known that if $\delta \leq N_1 B_1$, then

$$R(\delta) = B_1 \log_2 \frac{N_1 B_1}{\delta} \; \text{bits/sec}$$

The source is to be transmitted over an infinite-bandwidth white gaussian channel of two-sided noise spectral density $N_0/2$ and available average signaling power P_s.

(a) For $\delta \leq N_1 B_1$, how small must N_0 be to achieve average distortion no larger than δ, keeping the transmitted power at P_s?

(b) If fm is used, what does this say about the maximum frequency deviation f_d that can be used before breaking must occur no matter what fm demodulator we use? (See Section 4.7.)

(c) What is $(f_d)_{max}$ if $P_s = 10^{-18} \, \text{W}$, $B_1 = 3 \, \text{kHz}$, N_0 arises from thermal noise at temperature $T_0 = 20°K$, and N_1 is 6 W/Hz?

(d) What δ is attainable for this f_d?

(e) What is the rms deviation ratio $(f_d/B_1)(\overline{x^2})^{1/2}$?

HINT: show from Section 4.7.1 that the average distortion δ for fm is

$$\delta = B_1^3 N_0 / 3 P_s f_d^2$$

in the absence of breaking.

6. In the previous problem, the formula for $R(\delta)$ goes to 0 for $\delta = N_1 B_1$. Explain why this must be so.

7. In Problem 5, if all the parameters are given except for the allowable distortion δ, we see that for small δ, $R(\delta)$ behaves as

$$R(\delta) \sim B_1 \log_2 (1/\delta) \; \text{bits/sec}$$

Use sampling to show how we can attain this behavior. Conclude that sampling band-limited gaussian noise at the Nyquist rate, performing an analog-to-digital conversion, and ideal low-pass filtering to reconstruct the continuous-time waveform is nearly optimal for small distortions. HINT: review Section 5.6.

13.4. Speech Transmission and the Vocoder

In the digital transmission of speech, commonly a speech signal is sampled 8000 times per second and the sample amplitudes are encoded as 7- or 8-digit binary numbers. The quantizing levels, that is, the amplitudes that are encoded, are spaced increasingly farther apart as we go away from the zero-voltage axis. This is because the ear is sensitive to *fractional* errors in reproduced amplitude rather than to absolute errors.

Satisfactory telephone-quality transmission is thus attained with some 56,000 to 64,000 binary digits per second. Reference 2 is a good summary of the current state of the speech-coding art, including quality and implementation complexity.

By means of the vocoder, highly intelligible transmission of speech can be attained with as few as 2000 binary digits per second and there is every reason to believe that this can be improved upon by devices of similar nature.

The vocoder does not transmit or reproduce the waveform of the speech to be transmitted. Instead, what is transmitted is a *description* of the speech. From this description, intelligible and satisfactory speech can be reconstructed. Vocoders are so effective because they take full advantage of the physics of the speech generation mechanism. We could not do nearly as well if we simply tried to approximate the speech as an instantaneous waveform.

In the early thirties, long before Shannon's work on information theory, Homer Dudley of Bell Telephone Laboratories invented a form of efficient speech transmission which he named the *vocoder* for *voice coder*. The transmitting (*analyzer*) and receiving (*synthesizer*) units of a typical vocoder, the channel vocoder, are illustrated in Fig. 13.5.

In the analyzer, the speech signal is low passed to 3200 Hz and fed to 16 filters. Each of these filters determines the short-term energy in the speech signal in a particular 200-Hz band of frequencies and transmits a 20-Hz signal giving the power in the band. This signal is received by the corresponding synthesizer. In addition, an analysis is made to determine

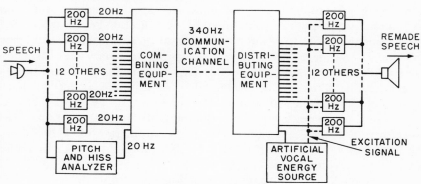

Figure 13.5. Channel vocoder concept.

whether the sound is *voiceless* or *voiced*. Here voicing means that the vocal cords are vibrated during sound generation. For example, *z*, *v* are the voiced versions of *s*, *f*, respectively. And, all vowels are voiced as well. (Reference 4 of Chapter 3 discusses the concepts of voicing, pitch, and related phonetic concepts of relevance to speech communication.) If the sound is voiced, the *pitch* or fundamental frequency of vocal cord vibration is determined by a further analysis. A 20-Hz signal is outputted to describe the pitch. The total bandwidth is $20 \times 17 = 340$ Hz.

At the synthesizer, when the sound is voiceless, a hissing, more or less white gaussian noise, is produced; when the sound is voiced, a sequence of electrical pulses at the proper pitch rate is produced. These pulses correspond to the puffs of the air passing the vocal cords of the speaker.

The hissing or pulses are fed to an array of filters, each passing the same 200-Hz band of frequencies that the corresponding filter in the analyzer passes. The short-term sound energy passing through a particular filter in the synthesizer is controlled by the output of the corresponding analyzer filter. Thus the short-term energy spectrum (see Section 3.11) of the sound produced by the synthesizer is made roughly the same as the short-term energy spectrum of the sound input to the analyzer.

The result is the reproduction of intelligible speech. In effect, the analyzer listens to and analyzes speech, and then instructs the synthesizer, which is an artificial speaking machine, how to say the words all over again with the very pitch and even accent of the speaker. Most vocoders have, however, had a strong and unpleasant electrical accent. Modern integrated circuits help to overcome this by accurate operation. Digital processing is perhaps more easily adapted to the *linear predictive coefficient* vocoder than to the channel vocoder (see Reference 3). Both are closely related.

The study of vocoder quality has led to new and important ideas concerning what determines and influences speech quality. But, even low-quality vocoders can sometimes be very useful. For instance, it is sometimes necessary to resort to enciphered or encrypted speech transmission. If one merely directly reduces speech to binary digits by pulse code modulation, 15,000 to 30,000 binary digits per second must be sent to give speech of acceptable intelligibility. By using a vocoder, comparable speech can be sent with around 2400 binary digits per second, or even less. Four-bit quantization of the samples of the 340-Hz transmitted waveform of Fig. 13.5 would result in an output rate of 2720 bits/sec. And, we can get a lower bit rate by not transmitting the high-frequency filter outputs with equal accuracy.

A channel vocoder sends information concerning from 10 to 30 frequency bands (16 in the example of Fig. 13.5). But, the short-term spectra of speech sounds actually have only a few very prominent frequency ranges called *formants*. These correspond to the resonances of the vocal tract. One can recreate intelligible speech by sending the location and intensity of two or three formants. Such a *formant-tracking vocoder* can be used to transmit speech with even fewer binary digits per second than the channel vocoder of Fig. 13.5 needs. But, the equipment is more complex and the quality somewhat lower.

In an even lower-rate but at present impractical device called the *phoneme vocoder*, the analyzer recognizes a number of basic voice sounds called *phonemes* and instructs the synthesizer to speak these.

For ordinary telephone use, vocoder quality has been barely adequate. A major reason is the unnatural sound of the output. In the channel vocoder, this unnaturalness appears to be associated with the failure of the sound generator of the synthesizer to adequately follow pitch, changes from voiced to voiceless sound, and more subtle qualities of the excitation of the speaker's vocal tract.

By sending in a band of its own an unmodified few hundred hertz of the lower frequencies of the speech to be recreated, and then distorting this at the synthesizer, we can obtain a more satisfactory source of sound with which to feed the synthesizer filters. Such a *voice-excited vocoder* sounds almost as good as regular telephone speech and takes around one-half as many bits per second to transmit as regular digital transmission of speech. It may be easier to obtain high quality with this kind of vocoder than with a channel vocoder, but the output bit rate is considerably higher.

The Huffman coding techniques of Section 13.1 or other source coding techniques can be used as a supplement to vocoding to further reduce the bit rate or bandwidth. The additional reduction does not come to even a factor of 2, whereas we have seen that the vocoder by itself can reduce the channel requirements by factors of 20 to 40 or more.

We note that the transmission of voice using even the lowest-rate vocoder takes many more bits per spoken word than the transmission of English text does. Partly, this because of the technical difficulties of analyzing and encoding speech as opposed to print. Partly, it is because in speech we are actually transmitting information about speaker identification, speech quality, pitch, stress, and accent as well as the information in text. We may say if we wish that the rate distortion function of speech is somewhat greater than the entropy of text, if the distortion is to be so low that the system sounds entirely natural.

That the vocoder does encode speech more efficiently than other methods depends on the fact that the configuration of the vocal tract changes less rapidly than the fluctuations of the sound waves which the vocal tract produces. Its effectiveness also depends on limitations of the human sense of hearing, which defines the bounds of relevant distortion measures. But, we have no good quantitative model for a true distortion measure corresponding to hearing, nor of the parameters of speaker identification and other nontextual information.

Can we nevertheless estimate what theoretical minimum number of bits per second might be required to transmit and reconstruct very high fidelity speech? We shall examine this problem by means of a simple model, assuming that the speech is intended for a single listener whose identity is known.

It is reasonable to suppose that an individual can recognize the voices of no more than about 100,000 (or 2^{17}) different individuals, and that it takes about a 2-sec sample of their speech to do so. This requires 17 bits in 2 sec, or about 9 bits/sec for speaker identification. There are perhaps 25 or 30 recognizable emotional states, and we will for simplicity assume 32 or 2^5 such states. The human reaction time is 300 msec. If we may have one new emotional state every 300 msec, to transmit emotional states requires 5 bits every 300 msec, or about 17 bits/sec. This 17-bit/sec rate accounts for all the information content in pitch, stress, accent, loudness, etc. The total for nontextual information estimated in this way is about 26 bits/sec.

What about textual information? A rapid talker can speak at most about 6 words/sec. The entropy of printed English text can be used to estimate the entropy of spoken words, even though the two sources are not exactly the same. With an entropy for printed text of about 10 bits/word, the entropy of spoken words can be estimated to be about 60 bits/sec. The total estimate for the entropy of spoken language, or for its rate distortion function at very low distortion is about 86 bits/sec.

We see that we have reason to believe that speech can be compressed to fewer than 100 bits/sec, and with extremely high quality. We may expect prohibitively complex terminal equipment if we do this. But, the quality would be far higher than normal telephone quality, and of course well above the minimal quality of present-day vocoders. The 100 bits/sec represents a factor-of-5 improvement over the most advanced experimental vocoders, and these have very poor quality.

In the next section, we shall consider how we might find "the message behind the message" for other data sources, so that we can greatly reduce their entropy. We shall discuss a technique called multidimensional scaling.

Problems

1. Estimate the average output bit rate of a vocoder such as in Fig. 13.5, with the following modification. The 16 band-pass filter outputs are analyzed 40 times

per second. The output of each is quantized to 4 bits. But, only the outputs of the two most significant filters are transmitted. Here significance is judged by the energy in each frequency band (differing weights may be used for each band, but this is not relevant in counting bits). The indices of the two most significant filters need to be transmitted as well. The pitch and hiss analyzer also outputs every 1/40 sec. This means that one bit is used to tell whether the sound is voiced or voiceless. If the sound is voiced, which occurs about half the time, an additional 8 bits are sent to describe the pitch to about $\frac{1}{2}$% accuracy.

2. Infinitely clipped or perfectly hard-clipped speech is speech turned into a constant-amplitude signal by sending $+v_0$ if the waveform is positive and $-v_0$ if it is negative. Its use is suggested by the fact that the pitch information in speech is contained in the zero crossings of the speech amplitude. Infinitely clipped speech that has first been low-pass filtered to 2 kHz is intelligible (but barely). To transmit this waveform digitally by sending a 0 if the speech waveform is positive and a 1 if it is negative requires about 12,000 bits/sec to retain its minimal intelligibility. Why is this higher than the Nyquist sampling rate of 4 kHz required to transmit a 2-kHz signal? Suggest a method based on Section 13.1 or Section 13.2 for further reducing the bit rate.

13.5. Multidimensional Scaling and the Message behind the Message

In speech transmission, it may appear that the message is the speech waveform, but behind this is another message, the act or process of speaking. In the vocoder, getting at the message behind the message reduces the apparent source entropy far more than it could be reduced by using any simplified source model and encoding the source as suggested by the model. The message behind the message is in this case the slowly changing configuration of the vocal tract and vocal cords. The vocoder transmits only such configurational changes, instead of blindly trying to sample or approximate the waveform as suggested by simplified models.

Consider reproducing a performance of marionettes, or string-operated dolls. We could use very high-quality television to transmit and reproduce a good image of the performance. Or, we could recreate the performance with far fewer bits per second by sending the string-end positions. The receiver which reproduces these string-end positions has a set of marionettes identical to those used in the performance being transmitted. Such a receiver is not like any in existence. It would prove very costly, and would serve a very narrow purpose. Nevertheless, by its use, we could transmit marionette shows over a very-low-capacity channel. And, we would get a much better representation of the show than would be provided by any possible television system. We have used the message behind the message to improve communication.

Suppose we want to transmit bar graphs as black and white pictures.

We might try to do this using facsimile or slow-scan TV, as in Section 13.2. But, facsimile can be applied to *any* two-level picture. If we take advantage of the structure of bar graphs, we can do much better. Here we should send only the heights and widths of the bars. Because this is the message behind the message, far fewer bits will be required than for facsimile, even for facsimile using run-length coding. But, the method would not work for other two-level pictures we might wish to send, such as weather maps. Here we see a design dilemma that often arises in communication. We can specialize communication to a particular source and reduce or compress the number of bits required. But, we will have a system that can be used only for the narrow purpose for which it was originally designed. The specialized system will also be more sensitive to errors. Which approach to take depends on particular circumstances and on economic factors.

Fingerprints are another data source for which there could be a message behind the message. Figure 13.6 shows typical fingerprints, and Reference 4 shows how fingerprints are classified by structure. Even if one were to transmit a fingerprint as a line drawing by describing the pencil positions as the lines are traced, the entropy would be several thousand bits per print. And, there would be little hope of using the tracings to search a computer file of fingerprints rapidly for a particular print.

It is said that no two sets of 10 fingerprints are alike. The total number of people in the next ten millennia (300 generations times 10 billion people per generation) is about 3×10^{12}. Any fingerprint characterization scheme that permits this many individuals to be distinguished will be satisfactory. Let us conservatively assume that fingerprint types do not occur independently from one finger to the next on a given individual. Rather, we will assume the opposite, in which all 10 prints are always the same. We see that we ought to be able to characterize a single fingerprint or a single individual in only

$$\log_2 (3 \times 10^{12}) = 41.4 \text{ bits/print}$$

We are a very long way from this.

Is there a message behind the fingerprint message that might put us within range of the 41.4-bits/print upper limit? This might involve specifying a few point and line *sources*, and defining a *gradient* which determines the spacing of the "epidermal ridges" surrounding or parallel to the source. This may be reasonable because the fingerprint-generation mechanism or message behind the message may lie in embryology.

We see that human voice as well as other data sources have a message behind their message. Is there any general or systematic way in which we can discover an underlying structure for data sources? There is

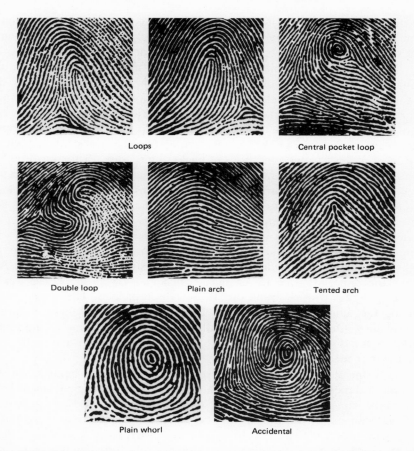

Figure 13.6. Typical fingerprints. (From *The Science of Fingerprints*, Federal Bureau of Investigation, U.S. Government Printing Office, 1963.)

a technique of some generality which can sometimes lead to such structure. It is called multidimensional scaling, developed for psychological measurement (Reference 5). We shall describe this algorithm.

The algorithm for Viterbi decoding that we studied in Section 12.2 is used for direct operation on the received signal. The algorithm for Huffman coding that we studied in Section 13.1 is not used in the transmitter or receiver, but is used in the design of the codes that are used in some transmitters and receivers. The multidimensional scaling algorithm which we now study is one step further removed from the communication system. It is used to discover principles behind a message source. It is these principles which are embodied in the design of a communication system, or in a part of the system analogous to the vocoder. We shall illustrate a potential kind of use of the multidimensional scaling algorithm by rediscovering some of the concepts of phonetics which were used in conceiving of the vocoder.

Multidimensional scaling starts with a set of n points or objects or signals. These may be 16 of the recognizable consonants in English [b, d, f, g, k, m, n, p, s, t, v, z, sh, th (as in think), th (as in that), and zh (as in azure)]. Estimates are made of the *discrepancies* between all pairs of signals. In this case, the discrepancies are the probabilities that one consonant of the pair will be distinguishable from, or not confused with, the other under various conditions of background noise, distortion, etc. In general, discrepancy is a measure of the dissimilarity between the signals. But, only the rank ordering (smallest to largest) of the values of the discrepancies is what matters in multidimensional scaling. A discrepancy measure need only be nonnegative and attain its minimum value when it measures the discrepancy of any signal from itself.

Multidimensional scaling attempts to represent or embed the n signals as points in a Euclidean space of dimension much less than n. At the same time, it tries to produce Euclidean distances whose rank ordering agrees with the rank ordering of discrepancies. This means that signals far apart as points of the Euclidean space will have a large discrepancy. Or, the rank order of discrepancies is the same as the rank order of the distances. We can, if we wish, use the rank order itself as a discrepancy measure.

In principle, multidimensional scaling finds the Euclidean space of minimum dimension in which the rank order of the distances, including ties, is the same as the rank order of the discrepancies. The configuration in Euclidean space will, it turns out, be practically unique. But, it may not be possible to make the rank orders of the discrepancies exactly match the rank orders of the distances in a space of sufficiently low dimension to be interesting or useful. This may not be too bad, because usually the discrepancies that we have come upon will only be noisy estimates of the true discrepancies anyway. We can then try to get an embedding into a space of sufficiently low dimension by trading preservation of rank order for lowering of dimensionality. We may often prefer to decrease the dimension if we can preserve most of the rank orders.

In performing the multidimensional scaling algorithm, we note that the rank order of the discrepancies corresponding to different signals can be matched exactly in $n - 1$ dimensions. We see that there are $n(n - 1)/2$ of these discrepancies if there are n signals. To do this matching, we start with a *regular simplex*, which is a regular geometric figure with n vertices such that all distances between different vertices are equal. An example of this is the vectors or codewords or vertices corresponding to a maximum-length shift register sequence as in Section 12.8. We center the simplex around the origin, so that the centroid or average of the signal vectors or average of their components along any axis will be 0. This provides an absolute location for the configuration. We also choose the length of the side of the simplex so that the "energy," or squared distance of the vertices from the origin, is 1, to provide an absolute scale for the configuration. We will maintain the location constraint that the centroid be 0 and the scale constraint throughout the multidimensional scaling process. For figures that are not regular simplices, the scale constraint is that the *average* energy, or mean-square distance from the centroid or origin, be 1. In taking this average, each vertex or signal vector has the same weight.

It is easy to show that we may make an infinitesimal change in the position of the n vertices of the regular simplex to obtain any rank order of the distances whatever. But, we would rarely be satisfied with a solution which is a small perturbation of the original simplex in $n - 1$ dimensions, even if all the rank orders are correct. Multidimensional scaling seeks to lower the dimensionality of the geometric figure, starting with the regular simplex, while at the same time making the rank order of the distances match the required rank order more and more closely. As explained above, the procedure may in some cases reduce the number of dimensions at the expense of a few changes in the rank order.

We finally reach a configuration of acceptably low dimensionality and acceptable rank order agreement, in which lowering the dimension still further changes too many rank orders. We then stop trying to lower the dimension, but continue to try to increase the agreement of rank orders. We finally can no longer improve the rank order agreement, and so we accept the configuration that we have derived. We note that the acceptability of the dimension or the rank order agreement may depend on physical or engineering insight.

Multidimensional scaling now produces a particular set of orthonormal coordinates in the minimum-dimensional space. The axes in this space are called the *principal axes*. The first principal axis is the direction along which there is the most average energy, or sum of the squares of the components of the n signals along that direction. The second principal axis is the orthogonal direction in which there is the next most energy, and so on. The principal axes are a characterization of the signal in terms of a few parameters that are naturally related to the original rank order of discrepancies. The parameters which multidimensional scaling uncovers may represent the message behind the message.

We will now describe an iterative algorithm taken from Reference 6 for simultaneously reducing the dimension and increasing rank order agreement. An improvement procedure is used which, by itself, would infinitesimally move the original simplex to another configuration with the exact desired rank order of distances. For this, we compute $n - 1$ vectors for each point of the configuration. Each vector falls along the line connecting the given point with one of the other $n - 1$ points of the configuration. A vector from one point is directed toward or away from another point depending on whether the distance between the points is too large or too small, as determined by a comparison of the rank order of the Euclidean distances of the configuration with the desired rank order as given by the discrepancies.

The length of the displacement vector is determined by the size of the rank deviation, with bigger deviations producing longer lengths. The vector is zero if the rank order is exactly right. We use the absolute value of the difference between the actual rank and the desired rank as the length of the displacement vector. And, we need to multiply the sum of these $n - 1$ displacement vectors by a small positive constant a to insure convergence of the iterative process. However, before we actually displace each of the n points by the overall rank-order-improving displacement vector from the point, we need to compute displacement vectors for dimension reduction as well.

How does dimension reduction work? It is based on a procedure for flattening a configuration so as to be closer to lying in a space of lower dimension. We observe that flattening tends to increase the *variance* of the distances between points. Here the variance is computed assuming each distance is equally weighted or equally likely. We can also say that increasing the variance tends to flatten the configuration and lower the number of dimensions. To increase the variance, we stretch those distances that are already large and shrink those that are already small, in comparison with the mean or average interpoint distance. This means that for each point, we construct $n - 1$ additional displacement vectors which point toward those points that are closer to the given point than the average distance, and away from points that are further than the average. The length of each such displacement vector is the magnitude of the difference of the distance from the average distance. A small multiplier b is also applied here to the sum to insure convergence. It has been found that b should be smaller than a for best convergence.

For each of the n points of the configuration, we have now computed $2(n - 1)$ small displacement vectors, one set of $n - 1$ to improve the rank order and the other set of $n - 1$ to lower the dimension. We add these $2(n - 1)$ vectors to get one final displacement vector for each point. When we displace each point by its displacement vector, we produce a new configuration which tends to have more rank orders correct and also to be flatter, or closer to being in a lower-dimensional subspace. Here we must remember to translate or center the configuration around its new centroid or mean to keep the vector sum of the new set of n signals equal to zero, and also to change the scale so that the average energy remains at 1.

We can now find the principal axes of the new configuration. We will delete those coordinates, if any, which are judged to contain sufficiently little variance or energy. By deleting coordinates, we have not only flattened the previous configuration, we have actually embedded the configuration into a lower-dimensional space.

The improvement procedure is now reapplied to this possibly lower-dimensional configuration. When the dimension is finally judged to be sufficiently low, the $n - 1$ dimension-reducing displacement vectors are not computed, and only the rank-improvement vectors used. After several more iterations, the procedure converges to a stable and we hope acceptable and useful configuration. The principal axes are found for the final low-dimensional configuration, and we can now try to find the message behind the message.

Let us see how this works for the 16 English consonants we had earlier. A discrepancy table or array is prepared from a large series of measurements of confusion between the 16 consonants

b, d, f, g, k, m, n, p, s, t, v, z, sh, th, t̲h̲, zh

Each discrepancy or entry in the 16×16 array is the probability or relative frequency that one consonant is not confused with another under varying conditions of noise and distortion. For example, we measure the probability that when p is spoken, the listener does not think it is a b, f, etc. The probabilities in the half array above the main diagonal and in the half below are averaged to obtain an average probability of not mistaking one letter of a pair for the other. Thus the probability of not mistaking b for p is averaged with the probability of not mistaking p for b. We put zeros down the diagonal of the array as the discrepancy between each consonant and itself. We now have the discrepancies between all pairs of letters in a symmetric array.

We note that the minimum discrepancy, 0, is obtained for identical consonants. These 16 zero discrepancies are given rank 0. The other discrepancies are ranked, starting with rank 1. In case of ties, ranks are repeated, but this is not likely for averages of measurements with large samples. We observe that the largest rank is

$$120 = \frac{16 \cdot 15}{2}$$

If there are ties, the largest rank would be lower than this.

Multidimensional scaling can now be applied to the 16 points. Reference 6 found that six dimensions were needed to represent the rank

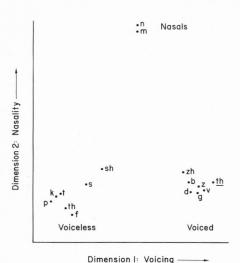

Figue 13.7. First two principal axes for 16 consonants.

ordering of the discrepancies adequately. A few adjacent ranks were interchanged to obtain a single last large drop in dimension. The improvement procedure was continued in six dimensions until the rank order of Euclidean distances matched the rank order of discrepancies as closely as possible. The configuration was then accepted, and the principal axes found.

There are six principal axes in six dimensions. The first principal axis is, as we saw, the one with the greatest average energy or variance for distinguishing the 16 consonants. This axis spans a plane, or two-dimensional subspace, together with the second axis. The projection of the final six-dimensional configuration onto that plane is shown in Fig. 13.7, which is taken from Reference 7. (The configuration is not centered around the origin in the figure.) Because dimension one neatly distinguishes the *voiceless* consonants (p, k, t, th, f, s, sh) from the *voiced* consonants (b, g, d, th, v, z, zh), the dimension is called the *voicing* dimension. Dimension two distinguishes the nasal consonants (m, n) from all others. It is called the *nasality* dimension.

The other four principal axes also have interpretations in classical phonetics, although somewhat more complicated than these two. But, even with knowledge of classical phonetics, we might not have known the particular two dimensions of voicing and nasality as the two most important ones for distinguishing consonants. We have gone quite a way toward discovering the message behind the message in transmitting the signals corresponding to consonants.

In the next section, which is the last in the book, we will apply what we have learned about sources and channels to the design of entire communication systems. We shall design a communication system in a highly idealized case so as to come as close as possible to the theoretically attainable performance when there is a cost limit preventing us from implementing everything specified by the theory.

Problems

1. We found in the text an *upper* bound on the number of bits required to characterize one of 3×10^{12} individuals uniquely based upon fingerprints. This bound assumed that all the 10 fingerprints of a person are the same. Derive a *lower* bound based on the most favorable assumption we can make about dependence. Does it seem reasonable that this bound can ever be attained?

2. In multidimensional scaling with three signals a, b, c, suppose $0 < d(ac) < d(ab) < d(bc)$. Here d is discrepancy. How many dimensions does multidimensional scaling require to represent the rank order of such signals exactly? How unique is the configuration?

3. Give an example of three points in two dimensions whose rank ordering of Euclidean distances is not representable by any three points in one dimension.

4. Write down the coordinates of a regular simplex with four vertices in three dimensions, using only 0's and 1's as coordinates. Relate this to a maximum-length shift register sequence.

5. Explain why it is true that by infinitesimally small changes in the coordinates of a three-dimensional regular simplex (one with four vertices), any rank ordering of the six distances can be obtained.

6. For a regular simplex in three dimensions, what is the variance of the distances? The simplex is flattened into two dimensions by forcing one vertex orthogonally into the plane of the other three vertices. What is the new variance of the distances? Assume that the original side is of length 1. What is the original average energy, if the simplex is centered around 0? (Problem 4 above may be used to help in this.)

7. The variance-increasing algorithm is applied in one dimension to the three points 0, 1, and 3 with the multiplier b set equal to $\frac{1}{2}$. By what factor is the variance of the interpoint distances increased? (First rescale the new configuration so as to have the same average energy with respect to its centroid as the original configuration had.)

13.6. The Ultimate Communication System

In this book, we have seen how antenna aperture affects communication performance. We have leaned how to represent noise and how to find error probabilities. We have evaluated the signal-to-noise ratio and frequency spectrum for various forms of modulation. We have discovered that it is possible to send information over a noisy channel with negligibly few errors. We have explored the encoding of particular signals into a small number of bits per second. We now know that there are many aspects of communication that the communications engineer must take into account.

There are important aspects of communication that were not covered in this book. Communication switching is the most important of these. And, economic and regulatory aspects can sometimes determine the form a communication system takes. In spite of these omissions, we shall discuss the process of communication system design and give a simple example involving tradeoffs in an idealized case.

Those who have read the Requests for Proposals (RFPs) and the Solicitation Orders and Awards issued by NASA and the Department of Defense may receive the impression that wonderful new communication systems start with a known and inflexible set of requirements. Actually, advances in communication begin either with new, higher-performance hardware that makes new systems possible, or with concepts for new functions that can be performed with already available hardware. Large-scale integrated circuits gave us microprocessors that can be used in complex communication modems (or modulator–demodulator terminals) and programmable switches. Communication satellites opened up a new use for existing microwave and space technology, and also created an impetus for new technology in these areas.

New hardware and new sorts of systems have both potentialities and limitations. Solid-state microwave power generation devices are light, long lived, and efficient, but microwave vacuum tubes give greater power. If a high-power device is specified, solid-state devices with their great advantages are ruled out. Communication satellites in synchronous orbit give communication paths with delays much longer than terrestrial microwave, coaxial cable, or optical fibers. If a small delay is specified, satellite circuits are ruled out. Existing vocoders that operate over a 2400-bit/sec circuit give a quality poor compared with an average telephone call, and do not transmit two voices at a time, nor a voice in a very noisy background.

New hardware or new systems will be useful and economical if we can find specified tasks that they can accomplish well and economically. If we specify too high a performance, or unnecessarily restrict details of a system (how it is to do the task rather than the task to be done), we may specify a useful system or technology right out of existence. Indeed, some RFPs insure a bad system (if the system is ever built) by demanding too high a performance, or too many features or uses, or by specifying in too great detail how the overall result shall be attained.

Large advances in communication lie in the hands of engineers and scientists who have a good deal of freedom. This can involve freedom to try to build new devices or freedom to search for new uses of new or existing devices. Various experimental systems must be built before sensible uses and sensible requirements can be arrived at. That has been the history of radio, television, frequency-division and time-division multiplex, vocoders, radar, communication satellites, and all other advances in communications.

Let us, however, imagine an unchanging world in which we know all the costs of performing various functions and the exact requirements that must be met. In this somewhat unreal world, how do we go about designing an economical communication system?

In this book, we have modeled the data source, but not the data sink. We do not know the uses to be made of the data being communicated. We expect that we will have to know the uses to find an appropriate fidelity criterion. We shall nevertheless assume that we have a fidelity criterion, as unrealistic as this may be.

We will assume that we know the message behind the message for the source, so that we know how many bits per second it takes to provide acceptable fidelity. Here we assume that the desired fidelity cannot be changed no matter what the tradeoff between cost and fidelity turns out to be. This is an unrealistic assumption, but we shall make it. We shall suppose that it takes at least 100 bits/sec for acceptable fidelity.

But, in order to attain this bit rate, we will need complex terminal equipment, such as vocoders, at both ends of the communication link. The closer to 100 bits/sec we want to get, the more we expect such equipment will cost. We may have a curve of such cost as a function of bit rate. The one we shall assume is

$$C_0(r) = C_0 e^{-r/100} \text{ dollars} \qquad (13.6\text{-}1)$$

Here r is the output bit rate, at least 100 bits/sec, and C_0 is in dollars. For $r = 100$ bits/sec, $C_0(100) = c_0/e$; for $r = 200$ bits/sec, $C_0(200) = C_0/e^2$; for $r = 1000$ bits/sec, $C_0(1000) = C_0/e^{10}$; and so on. The cost of the terminal equipment drops rapidly when we do not seek high degrees of compression, or extremely low bit rates. This seems very much like the real world.

The quality of the channel we will need will depend on how much we have compressed the source toward the 100-bit/sec limit. We expect that more compression means that we will need a channel of lower bit error probability. We note, for example, that a vocoder requires a lower bit error probability than ordinary digital telephony does, in order to be useful at all. Here we will assume that the required bit error probability $p(r)$ varies for r at least 100 bits/sec as

$$p(r) = p_0 \left[1 - \left(\frac{100}{r} \right)^{1/2} \right]^2 \qquad (13.6\text{-}2)$$

Here p_0 is the high-rate bit error probability, or the error probability required if the source is not compressed but outputs very many symbols per second. For $r = 100$ bits/sec, $p(100) = 0$, a noiseless channel; for $r = 200$ bits/sec, $p(200) = (0.086)p_0$; for $r = 1000$ bits/sec, $p(1000) = 0.47p_0$; and so on. We have assumed in (13.6-2) that only the average bit error probability is important in determining system performance. In practice the actual pattern of errors may also matter, but we shall ignore this.

We shall assume that unlimited bandwidth is available, although this is unreasonable if we are transmitting by radio, wire, or coaxial cable. We shall also assume thermal white gaussian receiver noise of power spectral density N_0. We have learned in Chapter 2 how this noise density depends upon the receiver noise temperature.

An approximate relation is known from coding theory, which gives the minimum attainable output bit error probability $p = p(r)$, in terms of the communications rate r, the length n of an appropriate block code, and the received signal power P_s and noise spectral density N_0. Note that we have restricted ourselves to block codes rather than convolutional codes. The approximate formula for large n is

$$p(r) \doteq 2^{-n(P_s/rN_0 \ln 2 - 1)} \tag{13.6-3}$$

$$n(r, P_s) = \frac{\ln[1/p(r)]}{(P_s/rN_0) - \ln 2} \tag{13.6-4}$$

In the above expression, r must be less than the capacity of the infinite bandwidth gaussian channel:

$$r < \frac{P_s}{N_0 \ln 2} \tag{13.6-5}$$

Otherwise a different formula holds, which is based on rate distortion theory. But, engineering insight tells us that it will usually not be economical to try to operate in the region where r exceeds channel capacity. So, we will assume that the bit rate r is less than the capacity, which means that (13.6-5) holds.

What does it cost to have a coding scheme of block length n? There is preliminary theoretical evidence that the cost, which includes the cost of encoders and decoders, may be proportional to the fourth power of the block length:

$$C_1(n) = C_1 n^4 \text{ dollars} \tag{13.6-6}$$

Here the constant C_1 is in dollars. We will assume that any decoding delay, or any n, is acceptable. This is unrealistic in the case of two-way communication such as telephony, but n could be rather large before the delay exceeded the 1/4 sec which has been to be the maximum acceptable.

We will assume preexisting microwave transmitting and receiving antennas at fixed frequency, with fixed noise temperature, a fixed distance apart in a free space link. In this design exercise, only the transmitter power is permitted to vary, and the received signal power P_s is directly proportional to this. In actual cases, we would want the antenna effective areas and receiver noise temperature to be subject to design tradeoffs as well.

We assume that the cost $C_2(P_s)$ of implementing a transmitter that results in received power P_s increases faster than linearly as a function of P_s. The formula is assumed to be

$$C_2(P_s) = C_2\left(\frac{P_s}{P_0}\right)^{3/2} \text{ dollars} \qquad (13.6\text{-}7)$$

Here C_2 is in dollars and P_0 is a reference received power.

We have now given all the implementation costs. But, we have not discussed the cost of maintaining and operating the communication equipment over its useful life. These life-cycle costs are an important consideration in the communication system design process, but they are harder to determine and we shall ignore them. So, the equation defining the total cost $C_{\text{TOT}}(r, P_s)$ if we operate at output rate r and received power P_s is

$$C_{\text{TOT}}(r, P_s) = C_0(r) + C_1(n) + C_2(P_s) \text{ dollars} \qquad (13.6\text{-}8)$$

We have assumed in (13.6-8) that the cost $C_0(r)$ of the source encoding–decoding equipment, the cost $C_1(n)$ of the channel encoding–decoding equipment, and the cost $C_2(P_s)$ of the transmitter which results in received power P_s are additive. But, interactions between the costs could arise from factors such as competition for space in equipment racks or for the operating power. We shall assume that such interactions are negligible here.

We must now choose r, the amount of source compression, and P_s, the power level, to minimize $C_{\text{TOT}}(r, P_s)$ as given by (13.6-8). Here the block length $n = n(r, P_s)$ is found from substituting (13.6-2), which determines the bit error probability $p(r)$ required if we communicate at bit rate r, into (13.6-4), which determines n from p. The encoding–decoding cost $C_1(n)$ becomes

$$C_1(n) = C_1\left(\frac{\ln\{p_0^{-1}[1 - (100/r)^{1/2}]^{-2}\}}{P_s/rN_0 - \ln 2}\right)^4 \text{ dollars} \qquad (13.6\text{-}9)$$

From (13.6-1), (13.6-9), and (13.6-7), (13.6-8) becomes

$$C_{TOT}(r, P_s) = C_0 e^{-r/100} + C_1 \left(\frac{\ln \{p_0^{-1}[1 - (100/r)^{1/2}]^{-2}\}}{P_s/rN_0 - \ln 2} \right)^4$$

$$+ C_2 \left(\frac{P_s}{P_0} \right)^{3/2} \text{ dollars} \quad (13.6\text{-}10)$$

The designer must now minimize the total cost given in (13.6-10) by choice of r and P_s. The noise power spectral density N_0, the reference received power level P_0, and the high-rate bit error probability p_0 are given. Here r must be at least 100 bits/sec, and P_s and r must satisfy the capacity inequality (13.6-5). Since we have observed that good designs would not operate at bit rates above channel capacity, we may usually ignore inequality (13.6-5) in finding the optimum r and P_s.

The communication system designer is now faced with an ordinary two-variable minimization, whose solution by numerical techniques is not difficult. We shall not work out any solutions here. In practice, such optima are rather broad, and intangible factors or nonengineering considerations usually are involved in the final choice.

In a less idealized world than that assumed above, there are many major communication system design issues that will face communication system engineers in the next decade. One issue involves the competition between communication satellites and fiber optics for long-distance communication of high volumes of traffic, including not only transcontinental links but also links across oceans. Another issue involves the potential for optical frequencies to replace microwaves for deep space to near earth or near-earth space-to-space relay communication. A third involves the proper choice of a modulation scheme for land mobile radio telephony. And there will of course be very many more.

The book does not resolve these issues. But, the student who has learned what is in this book will be able to frame the questions that must be answered before good decisions can be made. And, he or she will be able to begin to answer these questions and to provide the technical basis for sound implementation and investment decisions. In most cases, more specialized material will have to be mastered and more experience gained. But, this introduction provides the foundation on which to build that knowledge and experience.

This concludes our introduction to communication science and systems. There are many directions leading from this point. Some paths lead to communication networks linking nations and peoples. Some paths lead to a society which is decentralized but not fragmented. Some paths lead to the stars.

Problems

1. Show that (13.6-10) has a unique minimizing solution for P_s when the bit error rate r is fixed. HINT: differentiate and show that there is a unique P_s which makes the derivative 0.

2. Show that the total cost $C_{TOT}(r, P_s)$ must become infinite if we let r approach 100 bits/sec with P_s fixed. Explain this with as little calculation as you can.

3. What happens to the optimum design as we make C_0, the compression cost, large, with C_1 and C_2 fixed? The same if C_1 becomes large with the other two fixed? In each case, describe the effect on the following: source output symbol rate r, received signal power P_s, average bit error probability $p(r)$ out of the channel decoder, block length n, and total cost C_{TOT}. What phenomenon could we observe when C_2 becomes very large with the other two fixed?

4. Derive an equation similar to (13.6-10) when the option exists of increasing the diameter D of the receiving antenna. Assume the cost is

$$C_3(D) = C_3\left(\frac{D}{D_0}\right)^3$$

if the efficiency is fixed independent of diameter. Here D_0 is the reference diameter giving received power P_s. (An approximate equation such as this with a power slightly less than 3 has sometimes been used in comparing single large apertures with arrays when considering designs for ground microwave antennas to be used for satellite and deep space reception.)

References

Preface

1. *Telecommunications Transmission Engineering, Vol. I, Principles,* Bell System Center for Technical Evaluation, Murray Hill, New Jersey (1974).
2. Wozencraft, John M., and Irwin M. Jacobs, *Principles of Communication Engineering,* Wiley, New York (1965).
3. Carlson, A. Bruce, *Communications Systems,* McGraw-Hill, New York (1968).
4. Lucky, R., R. Salz, and E. Weldon, *Principles of Data Communication,* McGraw-Hill, New York (1968).
5. Ziemer, Rodger E., and William Tranter, *Principles of Communications: Systems Modulation and Noise,* Houghton Mifflin, Boston (1976).
6. Inose, Hiroshi, *An Introduction to Digital Integrated Communication Systems,* University of Tokyo Press, Tokyo (1979).

Chapter 1

1. Rusch, W. V. T., and P. D. Potter, *Analysis of Reflector Antennas,* Academic Press, New York (1970).
2. Pierce, John R., *Almost All About Waves,* MIT Press, Cambridge, Massachusetts (1974).
3. Skolnik, Merril I., *Introduction to Radar Systems,* McGraw-Hill, New York (1962).
4. *Transmission Systems for Communications,* Bell Telephone Laboratories, Murray Hill, New Jersey (1970).

Chapter 2

1. Robinson, F. N. H., *Noise and Fluctuations,* Clarendon Press, Oxford (1974).
2. Bell, D. A., *Electrical Noise—Fundamentals and Physical Mechanisms,* Van Nostrand, London (1960).
3. Oliver, B. M., ed., *Project Cyclops, A Design Study of a System for Detecting Extraterrestrial Intelligent Life,* NASA Ames Research Center, Moffet Field, California (1973).

4. Morrison, Phillip, John Billingham, and John Wolfe, eds., *The Search for Extra-Terrestrial Intelligence—SETI*, NASA Report FP-419, U.S. Government Printing Office, Washington, D.C. (1977).
5. Kraus, John D., *Radio Astronomy*, McGraw-Hill, New York (1966).
6. Gordon, J. P., W. H. Louisell, and L. R. Walker, Quantum fluctuations and noise in parametric processes, II, *Phys. Rev.* **129,** 481–485 (1963).
7. Menzies, R. T., Laser heterodyne detection techniques, Chapter 7 of *Laser Monitoring of the Atmosphere*, E. D. Hinkley, ed., Springer, Berlin (1976).
8. Ross, Monte, *Laser Receivers*, Wiley, New York (1966).

Chapter 3

1. Bose, Amor G., and Kenneth N. Stevens, *Introductory Network Theory*, Harper and Row, New York (1965).
2. Bracewell, Ron, *The Fourier Transform and its Applications*, McGraw-Hill, New York (1965).
3. Papoulis, Athanasios, *The Fourier Integral and its Applications* McGraw-Hill, New York (1962).
4. Flanagan, James, *Speech Analysis, Synthesis and Perception*, 2nd ed., Springer, Berlin (1972).

Chapter 4

1. Carson, John R., and Thorton C. Fry, Variable frequency electric circuit theory with application to the theory of frequency modulation, *Bell Syst. Tech. J.* **16,** 513–540 (1937).
2. Lindsey, W. C., *Synchronization Systems in Communication and Control*, Prentice-Hall, New Jersey (1972).
3. Gregg, W. David, *Analog and Digital Communication*, Wiley, New York (1977).

Chapter 5

1. Rabiner, Lawrence R., and Bernard Gold, *Theory and Application of Digital Signal Processing*, Prentice-Hall, Englewood Cliffs, New Jersey (1975).

Chapter 6

1. Papoulis, Athanasios, *Probability, Random Variables, and Stochastic Processes*, McGraw-Hill, New York (1965).

Chapter 7

1. Murnaghan, Francis P., *The Calculus of Variations*, Spartan Books, Washington, D.C. (1962).

2. Friis, H. T., Oscillographic observations on the direction of propagation and fading of short waves, *Proc. Inst. Radio Eng.* **16,** 658–665 (1928).
3. Friis, H. T., and C. B. Feldman, A multiple unit steerable antenna for short wave reception, *Proc. Inst. Radio Eng.* **25,** 841–917 (1937).
4. Cutler, C. C., R. Kompfner, and L. C. Tillotson, A self-steering array repeater, *Bell Syst. Tech. J.* **42,** 2013–2032 (1963).
5. Reudink, D. O., and Y. S. Yeh, A scanning spot-beam satellite system, *Bell Syst. Tech. J.* **56,** 1549–1560 (1977).

Chapter 8

1. Peirce, B. O., *A Short Table of Integrals*, 4th ed., Revised by R. M. Foster, Blaisdell, Waltham, Massachusetts (1957).
2. Feller, William, *An Introduction to Probability Theory and its Applications*, Vol. I, 3rd ed., Wiley, New York (1968).

Chapter 9

1. Weber, Charles L., *Elements of Detection and Signal Design*, McGraw-Hill, New York (1968).

Chapter 10

1. Lucky, R. W., J. Salz, and E. J. Weldon, Jr., *Principles of Data Communication*, McGraw-Hill, New York (1968).
2. Forney, G. David, Maximum likelihood sequence estimation of digital sequences in presence of intersymbol interference, *IEEE Trans. Inf. Theory* **IT-18,** 363–378 (1972), especially bibliography on pp. 377–378.
3. Leeper, David G., A universal digital data scrambler, *Bell Syst. Tech. J.* **52,** 1851–1865 (December 1973).
4. Golomb, S. W., *Shift Register Sequences*, Holden-Day, San Francisco (1967).
5. Cahn, Charles R., Performance of digital phase-modulation communication systems, *IRE Trans. Commun. Syst.* **CS-7,** 3–6 (1959).
6. Amoroso, F., Pulse and spectrum manipulation in the minimum (frequency) shift keying (MSK) format, *IEEE Trans. Commun.* **COM-24,** 381–384 (1976).

Chapter 11

1. Shannon, C. E., and Warren Weaver, *The Mathematical Theory of Communication*, The University of Illinois Press, Champaign–Urbana (1959).
2. Shannon, C. E., Communication in the presence of noise, *Proc. Inst. Rad. Eng.* **37,** 11–21 (1949).
3. Pierce, J. R., *Symbols, Signals and Noise*, Harper Brothers, New York (1961).
4. Helstrom, C. W., *Quantum Detection and Estimation*, Academic Press, New York (1976).

Chapter 12

1. McEliece, Robert J., *The Theory of Information and Coding*, Addison-Wesley, Reading, Massachusetts (1977).
2. Wolf, Jack K., A survey of coding theory 1967–72, *IEEE Trans. Inf. Theory* **IT-19**, 381–389 (1973).
3. Forney, G. David, The Viterbi algorithm, *Proc. IEEE* **61**, 268–278 (1973).
4. Helstrom, Carl W., Jane W. S. Liu, and James P. Gordon, Quantum mechanical communication theory, *Proc. IEEE* **58**, 1578–1598 (1970).
5. Pierce, J. R., Edward C. Posner, and Eugene R. Rodemich, The capacity of the photon counting channel, *IEEE Trans. on Inf. Theory* **IT-27** (to appear January 1981).
6. *IEEE Trans. Commun* **COM-25** (1977), entire special issue on Spread Spectrum Communications.

Chapter 13

1. Shannon, Claude E., Coding theorems for a discrete source with a fidelity criterion, in *Information and Decision Processes*, R. E. Machol, ed., McGraw-Hill, New York (1960), pp. 93–126.
2. Flanagan, J. L., M. R. Schroeder, B. S. Atal, R. E. Crochiere, N. S. Jayant, and J. M. Tribolet, Speech coding, *IEEE Trans. Commun.* **COM-27**, 710–737 (1979).
3. Atal, B. S., and S. L. Hanauer, Speech analysis and synthesis by linear prediction of the speech wave, *J. Acoust. Soc. Amer.* **50**, 637–655 (1971).
4. Eleccion, Marce, Automatic fingerprint identification, *IEEE Spectrum* **10**, 36–45 (1973).
5. Wish, Myron, and Joseph B. Kruskal, *Multidimensional Scaling*, Sage Publications, Beverly Hills, California (1978).
6. Shepard, Roger N., The analysis of proximities: multidimensional scaling with an unknown distance function, *Psychometrika* **27**, I–pp. 125–140; II–pp. 219–246 (1962).
7. Wish, Myron, and J. Douglas Carrol, Multi-dimensional scaling with differential weighting of dimensions, in *Mathematics in the Archaeological and Historical Sciences*, Edinburgh University Press, Edinburgh (1971), pp. 150–167.

Index

Viterbi, A., 299
Viterbi decoding, 299–306; *see also*
    Convolutional codes, Decoding
VLSI, 112, 356, 368
Vocoder, 354–359, 368–369
  channel, 354–356
  formant-tracking, 357
  linear predictive coefficient, 356
  phoneme, 357
  voice-excited, 357
Voice: *see* Speech
Voice-excited vocoder, 357
Voicing, 356, 362, 366

Voltage spectrum, 63; *see also* Fourier
    transform, Spectrum

Waveguide, 22, 25, 44
Waves: *see* Electromagnetic wave, Micro-
    waves, Millimeter waves
White noise, 12, 37, 94, 162, 173–174,
    209–210; *see also* Gaussian noise,
    Johnson noise, Shot noise
Window, 79–80
Wire pairs, 24–25, 210
Word error probability, 237–238, 254,
    297–298